Replacement of Neanderthals by Modern Humans Series

Edited by

Takeru Akazawa
Research Institute, Kochi University of Technology
Kochi 782-8502, Japan
akazawa.takeru@kochi-tech.ac.jp

Ofer Bar-Yosef
Department of Anthropology, Harvard University
Cambridge, Massachusetts 02138, USA
obaryos@fas.harvard.edu

The planned series of volumes will report the results of a major research project entitled "Replacement of Neanderthals by Modern Humans: Testing Evolutionary Models of Learning", offering new perspectives on the process of replacement and on interactions between Neanderthals and modern humans and hence on the origins of prehistoric modern cultures. The projected volumes will present the diverse achievements of research activities, originally designed to implement the project's strategy, in the fields of archaeology, paleoanthropology, cultural anthropology, population biology, earth sciences, developmental psychology, biomechanics, and neuroscience. Comprehensive research models will be used to integrate the discipline-specific research outcomes from those various perspectives. The series, aimed mainly at providing a set of multidisciplinary perspectives united under the overarching concept of learning strategies, will include monographs and edited collections of papers focusing on specific problems related to the goals of the project, employing a variety of approaches to the analysis of the newly acquired data sets.

For further volumes:
http://www.springer.com/series/11816

Alex Mesoudi • Kenichi Aoki
Editors

Learning Strategies and Cultural Evolution during the Palaeolithic

Springer

Editors
Alex Mesoudi
Department of Anthropology and Centre for the Coevolution of Biology and Culture
Durham University
Durham, UK

Kenichi Aoki
Organization for the Strategic Coordination of Research and Intellectual Properties
Meiji University
Tokyo, Japan

Replacement of Neanderthals by Modern Humans Series
ISBN 978-4-431-55362-5 ISBN 978-4-431-55363-2 (eBook)
DOI 10.1007/978-4-431-55363-2

Library of Congress Control Number: 2015938595

Springer Tokyo Heidelberg New York Dordrecht London

Printed on acid-free paper

Springer Japan KK is part of Springer Science+Business Media (www.springer.com)

Preface

This volume is being published as part of a 5-year research project, funded by the Japan Ministry of Education, Culture, Science, and Technology (Monbukagakusho), on the "Replacement of Neanderthals by Modern Humans (RNMH)." There are two basic premises of the RNMH project. First, Neanderthals were replaced or assimilated by modern humans (*Homo sapiens*). Second, the replacement or assimilation was driven by cultural differences between competing Neanderthal and modern human groups, potentially due to some advantage(s) associated with the culture(s) of modern humans.

The current volume focuses on how differences in the cultures of Palaeolithic or Stone Age hominin societies might arise as a result of differences in learning strategies, social and demographic factors, and so on. This includes the knotty inverse problem of inferring learning strategies from actual trajectories of cultural change. With the exception of one chapter, the replacement process itself is not addressed.

The majority of contributors to this volume are not members of the RNMH project. Nevertheless, we have invited them to submit chapters, because they are leading anthropologists, archaeologists, biologists, and psychologists who are directly involved in the effort to decipher hominin cultural change and cultural diversity during the Palaeolithic (see list of contributors).

In addition to the contributors, we wish to thank Jelmer Eerkens, Yasuo Ihara, Jeremy Kendal, Steven Kuhn, Charles Perreault, Katsuhiro Sano, Jonathan Scholnick, Pontus Strimling, Jamie Tehrani, Claudio Tennie, and Taro Yoshida, for valuable comments on and constructive reviews of the chapters. Their input is gratefully acknowledged.

This volume was made possible by Monbukagakusho grant 22101004 and indirectly by the Japanese taxpayer. We are deeply grateful for their support and interest.

Finally, we thank Ken Kimlicka and Taeko Sato of Springer, Japan for their guidance through the intricacies of the publication process.

Durham, UK — Alex Mesoudi
Tokyo, Japan — Kenichi Aoki

Contents

Contributors

Kenichi Aoki Organization for the Strategic Coordination of Research and Intellectual Properties, Meiji University, Tokyo, Japan

R. Alexander Bentley Department of Archaeology and Anthropology, University of Bristol, Bristol, UK

Matthew T. Boulanger Department of Anthropology, University of Missouri, Columbia, MO, USA

Briggs Buchanan Department of Anthropology, University of Tulsa, Tulsa, OK, USA

Christine A. Caldwell Psychology, School of Natural Sciences, University of Stirling, Stirling, UK

Mark Collard Human Evolutionary Studies Program and Department of Archaeology, Simon Fraser University, Burnaby, Burnaby, BC, Canada

Department of Archaeology, University of Aberdeen, Aberdeen, UK

Francesco d'Errico CNRS (Centre National de la Recherche Scientifique), UMR (Unité Mixte de Recherche) 5199, PACEA (De la Préhistoire à l'Actuel: Culture, Environement et Anthropologie), Université Bordeaux 1, Talence, France

Department of Archaeology, History, Cultural Studies and Religion, University of Bergen, Bergen, Norway

Metin I. Eren Department of Anthropology, University of Missouri, Columbia, MO, USA

Marcus W. Feldman Department of Biology, Stanford University, Stanford, CA, USA

Laurel Fogarty Department of Biology, Stanford University, Stanford, CA, USA

Seiji Kadowaki University Museum, Nagoya University, Nagoya, Japan

Anne Kandler Department of Mathematics, City University London, London, UK

Marius Kempe Department of Anthropology and Centre for the Coevolution of Biology and Culture, Durham University, Durham, UK

Yutaka Kobayashi Department of Management, Kochi University of Technology, Kami-city, Japan

Mirna Kovacevic CoMPLEX (Centre for Mathematics and Physics in the Life Sciences and Experimental Biology), University College London, London, UK

Research Department of Genetics, Evolution and Environment, University College London, London, UK

Carl P. Lipo Department of Anthropology and IIRMES, California State University at Long Beach, Long Beach, CA, USA

Stephen J. Lycett Department of Anthropology, University at Buffalo SUNY, Buffalo, NY, USA

R. Lee Lyman Department of Anthropology, University of Missouri, Columbia, MO, USA

Mark E. Madsen Department of Anthropology, University of Washington, Seattle, WA, USA

Alex Mesoudi Department of Anthropology and Centre for the Coevolution of Biology and Culture, Durham University, Durham, UK

Masaki Naganuma Center for Ainu & Indigenous Studies, Hokkaido University, Sapporo, Japan

Wataru Nakahashi School of Advanced Sciences, SOKENDAI (The Graduate University for Advanced Studies), Hayama, Japan

Michael J. O'Brien Department of Anthropology, University of Missouri, Columbia, MO, USA

Adam Powell Palaeogenetics Group, Institute of Anthropology, University of Mainz, Mainz, Germany

L.S. Premo Department of Anthropology, Washington State University, Pullman, WA, USA

Department of Human Evolution, Max Planck Institute for Evolutionary Anthropology, Leipzig, Germany

Kerstin Schillinger Department of Anthropology, University of Kent, Canterbury, UK

Stephen Shennan Institute of Archaeology, University College London, London, UK

Mark G. Thomas Research Department of Genetics, Evolution and Environment, University College London, London, UK

Marian Vanhaeren CNRS (Centre National de la Recherche Scientifique), UMR (Unité Mixte de Recherche) 5199, PACEA (De la Préhistoire à l'Actuel: Culture, Environement et Anthropologie), Université Bordeaux 1, Talence, France

Joe Yuichiro Wakano School of Interdisciplinary Mathematical Sciences, Meiji University, Tokyo, Japan

1 Introduction to "Learning Strategies and Cultural Evolution During the Palaeolithic"

Kenichi Aoki and Alex Mesoudi

Abstract

In this introductory chapter, we first provide some background on the two major recurrent themes of the volume, i.e. learning strategies of individuals, and social and demographic characteristics of populations. This is followed by a brief summary of each chapter. Then, we conclude with some thoughts on why and how the methods and findings presented in this volume are relevant to, and might inform our understanding of, the replacement of Neanderthals by modern humans (*Homo sapiens*).

Keywords

Learning strategy • Demographic factors • Cultural change • Cultural diversity

This volume provides up-to-date coverage on the theory of cultural evolution as is being used by anthropologists, archaeologists, biologists, and psychologists to decipher hominin cultural change and cultural diversity during the Palaeolithic. The contributing authors are directly involved in this effort, and the material presented includes novel approaches and findings. The common theoretical framework of the volume is that cultural change constitutes a dynamic evolutionary system, which can be analyzed using tools and methods derived from the theory of biological evolution (Cavalli-Sforza and Feldman 1981; Boyd and Richerson 1985).

Various chapters show how learning strategies in combination with social and demographic factors (e.g. population size and mobility patterns) predict cultural evolution in a world without the printing press, radio, or the internet—which is to say that cultural traits can be acquired from others only by directly observing their actions or the results of these actions. Also addressed is the inverse problem of how learning strategies may be inferred from actual trajectories of cultural change, for example as seen in the North American Palaeolithic. Mathematics and statistics, a sometimes necessary part of theory, are explained in elementary terms where they appear, with details relegated to appendices. Full citations of the relevant literature will help the reader to further pursue any topic of interest.

1.1 Learning Strategies

Before proceeding it will be useful to briefly explain what the contributing authors and the editors mean by a "learning strategy." A learning strategy is the way in which an organism combines individual learning and social learning, either simultaneously or sequentially, and its relative dependence on each. Here, individual learning occurs when the organism depends on personal experience to gather information, e.g. by trial-and-error. Social learning refers to obtaining information from other organisms, e.g. by imitation. Biases associated with social learning in the choice of whom to copy are also an integral part of a learning strategy.

K. Aoki (✉)
Organization for the Strategic Coordination of Research and Intellectual Properties, Meiji University, Nakano 4-21-1, Nakano-ku, Tokyo 164-8525, Japan
e-mail: kenaoki@meiji.ac.jp

A. Mesoudi
Department of Anthropology and Centre for the Coevolution of Biology and Culture, Durham University, Durham DH1 3LE, UK
e-mail: a.a.mesoudi@durham.ac.uk

A. Mesoudi and K. Aoki (eds.) *Learning Strategies and Cultural Evolution during the Palaeolithic*, Replacement of Neanderthals by Modern Humans Series, DOI 10.1007/978-4-431-55363-2_1,

Much theoretical work has been directed toward examining the adaptiveness of various social learning biases (Boyd and Richerson 1985; Laland 2004), such as "success bias" and "prestige bias" which entail preferentially copying a successful or a prestigious individual, respectively (Henrich and Gil-White 2001; Nakahashi et al. 2012), and "conformist bias" which entails copying the majority cultural behavior of one's group (Henrich and Boyd 1998; Wakano and Aoki 2007; Nakahashi 2007; Kendal et al. 2009; see Aoki and Feldman 2014 for a comprehensive review). "Teaching" represents a powerful adjunct to social learning, where the individual being copied (the "teacher") modifies his/her behavior to facilitate social learning by a naïve individual (the "pupil") (Caro and Hauser 1992). Still another aspect of learning—in particular the learning of complex technical knowledge comprising various different skills—is that cultural traits may be interdependent, some serving as prerequisites for the acquisition of others. Several chapters in this volume tackle the question of how different learning strategies might structure population-level cultural change and variation, and the even more difficult problem of how to identify these population-level signatures in the often sketchy archaeological record.

1.2 Social and Demographic Factors

Many archaeologists and anthropologists currently emphasize social and demographic factors in interpreting "sudden" and "dramatic" changes in stone tools or other cultural artefacts during the Late Pleistocene (between 130,000 and 10,000 years ago), in particular the "creative explosions" (Kuhn 2012) of the African late Middle Stone Age and the European Upper Palaeolithic (Shennan 2001; Henrich 2004; Kline and Boyd 2010; Zilhão et al. 2010; Mesoudi 2011; Clark 2011; Kuhn 2013). In fact, theoretical studies have repeatedly shown that population size can have a large effect on cultural evolutionary rate and cultural diversity (Shennan 2001; Henrich 2004; Strimling et al. 2009; Mesoudi 2011; Lehmann et al. 2011; Aoki et al. 2011; Kobayashi and Aoki 2012; Aoki 2013), as can interconnectedness of subpopulations (Powell et al. 2009; Perreault and Brantingham 2011). Transmission chain experiments conducted in the laboratory also provide some support for a link between population (or group) size and cultural complexity (Derex et al. 2013; Muthukrishna et al. 2014; Kempe and Mesoudi 2014; but see Caldwell and Millen 2010).

However, archaeological evidence on the role of demographic factors is inconclusive or even contradictory. Two recent studies of Late Pleistocene South Africa are particularly relevant. Clark (2011) looked for signatures of population growth and/or demographic stress in an increase of diet breadth (e.g. the use of non-preferred prey animals), obtaining some support for an association with the heightened creativity of Howieson's Poort. But, as Clark (2011) is careful to note, this association is open to an alternative interpretation, namely that rapid cultural change produced new tools, which were used to exploit novel resources. Klein and Steele (2013) (see also Klein 2008, Box 1) observed that edible shellfish remains from Middle Stone Age middens are significantly larger than those from Later Stone Age middens. If shellfish size reflects human collection intensity, then this finding suggests that the precocious appearance of modern behaviors in the Still Bay and Howieson's Poort may not have been associated with population growth.

The claim that pre-contact Neanderthals in Spain used necklaces made of shells strung together as body ornamentation 50,000 years ago is also laden with ambiguity, in more ways than one. Zilhão et al. (2010) regards this as evidence for the cognitive equality of Neanderthals and modern humans, "support[ing] models of the emergence of behavioral modernity as caused by technological progress, demographic increase." However, perforation may not have been anthropogenic, and shells with naturally-formed holes of appropriate size for threading may have been selectively collected. Moreover, according to Prüfer et al. (2014), Neanderthal population size in the Altai region as estimated from genetic data shows a continual decrease after one million years ago, which is not true of various current modern humans. Similarly, Mellars and French (2011) argue for small population size in pre-contact European Neanderthals (MTA) compared to the Aurignacian. By implication, both Neanderthals and modern humans achieved the same cultural level, in spite of a difference in population size. Note, however, modern human beads occur much earlier—as early as 100,000 years ago in the Levant (Vanhaeren et al. 2006)—so perhaps population size did play a role.

In addition, statistical analyses of ethnographic hunter-gatherers have failed to detect an association between population size and the number of food-getting tools (Collard et al. 2005; Read 2006). On the other hand, ethnographic food-producing societies (e.g. small-scale farmers and herders) do conform to the theoretical prediction that population size and the number of food-getting tools should be positively correlated (Kline and Boyd 2010; Collard et al. 2013). Possible explanations for these contrasting results have been suggested, including higher degrees of specialization in the latter societies.

A fundamental problem in human evolution is how to account for an apparently abrupt cultural change, without invoking a major genetic change in cognition (e.g. innovativeness), for which there is at present no strong evidence (Klein 2008). Needless to say, absence of evidence does not constitute evidence of absence, and we are obliged to keep an open mind (Akazawa et al. 2013). Richerson et al. (2009)

(see also Richerson and Boyd 2013) discuss the possibility of spontaneous transitions between stable regimes—a small population at a low cultural level and a large population at a high cultural level. Developing this idea further and based on an explicit mathematical model, Aoki (2015) shows that a saltatory cultural change can be triggered by a gradual evolutionary change in the genetic basis for innovativeness. This scenario is not inconsistent with the "neural hypothesis," a recent version of which invokes "a neural change that promoted the extraordinary modern human ability to innovate" (Klein 2008, p. 271). However, this neural change would not be attributable to just one "fortuitous mutation" in a major gene 50,000 years ago.

1.3 Summary of the Chapters

This volume comprises ten chapters, which use a range of methods to address different aspects of cultural evolution during the Palaeolithic.

In Chap. 2, Fogarty et al. present a theoretical analysis examining the modes and pathways of social learning, and how they affect the expected number of cultural traits maintained in a population. Specifically, they compare random oblique, best-of-K (an example of direct bias, which entails a preference for a particular variant of a cultural trait), success bias, and one-to-many. Given the current emphasis among archaeologists and anthropologists on demographic factors, the effect of population size is also investigated, as is the less acknowledged role of innovation.

Fogarty et al. classify cultural traits as simple or complex, depending on the ease or difficulty of acquisition by social learning and innovation. Assuming an innate upper limit to the number of cultural traits that can be imagined—a limitation that may possibly be overcome by a mechanism analogous to "embedding" in linguistics—they show that the number of simple cultural traits may saturate as population size increases, in which case a statistical association between the two variables is not predicted. At smaller population sizes, there is a major effect of the mode of social learning. By contrast, the relation between the number of complex cultural traits and population size is approximately linear and almost identical for all modes of social learning investigated. This is because most of the complex cultural traits that are maintained in the population can be accounted for by innovation alone, which raises the question of whether such traits qualify as "cultural" (Whiten et al. 1999).

In Chap. 3, Nakahashi describes and analyzes a new mathematical model for the evolution of teaching that is culturally transmitted rather than genetically determined. Teaching is here defined sensu Caro and Hauser (1992) as a knowledgeable individual (the teacher) altering its behavior in the presence of a naïve individual (the pupil), suffering a cost to do so, and thereby promoting social learning by that naïve individual. In this model, there are an infinite number of cultural traits, which are acquired by either individual learning or social learning, and where their acquisition entails a viability cost. Moreover, cultural traits are either beneficial or neutral, and only the former are assumed to contribute to fertility.

Nakahashi shows that teaching behavior can evolve culturally—i.e. teachers can invade and exist at a stable positive equilibrium—if a teacher can socially transmit more cultural traits than a non-teacher. However and surprisingly, it cannot evolve if teaching merely improves the accuracy of social learning by pupils. This latter result differs from the predictions of previous theoretical work that assume genetic determination of teaching behavior (Fogarty et al. 2011).

The next three chapters deal with structured populations. Kobayashi et al. (Chap. 4) directly address the cultural correlates of the replacement of Neanderthals (and other archaic humans) by modern humans. Their chapter begins with a detailed review of the archaeology of the Middle to Upper Palaeolithic transition in various parts of Eurasia, which suggests varying degrees of cultural continuity during/after the arrival of modern humans. In particular, China is apparently characterized by the late persistence of primitive core-and-flake industries (Norton and Jin 2009; Bar-Yosef and Wang 2012). Several Upper Palaeolithic industries in western Eurasia, e.g. the Emiran in the Levant and the Early Baradostian in the Zagros, may also exhibit recognizable elements of the preceding Middle Palaeolithic.

Kobayashi et al. describe a new model in which an invading modern human population has a demographic advantage (a higher relative growth rate), but receives unidirectional cultural influences from the indigenous archaic population. The cultural traits that the modern humans acquire from the archaics are assumed to be of a different kind from those that may be contributing to the demographic advantage of the former. Using approximate analytical methods and agent-based simulations, these authors show that biological replacement can be associated with either the rapid disappearance, the gradual disappearance, or the persistence of these autochthonous cultural traits. Gradual disappearance or persistence, i.e. cultural continuity, is predicted when a small modern human population invades a region with a relatively unfavorable physical environment. Importantly, cultural continuity is not an indicator of biological continuity.

The pattern of mobility within a geographically-structured population is recognized to be an important demographic factor in cultural evolution, through its effect on the variety of social learning opportunities (Powell et al. 2009). In addition, mobility may place a limit on the number of portable artefacts (Torrence 1983; Shott 1986). Premo (Chap. 5) gives an excellent introduction to residential mobility and logistical mobility (Binford 1980; Kelly 1983), where the former refers to the relocation of a hunter-gatherer residential base and the latter to the movement of a subgroup on task-specific forays.

Premo conducts agent-based simulations of a spatially-distributed metapopulation to obtain its effective size. Each agent is a culturally monomorphic group of hunter-gatherers that relocates its residential base if a logistical foray proves unsuccessful in finding food. All agents initially carry different variants of a cultural trait, but agents within a certain interaction radius influence each other culturally, and the mean time until the metapopulation is fixed for one of these variants is used as a proxy measure of effective population size. The main result of this chapter is that the effective size of the metapopulation increases—sometimes quite substantially—as the frequency of residential moves decreases. This prediction has implications for the cultural diversity that is expected to be maintained in the metapopulation.

Madsen and Lipo (Chap. 6) describe and analyze a new agent-based simulation model for the cultural evolution of hierarchically-structured cultural traits (e.g. knowledge, skills), where some cultural traits are prerequisites for the acquisition of others. That is, while most models assume the transmission of independent cultural traits, Madsen and Lipo build on previous efforts (Mesoudi and O'Brien 2008; Mesoudi 2011; Enquist et al. 2011) to explore the more realistic situation where acquiring a cultural trait is dependent on already possessing other cultural traits. Cultural interactions, providing opportunities for social learning, can occur between two adjacent agents on a square lattice, as in the model proposed by Axelrod (1997). Agents may also innovate. The major difference between the current model and the original Axelrod model is that the focal agent can acquire a cultural trait from its neighbor only if it already has the prerequisite cultural traits. Alternatively, the neighbor may structure the learning environment of the focal agent by supplying the latter with the necessary prerequisite cultural traits. Madsen and Lipo refer to this behavior by the neighbor as teaching.

In their simulations, Madsen and Lipo examine the effects of the fidelity of teaching, size of design space (maximum possible number of cultural traits), innovation rate, and population size on cultural diversity (number of different repertoires of cultural traits segregating in the population), "knowledge depth" of cultural traits (average number of prerequisites per extant cultural trait, which can be regarded as a measure of cultural complexity or cumulativeness), etc. Cultural diversity is found to increase with the fidelity of teaching and with the size of design space, but only when innovation is allowed. Knowledge depth increases with the fidelity of teaching, but again only when innovation is allowed.

Of particular interest is the possibly counterintuitive prediction that knowledge depth is minimally affected by population size. This result is likely a consequence of the assumption inherited from Axelrod (1997) that cultural interactions are spatially localized and moreover occur preferentially between culturally similar agents (homophily). If these assumptions apply to Palaeolithic hunter-gatherers—they may not to some present-day hunter-gatherers that travel long distances (Hewlett et al. 1982)—then the predictive value of population size per se should be viewed with caution.

Chapters 7 and 8 explore novel Bayesian methods for detecting signatures of learning strategies in archaeological data. Summary statistics such as the expected number of cultural traits are useful, but frequency distributions (spectrums) of variants of cultural traits are more informative. Kandler and Powell (Chap. 7) explain a powerful new method—approximate Bayesian computation (ABC) (Beaumont et al. 2002)—for identifying learning strategies that produce the observed population level data on such frequency distributions. Very briefly, ABC entails simulating a model with parameter values chosen from a prior distribution and retaining those parameter values that give the closest fit of the simulated data to the observed data; these retained parameter values approximate the posterior distribution.

As a concrete example of the application of this method, they first generate "observed" data by simulating a hypothetical model for the evolution of a cultural trait with known parameter values but with noise added. The parameters that define the learning strategy are the degrees of reliance on individual learning (i.e. innovation), directly-biased social learning, and conformist social learning. Then, the frequencies of the variants are sampled at various times, and the parameters of this model are estimated by ABC. It is shown that the original parameter settings are faithfully recovered by this estimation procedure.

Kovacevic et al. (Chap. 8) apply approximate Bayesian computation to empirical data on the geographical distribution of bead types in European Aurignacian sites. The goal of this chapter is to test the hypothesis, proposed by Vanhaeren and d'Errico (2006) based on this data, that these bead types had a symbolic meaning and served as markers of ethnic identity. Agents in the simulation model of Kovacevic et al. are mobile Aurignacian groups, which can undergo pairwise cultural interactions when in geographical proximity. Two cultural interaction processes are considered, "conflict" and "sharing." In the case of conflict, the bead types of the losing group are completely replaced by those of the winning group. Sharing entails the pooling and swapping of some bead types between the two groups.

If bead types are indeed ethnic markers, then—as Kovacevic et al. argue—two interacting groups that are relatively similar for bead types are more likely to share, whereas those that are relatively different are more likely to experience conflict (culture-dependent interaction model). On the other hand, if bead types have no such meaning, then the nature of the cultural interaction should not depend on these similarities or differences (null model). Kovacevic et al. find that the best fits of the simulated to the observed data are equally likely under the null model as the culture-dependent model. Thus, their study does not support the hypothesis that the Aurignacian was ethnically structured, at least in the sense that different bead types were used symbolically to mark ethnic identity.

In Chap. 9, O'Brien et al. provide an excellent summary of the variety of learning strategies discussed in the literature and of the well-studied archaeology of Palaeolithic North America. Among these various learning strategies, they focus on "guided variation" and "indirectly-biased" social learning as likely candidates for the Early Paleoindian period. Guided variation is a learning strategy in which unbiased social learning is followed by individual learning that targets the environmentally optimal behavior (Boyd and Richerson 1985). Indirect bias entails that an individual perceived to be successful or prestigious is preferentially copied. Whereas the former results in adaptive cultural change, the latter may not do so unless success or prestige is correlated with biological fitness.

O'Brien et al. review several recent studies (Morrow and Morrow 1999; Buchanan and Hamilton 2009; Hamilton and Buchanan 2009; Sholts et al. 2012; Buchanan et al. 2014) that ask whether the geographical variation in Clovis points is due to regional adaptation by guided variation or other factors such as random drift associated with indirectly-biased social learning. They argue that different learning strategies may be applied to different aspects of stone-tool production, specifically that patterns of flake removal may have been determined by prestige-biased social learning from skilled craftsmen, whereas point shape was determined by guided variation. The greater regional variation observed for point shape is explained by this difference in learning strategies.

Our final two chapters focus on laboratory experiments as a means of better understanding learning strategies, beginning with Caldwell (Chap. 10). The distinguishing feature of the culture of modern humans as opposed to non-human animal cultures—to a certain extent, perhaps also Neanderthal culture—is that it is cumulative. Thus, as Tomasello (1999, p. 512) explains, "[t]he most distinctive characteristic of human cultural evolution is the way that modifications to a cultural artifact or a social practice made by one individual or group of individuals often spread within the group, and then stay in place until some future individual or individuals make further modifications."

Caldwell reviews her work on transmission-chain laboratory experiments (e.g. Caldwell and Millen 2008a, b, 2009, 2010; Caldwell et al. 2012), which ask what unique aspects of cognition and social learning in present-day humans make cumulative cultural change possible. One prevailing view holds that only imitation, i.e. "process-oriented" or "action-copying" social learning, can support high-fidelity social learning, which is a necessary condition for cumulative cultural change. Against this, the experiments tested whether emulation, i.e. "goal-oriented" or "results-feedback" social learning, might also qualify.

Participants were given two tasks, building a paper aeroplane scored for flight distance and building a spaghetti tower scored for height. The experimental conditions differed in the type of information made available to the participants, specifically the actions used by the antecedents in the transmission chain to execute the tasks and/or the results (i.e. products) of these actions. Caldwell and coworkers found that high-fidelity social learning occurred even when participants were permitted to observe only the results. Moreover and consequently, performance was improved over the "generations" of the transmission chain. Teaching was also found to be effective, independently of imitation and emulation. However, as Caldwell notes, these experiments do not address the role of innovation, which is clearly also a necessary condition for cumulative cultural change (Borenstein et al. 2008). Furthermore, the tasks used are rather simple compared to even very early hominin technology such as flint-knapped handaxes, for which imitation may well be more important.

Lycett et al. (Chap. 11) review their work on transmission-chain experiments conducted in the laboratory (Kempe et al. 2012; Schillinger et al. 2014). They emphasize the importance of laboratory experiments in understanding how the dynamics of micro-evolutionary processes affect artefactual variation, to produce the macro-patterns seen in the archaeological record. Specifically, their interest focuses on elucidating the relation between copying error, either deliberate or unintentional, and cultural variation among populations.

In the first experiment (Kempe et al. 2012), each participant was asked to view an image of a handaxe drawn by his/her antecedent in the transmission chain and to faithfully copy its size. Observed copying error averaged 3.43 %, which is consistent with the value of 3 % reported in the psychophysical literature. Moreover, the variance among transmission chains increased as expected over the ten generations of the experiment. Then, individual based simulations were conducted to predict the variance of handaxe length and breadth after 200 generations. The predicted variance was far greater than that observed in a sample of more than 2,000 Acheulean handaxes from 21 sites spanning 1.2 million years, suggesting certain undetermined factors countermanding the effects of copying error.

In the second experiment (Schillinger et al. 2014), participants were asked to make a replica handaxe from plasticine using a knife. Two experimental conditions were compared: reductive only in which material may be removed but not added as is usually the case in stone-tool knapping, and additive-reductive in which both actions are permitted as in the manufacture of pottery. As expected, copying error was statistically greater under the former condition. These simple experiments show, using a "model artifact" for culture—in this case, handaxes—equivalent to "model organisms" in experimental population genetics, how features of perceptual systems and manufacturing techniques can generate systematic patterns in cultural datasets.

1.4 Concluding Remarks

The goal of this volume is to show how learning strategies, in combination with social and demographic factors, predict cultural change and cultural diversity during the Palaeolithic. Also addressed is the more challenging inverse problem of how learning strategies may be inferred from the sketchy archaeological record. Towards this end, our contributors have applied a diverse range of scientific methodologies, including mathematical models derived from population genetics, spatially explicit agent-based models, approximate Bayesian computation, and transmission chain laboratory experiments.

Our underlying motivation is to explain why and how Neanderthals were replaced (or assimilated) by modern humans, in terms of cultural differences between the two (sub-)species. This is an entirely reasonable premise, given that replacements in historical times were likely driven by cultural differences—and the demographic or disease-related corollaries of such cultural differences—between competing ethnic groups (Diamond 1997). Moreover, as recently as 2003, leading anthropologists noted that "[t]he ultimate mechanism for this replacement is widely considered to be a behavioral difference between non-modern and modern populations that lent an adaptive advantage to moderns" (Henshilwood and Marean 2003, p. 627). The theory described in this volume helps us to understand how various factors—innovativeness, biases and error associated with social learning, population size and structure, residential and logistical mobility, to name a few—influence the amount, complexity, and geographic variation of culture.

However, based on intensive archaeological work during the past decade, it has been suggested that contemporaneous Neanderthals and modern humans—in particular in Europe where the two (sub-)species apparently overlapped for several thousand years (Higham et al. 2014)—both exhibited most, perhaps all, "modern behaviors" including symbolic behavior (d'Errico and Stringer 2011; Zilhão 2013). See, for example, Table 3 of McBrearty and Brooks (2000) for a comprehensive list of modern behaviors, initially believed to constitute the distinguishing features of modern humans. From the standpoint of the cognitive equality of Neanderthals and modern humans, the possible "nail in the coffin" is the recent report of a rock engraving made by pre-contact Neanderthals in Gibraltar (Rodríguez-Vidal et al. 2014).

On the other hand, we must not forget that there was much variability both among Neanderthal regional groups and among modern human regional groups. Competition resulting in replacement would likely have occurred on a circumscribed geographical scale, not between the "champions" on either side. Hence, if Neanderthal regional cultures were *on average* slightly inferior to modern human regional cultures—perhaps reflecting a small (sub-)specific difference in cognition or in demography—then the premise that cultural differences contributed to replacement, which we subscribe to, would still be tenable.

References

Akazawa T, Nishiaki Y, Aoki K (2013) Dynamics of learning in Neanderthals and modern humans vol 1 cultural perspectives. Springer, Tokyo

Aoki K (2013) Determinants of cultural evolutionary rates. In: Akazawa T, Nishiaki Y, Aoki K (eds) Dynamics of learning in Neanderthals and modern humans vol 1 cultural perspectives. Springer, Tokyo, pp 199–210

Aoki K (2015) Modeling abrupt cultural regime shifts during the Palaeolithic and Stone Age. Theor Popul Biol 100:6–12

Aoki K, Feldman MW (2014) Evolution of learning strategies in temporally and spatially variable environments: a review of theory. Theor Popul Biol 91:3–19

Aoki K, Lehmann L, Feldman MW (2011) Rates of cultural change and patterns of cultural accumulation in stochastic models of social transmission. Theor Popul Biol 79:192–202

Axelrod R (1997) The dissemination of culture: a model with local convergence and global polarization. J Confl Resolut 41:203–226

Bar-Yosef O, Wang YP (2012) Paleolithic archaeology in China. Annu Rev Anthropol 41:319–335

Beaumont MA, Zhang W, Balding DJ (2002) Approximate Bayesian computation in population genetics. Genetics 162:2025–2035

Binford LR (1980) Willow smoke and dogs' tails: hunter-gatherer settlement systems and archaeological site formation. Am Antiq 45:4–20

Borenstein E, Feldman MW, Aoki K (2008) Evolution of learning in fluctuating environments: when selection favors both social and exploratory individual learning. Evolution 62:586–602

Boyd R, Richerson PJ (1985) Culture and the evolutionary process. University Chicago Press, Chicago

Buchanan B, Hamilton MJ (2009) A formal test of the origin of variation in North American early Paleoindian projectile points. Am Antiq 74:279–298

Buchanan B, O'Brien MJ, Collard M (2014) Continent-wide or region-specific? A geometric-morphometrics-based assessment of variation in Clovis point shape. Archaeol Anthropol Sci 6:145–162

Caldwell CA, Millen AE (2008a) Studying cumulative cultural evolution in the laboratory. Philos Trans R Soc B 363:3529–3539

Caldwell CA, Millen AE (2008b) Experimental models for testing hypotheses about cumulative cultural evolution. Evol Hum Behav 29:165–171

Caldwell CA, Millen AE (2009) Social learning mechanisms and cumulative cultural evolution: is imitation necessary? Psychol Sci 20:1478–1483

Caldwell CA, Millen AE (2010) Conservatism in laboratory microsocieties: unpredictable payoffs accentuate group-specific traditions. Evol Hum Behav 31:123–130

Caldwell CA, Schillinger K, Evans CL, Hopper LM (2012) End state copying by humans (*Homo sapiens*): implications for a comparative perspective on cumulative culture. J Comp Psychol 126:161–169

Caro TM, Hauser MD (1992) Is there teaching in nonhuman animals? Q Rev Biol 67:151–174

Cavalli-Sforza LL, Feldman MW (1981) Cultural transmission and evolution. Princeton University Press, Princeton

Clark JL (2011) The evolution of human culture during the later Pleistocene: using fauna to test models on the emergence and nature of "modern" human behavior. J Anthropol Archaeol 30:273–291

Collard M, Kemery M, Banks S (2005) Causes of toolkit variation among hunter-gatherers: a test of four competing hypotheses. Can J Archaeol 29:1–19

Collard M, Ruttle A, Buchanan B, O'Brien MJ (2013) Population size and cultural evolution in nonindustrial food-producing societies. PLoS One 8:e72628

Derex M, Beugin M-P, Godelle B, Raymond M (2013) Experimental evidence for the influence of group size on cultural complexity. Nature 503:389–391

d'Errico F, Stringer CB (2011) Evolution, revolution or saltation scenario for the emergence of modern cultures? Philos Trans R Soc B 366:1060–1069

Diamond J (1997) Guns, germs, and steel. Norton, New York

Enquist M, Ghirlanda S, Errikson K (2011) Modelling the evolution and diversity of cumulative culture. Philos Trans R Soc B 366: 412–423

Fogarty L, Strimling P, Laland KN (2011) The evolution of teaching. Evolution 65:2760–2770

Hamilton MJ, Buchanan B (2009) The accumulation of stochastic copying errors causes drift in culturally transmitted technologies: quantifying Clovis evolutionary dynamics. J Anthropol Archaeol 28:55–69

Henrich J (2004) Demography and cultural evolution: how adaptive cultural processes can produce maladaptive losses—the Tasmanian case. Am Antiq 69:197–214

Henrich J, Boyd R (1998) The evolution of conformist transmission and the emergence of between-group differences. Evol Hum Behav 19:215–241

Henrich J, Gil-White FJ (2001) The evolution of prestige: freely conferred deference as a mechanism for enhancing the benefits of cultural transmission. Evol Hum Behav 22:165–196

Henshilwood CS, Marean CW (2003) The origin of modern human behavior. Curr Anthropol 44:627–651

Hewlett B, van de Koppel JMH, Cavalli-Sforza LL (1982) Exploration ranges of Aka Pygmies of the Central African Republic. Man 17:418–430

Higham T, Douka K, Wood R, Bronk Ramsey C et al (2014) The timing and spatiotemporal patterning of Neanderthal disappearance. Nature 512:306–309

Kelly RL (1983) Hunter-gatherer mobility strategies. J Anthropol Res 39:277–306

Kempe M, Mesoudi A (2014) An experimental demonstration of the effect of group size on cultural accumulation. Evol Hum Behav 35:285–290

Kempe M, Lycett SJ, Mesoudi A (2012) An experimental test of the accumulated copying error model of cultural mutation for Acheulean handaxe size. PLoS One 7:e48333

Kendal J, Giraldeau L-A, Laland K (2009) The evolution of social learning rules: payoff-biased and frequency-dependent biased transmission. J Theor Biol 260:210–219

Klein RG (2008) Out of Africa and the evolution of human behavior. Evol Anthropol 17:267–281

Klein RG, Steele TE (2013) Archaeological shellfish size and later human evolution in Africa. Proc Natl Acad Sci U S A 110:10910–10915

Kline MA, Boyd R (2010) Population size predicts technological complexity in Oceania. Proc R Soc B 277:2559–2564

Kobayashi Y, Aoki K (2012) Innovativeness, population size and cumulative cultural evolution.Theor Popul Biol 82:38–47

Kuhn SL (2012) Emergent patterns of creativity and innovation in early technologies. In: Elias S (ed) Origins of human innovation and creativity. Elsevier, Amsterdam, pp 69–86

Kuhn SL (2013) Cultural transmission, institutional continuity and the persistence of the Mousterian. In: Akazawa T, Nishiaki Y, Aoki K (eds) Dynamics of learning in Neanderthals and modern humans vol 1 cultural perspectives. Springer, Tokyo, pp 105–113

Laland KN (2004) Social learning strategies. Learn Behav 32:4–14

Lehmann L, Aoki K, Feldman MW (2011) On the number of independent cultural traits carried by individuals and populations. Philos Trans R Soc B 366:424–435

McBrearty S, Brooks AS (2000) The revolution that wasn't: a new interpretation of the origin of modern human behavior. J Hum Evol 39:453–563

Mellars P, French JC (2011) Tenfold population increase in western Europe at the Neandertal-to-modern human transition. Science 333:623–627

Mesoudi A (2011) Variable cultural acquisition costs constrain cumulative cultural evolution. PLoS One 6:1–10

Mesoudi A, O'Brien MJ (2008) The learning and transmission of hierarchical cultural recipes. Biol Theory 3:63–72

Morrow JE, Morrow TA (1999) Geographic variation in fluted projectile points: a hemispheric perspective. Am Antiq 64:215–231

Muthukrishna M, Shulman BW, Vasilescu V, Henrich J (2014) Sociality influences cultural complexity. Proc R Soc B 281: 20132511

Nakahashi W (2007) The evolution of conformist transmission in social learning when the environment changes periodically. Theor Popul Biol 72:52–66

Nakahashi W, Wakano JY, Henrich J (2012) Adaptive social learning strategies in temporally and spatially varying environments. Hum Nat 23:386–418

Norton CJ, Jin JH (2009) The evolution of modern human behavior in east Asia: current perspectives. Evol Anthropol 18:247–260

Perreault C, Brantingham PJ (2011) Mobility-driven cultural transmission along the forager-collector continuum. J Anthropol Archaeol 30:62–68

Powell A, Shennan S, Thomas MG (2009) Late Pleistocene demography and the appearance of modern human behavior. Science 324:1298–1301

Prüfer K, Racimo F, Patterson N, Jay F et al (2014) The complete genome sequence of a Neanderthal from the Altai mountains. Nature 505:43–49

Read D (2006) Tasmanian knowledge and skill: maladaptive imitation or adequate technology? Am Antiq 71:164–184

Richerson PJ, Boyd R (2013) Rethinking paleoanthropology: a world queerer than we supposed. In: Hatfield G, Pittman H (eds) Evolution of mind, brain and culture. University of Pennsylvania Press, Philadelphia, pp 263–302

Richerson PJ, Boyd R, Bettinger RL (2009) Cultural innovations and demographic change. Hum Biol 81:211–235

Rodríguez-Vidal J, d'Errico F, Pacheco FG, Blasco R et al (2014) A rock engraving made by Neanderthals in Gibraltar. Proc Natl Acad Sci U S A 111:13301–13306

Schillinger K, Mesoudi A, Lycett SJ (2014) Copying error and the cultural evolution of "additive" versus "reductive" material traditions: an experimental assessment. Am Antiq 79:128–143

Shennan S (2001) Demography and cultural innovation: a model and its implications for the emergence of modern human culture. Camb Archaeol J 11:5–16
Sholts SB, Stanford DJ, Flores LM, Wärmländer SKTS (2012) Flake scar patterns of Clovis points analyzed with a new digital morphometrics approach: evidence for direct transmission of technological knowledge across early North America. J Archaeol Sci 39:3018–3026
Shott M (1986) Technological organization and settlement mobility: an ethnographic examination. J Anthropol Res 42:15–51
Strimling P, Sjöstrand J, Enquist M, Errikson K (2009) Accumulation of independent cultural traits. Theor Popul Biol 76:77–83
Tomasello M (1999) The human adaptation for culture. Ann Rev Anthropol 28:509–529
Torrence R (1983) Time budgeting and hunter-gatherer technology. In: Bailey G (ed) Hunter-gatherer economy in prehistory. Cambridge University Press, Cambridge, pp 11–22
Vanhaeren M, d'Errico F (2006) Aurignacian ethno-linguistic geography of Europe revealed by personal ornaments. J Archaeol Sci 33:1105–1128
Vanhaeren M, d'Errico F, Stringer C, James SL et al (2006) Middle Paleolithic shell beads in Israel and Algeria. Science 312:1785–1788
Wakano JY, Aoki K (2007) Do social learning and conformist bias coevolve? Henrich and Boyd revisited. Theor Popul Biol 72:504–512
Whiten A, Goodall J, McGrew WC, Nishida T et al (1999) Cultures in chimpanzees. Nature 399:682–685
Zilhão J (2013) Neandertal-modern human contact in western Eurasia: issues of dating, taxonomy, and cultural associations. In: Akazawa T, Nishiaki Y, Aoki K (eds) Dynamics of learning in Neanderthals and modern humans vol 1 cultural perspectives. Springer, Tokyo, pp 21–57
Zilhão J, Angelucci DE, Badal-Garcia E, d'Errico F et al (2010) Symbolic use of marine shells and mineral pigments by Iberian Neandertals. Proc Natl Acad Sci U S A 107:1023–1028

Factors Limiting the Number of Independent Cultural Traits That Can Be Maintained in a Population

2

Laurel Fogarty, Joe Yuichiro Wakano, Marcus W. Feldman, and Kenichi Aoki

Abstract

We obtain the theoretically expected number of independent cultural traits at equilibrium in a population where one of four modes of social learning—random oblique, best-of-*K*, success bias, or one-to-many—is practiced by its members. Cultural traits can be classified as simple or complex, depending on the ease or difficulty of acquisition by social learning and innovation. The number of simple cultural traits may saturate as population size increases, in which case a statistical association between the two variables is not predicted. At smaller population sizes, there is a major effect of the mode of social learning on the number of simple cultural traits. By contrast, the relation between the number of complex cultural traits and population size is approximately linear and almost identical for all four modes of social learning. We suggest that empirical studies of statistical association between number of cultural traits and population size should distinguish between simple and complex cultural traits.

Keywords

Cultural Moran model • Social learning • Innovation • Population size • Imagination space • Simple cultural trait • Complex cultural trait

2.1 Introduction

Beginning with the work of Cavalli-Sforza and Feldman (1973a, b, 1981), numerous theoretical models have been proposed for the quantitative study of cultural evolution. These models differ in how cultural variation among individuals is represented. Recent theoretical investigations of *long-term* cultural evolution have employed three approaches. The first measures the "skill" of an individual by one continuous variable, which may be regarded as the cumulative effect of the variants of many cultural traits carried by that individual (e.g. Henrich 2004; Powell et al. 2009; Kobayashi and Aoki 2012). The second assigns to each individual a "vector" (a one-dimensional array), the elements of which represent different cultural traits, and where any number of variants is permitted for each cultural trait (e.g. Shennan 2001; Mesoudi 2011). The third is a simplified version of the second, in which only the possession or non-possession of a cultural trait is noted (e.g. Strimling et al. 2009; Lehmann et al. 2011; Aoki et al. 2011; Aoki 2013).

Ethnography provides us with comparative data from numerous societies on the presence or absence of various cultural traits, including tool types and linguistic features. Such data have been used in recent statistical analyses of

L. Fogarty (✉) • M.W. Feldman
Department of Biology, Stanford University, Stanford, CA 94305, USA
e-mail: lfogarty@stanford.edu; mfeldman@stanford.edu

J.Y. Wakano
School of Interdisciplinary Mathematical Sciences, Meiji University, Nakano 4-21-1, Nakano-ku, Tokyo 164-8525, Japan
e-mail: joe@meiji.ac.jp

K. Aoki
Organization for the Strategic Coordination of Research and Intellectual Properties, Meiji University, Nakano 4-21-1, Nakano-ku, Tokyo 164-8525, Japan
e-mail: kenaoki@meiji.ac.jp

A. Mesoudi and K. Aoki (eds.) *Learning Strategies and Cultural Evolution during the Palaeolithic*, Replacement of Neanderthals by Modern Humans Series, DOI 10.1007/978-4-431-55363-2_2,

cultural evolution to estimate the cultural evolutionary rate (e.g. Rogers and Ehrlich 2008; Rogers et al. 2009), to determine whether certain cultural traits are co-transmitted (e.g. Brown and Feldman 2009; Jordan and Shennan 2009; Jordan and O'Neill 2010), and to examine the possibility of the serial founder effect (e.g. Atkinson 2011). Other statistical studies have investigated the effects of various environmental, demographic, and behavioral factors on the number of food-getting tool types in use in subsistence societies (Collard et al. 2005, 2011, 2012, 2013a, b, c; Read 2006). In these studies, the "cultural state" of a society is represented, explicitly or implicitly, in terms of a vector of 1's and 0's which denote the presence and absence, respectively, of the relevant cultural traits in that society. This is consistent with the third modeling approach described above, although variation among individuals within a society is not distinguished in these empirical studies. We note here an ambiguity in the concept of a "cultural trait," a matter we take up briefly in the Discussion.

The objective of this paper is to further develop a framework for modeling the evolution of culture-as-a-0, 1-vector. We show how processes at the individual level such as social learning and innovation, in conjunction with demographic factors such as population size, determine the extent of cultural diversity within a population. In particular, we wish to compare the effects of different modes of social learning (transmission)—specifically, random oblique, direct bias, success bias, and one-to-many—which have been argued to be important (Guglielmino et al. 1995; Henrich 2004; Lycett and Gowlett 2008; Powell et al. 2009; Mesoudi 2011). Various measures of within-population cultural diversity exist, e.g. the total number of cultural traits in the population, the average number of cultural traits carried by an individual, the fraction of cultural traits that are shared between individuals, etc. Here, we focus on the first measure, because it is directly relevant to the statistical studies on the number of tool types mentioned above. It may perhaps also be useful for our ultimate goal of interpreting the archaeological record of Palaeolithic technologies (e.g. Perreault et al. 2013) and of explaining the demise of the Neanderthals and their replacement by modern humans.

Several theoretical studies have already addressed this question (Strimling et al. 2009; Lehmann et al. 2011; Aoki et al. 2011), but they do not fully explore the detailed relationships among population size, innovation regimes, and trait simplicity/complexity. In addition, these papers may contain an unnatural assumption with regard to who innovates—empirically speaking, innovations are attributable to adults or juveniles (e.g. Reader and Laland 2001; Morand-Ferron et al. 2011; Hewlett 2013), but not newborn infants—which we rectify (see below). On the other hand, we continue, as in these previous studies, to ignore natural selection and to assume that cultural traits are independently transmitted. These simplifying conditions clearly restrict the applicability of our results and moreover have been shown in some instances not to apply. For example, Brown and Feldman (2009), Jordan and Shennan (2009), and Jordan and O'Neill (2010) provide evidence for the associated transmission of some, but not all, cultural traits. Contingent transmission of cultural traits has been modeled in a theoretical paper by Mesoudi (2011).

In the present paper, we first adapt the model of Strimling et al. (2009) to analytically derive a recursion for the total number of cultural traits in the population, which can be iterated numerically to yield a solution in the case of direct bias as well as for random oblique social learning. Then, we conduct extensive Monte Carlo individual-based simulations. Some of the results we obtain are at first sight counterintuitive or inconsistent with established views. For example, we find that the proposal by Henrich (2004) that "complex skills" are more easily lost than "simple skills" when population size declines is only partially supported.

In implementing the simulations, we have set an upper limit to the possible number of cultural traits, which we call the "imagination space." Although introduced for the sake of convenience, the imagination space of cultural traits may in reality be bounded due to innate constraints, or alternatively be dependent on the cultural level that has already been attained—an example of cultural niche construction (Odling-Smee et al. 2003; Ihara and Feldman 2004). Not surprisingly, we find that imagination space has a profound effect on cultural diversity.

2.2 Model

Let us assume a finite population comprising N individuals and in which a potentially infinite number of cultural traits may occur. A particular cultural trait may or may not be found in the population at any one time. Moreover, even when it is, an individual member of this population may or may not carry it. We will use the terms 1-variant and 0-variant to denote the presence and absence, respectively, of a cultural trait at the individual and population levels. Each individual can then be represented by a vector of 1's and 0's, and the population by a matrix formed by aligning N such vectors. Thus, the ijth element of this matrix, which we write C_{ij}, is equal to 1 if the jth cultural trait carried by the ith individual is the 1-variant, and 0 otherwise.

The cultural dynamics are defined by events occurring during one time step, which are innovation, birth, social learning (SL), death and replacement, in that order. First, the N individuals alive at the beginning of a time step, whom we call adults, innovate according to rules described later. Second, one individual is born and is considered to have the 0-variant of each cultural trait. Third, this newborn acquires either the 1-variant or the 0-variant of each extant cultural trait by SL from the adults. Fourth, a random adult, i.e. an individual other than the newborn, dies and is replaced by the newborn (except in the case of one-to-many SL, as described below, when it is the expert who dies). The second through fourth events constitute the birth-death version of the cultural Moran model (Strimling et al. 2009; Lehmann et al. 2011; Aoki et al. 2011; Aoki 2013).

Since deaths occur at random with probability $1/N$ per adult per time step, the life expectancy of a newborn is N time steps. Hence, we can regard N time steps as equivalent in duration to one generation (Gale 1990). This does not mean that the lifespan of an individual increases linearly with population size. Rather, the correct interpretation is that the interval between successive births decreases in inverse proportion to the population size. One virtue of the cultural Moran model is that the generations are naturally overlapping—e.g. an individual can be alive at the same time as his/her parents and/or children—which provides a closer approximation to hominin demography than earlier models of cultural transmission that assumed discrete non-overlapping generations.

Innovation of a cultural trait is defined as the spontaneous acquisition of the 1-variant by an individual who carries the 0-variant. We assume that all N adults may innovate during a time step. This differs from most previous models in which only the one newborn was allowed to innovate (e.g. Strimling et al. 2009; Lehmann et al. 2011). In addition, for the Monte-Carlo individual-based simulations (IBS, see below), we consider two different types of innovation processes, infinite-sites and recurrent. Infinite-sites innovation entails that an innovating adult may only acquire the 1-variant of a cultural trait that is absent from the population and that has not been acquired by another innovating adult during the same time step. Hence, after all N adults have had a change to innovate, each new 1-variant necessarily exists in single copy. With recurrent innovation, on the other hand, an innovating adult may acquire any 1-variant he/she does not currently carry, regardless of whether that 1-variant is carried by other adults.

The infinite-sites assumption may appear to be a restrictive way of modeling innovation. However, it is useful in deriving analytical results, and we make this assumption to do so (see Aoki et al. 2011 and below). Moreover, provided the innovation rate is low, the IBS suggest that the cultural dynamics apparently do not differ greatly whether we assume infinite-sites or recurrent innovation. On the other hand, we observe some interesting differences in the outcomes when the innovation rate is high.

The innovation rate, which we denote by μ, is defined per individual per generation/lifespan. Since the life cycle is described per time step, we also require the innovation rate per time step, which is μ/N for each individual and μ for the population. The reason we choose the generation rather than the time step to be our basic unit of time is that the former is biologically relevant whereas the latter is a mathematical construct.

We consider four different modes of SL by the newborn. In a "real-life" SL situation, these different modes of SL are likely to be combined. For example, the "expert" in one-to-many SL may also be the most "successful" individual who carries the greatest number of 1-variants. Nevertheless, we believe it is useful to consider the separate effect(s) of each mode of SL in isolation. See Aoki et al. (2011) and Aoki (2013) for details.

Random Oblique The newborn samples one adult at random from among the N adults. The 1-variant of each cultural trait is independently acquired from this exemplar adult with probability β, which we also refer to as the efficiency of SL.

Best-of-K The newborn samples K adults at random without replacement. The 1-variant of each cultural trait is independently acquired with probability β, provided at least one of the K sampled exemplar adults carries the 1-variant.

Success Bias The newborn samples K adults at random without replacement and then chooses one adult with the greatest number of 1-variants from among these K adults to be his/her exemplar. That is, we assume that the possession of cultural traits (some of which may be material goods) is regarded as an indicator of success, whether or not they contribute to reproductive success. The 1-variant of each cultural trait is independently acquired with probability β from this exemplar adult.

One-to-Many One adult has the special status of expert and retains this special status until death. The newborn acquires the 1-variant of each cultural trait independently with probability β from this expert. When the expert dies, which occurs with probability $1/N$ per time step, he/she is replaced by another adult or by the newborn. When a non-expert dies, he/she is replaced by the newborn, as with the other modes of SL.

Finally, for all four modes of SL, the default trait variant is the 0-variant.

In the above description of SL, it should be noted that the probability of acquisition of the 1-variant of a cultural trait, β, is defined per time step, in contrast to the innovation rate, μ, which has been defined per generation. If the generation length is fixed, then the duration of a time step, and hence the time available for SL, is expected to decrease as the population size, N, increases. However, our model does not take this into account and in effect assumes that SL is instantaneous. We assume $0 < \beta < 1$.

The variable of interest for us in this paper is the number of *different* cultural traits that can be found in the population at any one time, i.e. the number of cultural traits for which at least one individual carries the 1-variant, which we denote by C_{pop}. In terms of the matrix elements, C_{ij}, we can define C_{pop} as follows. Let $\chi_j = 1$ if $\sum_{i=1}^{N} C_{ij} > 0$, and $\chi_j = 0$ if $\sum_{i=1}^{N} C_{ij} = 0$. Then, $C_{pop} = \sum_j \chi_j$. We have followed previous theoretical work (e.g. Strimling et al. 2009; Lehmann et al. 2011; Aoki et al. 2011) in adopting this definition of C_{pop}, which is also implicit in the models of Shennan (2001) and Mesoudi (2011). However, it may not agree with the operational definition for identifying a cultural behavior in non-humans primates, which requires that it occur in several individuals (e.g. Whiten et al. 1999). We take this point up again very briefly in the Discussion.

We investigate how the expected value of this variable at equilibrium, $\hat{\overline{C}}_{pop}$ (the bar over the variable indicates that it is the expected value, whereas the hat indicates that it is evaluated at equilibrium), depends on the mode and efficiency of SL, the innovation rate, the imagination space (see below), and the population size. Other important measures of cultural diversity are the number of cultural traits carried by an individual, C_{ind}, and the number of cultural traits that are shared between individuals (see companion paper by Fogarty et al.). The reason we focus here on C_{pop} is that it is empirically most relevant to the ethnographic and archaeological records.

2.3 An Analytical Result

We adapt the model of Strimling et al. (2009) to develop an alternative derivation of $\hat{\overline{C}}_{pop}$ that allows us to obtain $\hat{\overline{C}}_{pop}$ for best-of-*K* as well as random oblique SL.

We make two independence assumptions. First, there is no association among the cultural traits carried by an individual; for example, the possession of the 1-variant of one cultural trait does not predict possession of the 1-variant of another cultural trait (analogous to the assumption of "linkage equilibrium" in genetics). Second, the cultural traits carried by an exemplar, or exemplars, are transmitted independently of each other to the newborn (analogous to "free recombination").

Let the variable P_i be the expected number of cultural traits at the beginning of a time step for which the 1-variant has "popularity" i in the population of N adults—i.e. has frequency i/N, where $1 \leq i \leq N$. Then $\overline{C}_{pop} = \sum_{i=1}^{N} P_i$. Immediately after infinite-sites innovation, this expected number becomes

$$P_i^* = P_i + \mu\delta_{i1} \tag{2.1}$$

where $\delta_{i1} = 1$ if $i = 1$ and 0 otherwise. For each such cultural trait, let b_i be the probability that the newborn acquires the 1-variant, and let d_i be the probability that death then strikes an adult carrying the 1-variant. Then, at the end of the time step, we have

$$P_i' = P_{i-1}^* b_{i-1}(1-d_{i-1}) + P_i^*\left[b_i d_i + (1-b_i)(1-d_i)\right] + P_{i+1}^*(1-b_{i+1})\, d_{i+1}. \tag{2.2}$$

A partial proof of Eq. (2.2) is given in the Appendix 1.

For random oblique SL (see Aoki et al. 2011), we set

$$b_i = \tfrac{i}{N}\beta,\ d_i = \tfrac{i}{N}. \tag{2.3}$$

For best-of-*K*, we set

$$b_i = \beta\left\{1 - \frac{\binom{N-i}{K}}{\binom{N}{K}}\right\},\ d_i = \tfrac{i}{N} \text{ for } K \leq N-i, \tag{2.4}$$

where $\binom{x}{y} = \dfrac{x(x-1)\cdots(x-y+1)}{y(y-1)\cdots 1}$, and $b_i = \beta$ for $K > N-i$ (see Aoki et al. 2011).

Setting $P_i = P_i' = \hat{P}_i$ in Eqs. (2.1) and (2.2), we obtain the nonhomogeneous second order difference equation that holds at equilibrium

$$\begin{aligned}&\hat{P}_{i-1}b_{i-1}(1-d_{i-1}) + \hat{P}_i(2b_i d_i - b_i - d_i) \\ &\quad + \hat{P}_{i+1}(1-b_{i+1})\, d_{i+1} \\ &= -\mu\delta_{i-1,1}b_{i-1}(1-d_{i-1}) - \mu\delta_{i1} \\ &\quad \left[b_i d_i + (1-b_i)(1-d_i)\right]\end{aligned} \tag{2.5}$$

where $b_0 = 0$ and $\hat{P}_{N+1} = 0$. We have been unable to derive the analytical solution of Eq. (2.5). Hence, we compute $\hat{P}_i$ by numerical iteration of Eqs. (2.1) and (2.2), with either Eqs. (2.3) or (2.4), from the initial conditions $P_i = 0$ for $1 \leq i \leq N$ until equilibrium is reached. Then, $\hat{\overline{C}}_{pop} = \sum_{i=1}^{N} \hat{P}_i$.

In the limiting case of $\beta = 0$, where SL is absent, Eq. (2.5) can be solved explicitly. We obtain $\hat{P}_1 = (N-1)\mu$ and $\hat{P}_i = 0$ for $2 \leq i \leq N$ (Appendix 2). Hence, in this case

$$\hat{\overline{C}}_{pop} = (N-1)\mu, \tag{2.6}$$

which is linear in N. When the efficiency of SL is low, i.e. when β is small, we will find below that Eq. (2.6) provides a good approximation.

2.4 Monte Carlo Individual-Based Simulations

Our principal results are obtained by Monte Carlo individual-based simulations (IBS). Implementing the IBS is for the most part straightforward, with three exceptions. First, whereas theory allows a potentially infinite number of cultural traits, we must set an upper limit, M, to this number in the IBS. We call M the "imagination space"—more precisely the size of the imagination space. Although parameter M is introduced for the sake of convenience, it may in fact have an important cognitive implication. For example, it may represent an innate bound to the number of cultural traits that can be imagined. We will find that assuming a finite number of cultural traits in the IBS may have a large effect on $\hat{C}_{pop}$.

Second, we need to carefully specify the details of the innovation process. The algorithm for infinite-sites innovation per time step is: (1) Choose an adult in random order who has not been chosen before. If all N adults have already been chosen, then exit loop. Else proceed. (2) Ask whether this adult innovates by comparing a uniformly-distributed pseudo-random number to the innovation rate. If innovation does not occur, return to step 1. If it does, proceed. (3) Choose a cultural trait at random that is fixed for the 0-variant in the population. If all such cultural traits have already been chosen, then exit loop. Else proceed. (4) Change this cultural trait in this adult from the 0-variant to the 1-variant. Return to step 1.

A slight modification of the above yields the algorithm for recurrent innovation per time step: (1) Choose an adult in random order who has not been chosen before. If all N adults have already been chosen, then exit loop. Else proceed. (2) Ask whether this adult innovates by comparing a uniformly-distributed pseudo-random number to the innovation rate. If innovation does not occur, return to step 1. If it does, proceed. (3) Choose a cultural trait at random for which this adult carries the 0-variant. If this adult carries the 1-variant of all cultural traits, then return to step 1. Else proceed. (4) Change this cultural trait in this adult from the 0-variant to the 1-variant. Return to step 1.

Third, we wish to estimate $\hat{C}_{pop}$, which is the mean of the stationary distribution, but this is difficult to do because of stochastic fluctuations. For each set of parameters, the IBS was run for 2,000 generations (i.e. $2{,}000 \times N$ time steps) from the initial conditions $C_{ij} = 0$ for $1 \leq i \leq N, 1 \leq j \leq M$. By visualizing the output of C_{pop}, we were able to determine that 1,000 generations was sufficient for directional changes to cease. Thus, the value of $\hat{C}_{pop}$ that we report is the average of C_{pop}, measured once every generation, between the 1,001st and 2,000th generations

2.5 Results for Random Oblique Social Learning

In Fig. 2.1a–d, we plot $\hat{C}_{pop}$ against N for random oblique SL.The solid line in each panel gives the analytical values—rigorously speaking, the values obtained by numerical iteration of Eqs. (2.1) and (2.2) with Eq. (2.3). The short-dash broken line gives the IBS values with infinite-sites innovation, and the long-dash-and-dot broken line gives the IBS values with recurrent innovation. In general, $\hat{C}_{pop}$ increases as N, β, or μ increase. In the IBS, we always set the size of the imagination space to $M = 500$.

We are particularly interested in two panels: Fig. 2.1a with parameter settings $\beta = 0.9$ and $\mu = 1$, and Fig. 2.1d with $\beta = 0.1$ and $\mu = 0.04$. The motivation for considering these two combinations of the efficiency of SL (β) and the innovation rate (μ) is that we are interested in the effect of population size (N) on the number of "simple" cultural traits and "complex" cultural traits that can be maintained at equilibrium. These two classes of cultural traits, which are analogous to the simple skills and complex skills of Henrich (2004), can be operationally defined by the efficiency/fidelity of SL and the ease of innovation, which are both expected to be high for the former ($\beta = 0.9$ and $\mu = 1$ as in Fig. 2.1a) and low for the latter ($\beta = 0.1$ and $\mu = 0.04$ as in Fig. 2.1d).

But before proceeding with this comparison, we note some important conclusions illustrated by these panels, some of them specific to random oblique SL. First, the analytical values of $\hat{C}_{pop}$ for random oblique SL are approximately linear in N. This property was previously noted and proved by Strimling et al. (2009) and by Lehmann et al. (2011), based on models of cultural evolution that differ slightly from the present one. The largest difference is the assumption by these authors that only the one newborn innovates. Let u be the innovation rate per time step for the newborn. Then, Strimling et al. (2009) show that (in our notation)

$$\hat{C}_{pop} \approx \frac{Nu}{\beta} \ln \frac{1}{1-\beta} + \frac{u}{1-\beta}, \tag{2.7}$$

which holds for large N and large β. Equation (2.7) would appear to give a slight overestimate relative to Eqs. (2.1) and (2.2) with Eq. (2.3), when we set $u = \mu$.

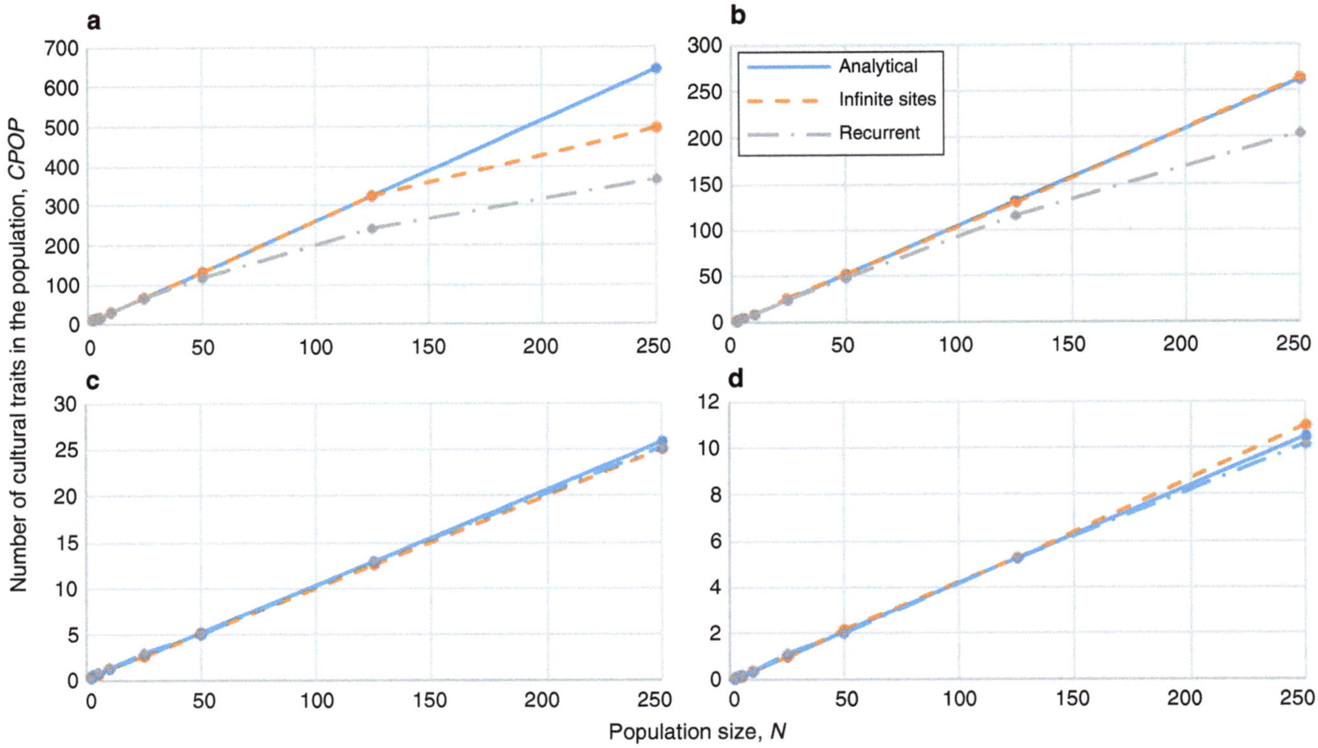

Fig. 2.1 Expected number of cultural traits at equilibrium, $\hat{\overline{C}}_{pop}$, is plotted against the population size, N, for the case of random oblique social learning. Solid (*blue*) line: analytical values. Short-dash broken (*orange*) line: individual-based simulation values with infinite-sites innovation. Long-dash-and-dot broken (*grey*) line: individual-based simulation values with recurrent innovation. **(a)** $\beta = 0.9, \ \ \mu = 1$. **(b)** $\beta = 0.1, \ \ \mu = 1$. **(c)** $\beta = 0.9, \ \ \mu = 0.04$. **(d)** $\beta = 0.1, \ \ \mu = 0.04$

Second, the IBS values agree well with the analytical values, obtained from Eqs. (2.1) and (2.2) with Eq. (2.3), when μ is small (Fig. 2.1c, d). For large values of μ, there is good agreement provided N is small (Fig. 2.1a, b). Recall that we assume independent transmission of the cultural traits, but not linkage equilibrium, in the IBS. Nevertheless, the IBS values are consistent with the analytical values. Thus, we suggest that the assumption of linkage equilibrium made to derive Eq. (2.2) may be justified in the case of random oblique SL.

Third, we observe a downward deviation of the IBS values from the analytical values, which is especially pronounced when β, μ, and N are all large (Fig. 2.1a). We attribute this to what we call the "saturation effect." For such combinations of parameter values, $\hat{\overline{C}}_{pop}$—also $\hat{\overline{C}}_{ind}$, the expected value of the average number of cultural traits carried by an individual at equilibrium—will approach the upper limit M. This entails that the number of cultural traits that are available for innovation will decrease, and hence the effective innovation rate will be smaller than the nominal value, μ. Interestingly, a close look at Fig. 2.1a shows that the saturation effect is more pronounced with recurrent innovation than with infinite-sites innovation; this is apparently true for all modes of SL considered in this paper, except perhaps one-to-many (below). This result is at first sight counterintuitive, because recurrent innovation can occur under more permissive conditions than infinite-sites innovation. Our explanation is that infinite-sites innovation, when it occurs, always adds to the number of segregating cultural traits, whereas recurrent innovation may not.

Let us now consider the claim made by Henrich (2004) that a decrease in population size should tend to result preferentially in the loss of complex skills. A comparison of Fig. 2.1a, d shows that this prediction is not upheld for the case of random oblique transmission. That is, the reduction in $\hat{\overline{C}}_{pop}$ when the population size decreases from $N = 100$ to $N = 50$, say, is larger for the simple cultural traits (Fig. 2.1a) than for the complex cultural traits (Fig. 2.1d), even when the saturation effect is taken into account.

2.6 Results for Best-of-2 Social Learning

Best-of-K is an example of direct bias. Here, the preference is assumed to be for having a cultural trait (carrying the 1-variant) rather than not having it (carrying the 0-variant).We find that there is a large increase in $\hat{C}_{pop}$ relative to random oblique SL, when a newborn is able to sample just 2 adults. Hence, in what follows, we report our results for best-of-2 SL.

In Fig. 2.2a–d, we plot $\hat{C}_{pop}$ against N for best-of-2 SL. The solid line in each panel gives the analytical values obtained from Eqs. (2.1) and (2.2) with Eq. (2.4), the short-dash broken line gives the IBS values with infinite-sites innovation, and the long-dash-and-dot broken line gives the IBS values with recurrent innovation. As before, we focus on the two panels, Fig. 2.2a with $\beta = 0.9$ and $\mu = 1$ illustrating the effect of population size on simple cultural traits, and Fig. 2.2d with $\beta = 0.1$ and $\mu = 0.04$, for complex cultural traits.

A striking feature of Fig. 2.2a is that the analytical values of $\hat{C}_{pop}$ "explode." That is, provided the imagination space is unlimited, best-of-2 SL—and a fortiori, best-of-K SL for $K \geq 3$—can support an extremely large number of simple cultural traits in the population. This property of best-of-K SL was previously noted by Aoki et al. (2011, see their Fig. 2) for a slightly different model. By contrast, the IBS values of $\hat{C}_{pop}$ are seen to rapidly approach the upper bound of $M = 500$, after which they remain independent of population size. Hence, the existence of a limit to imagination has a profound effect on the number of simple cultural traits that can be maintained by best-of-2 SL.

The inset shows that the IBS values with infinite-sites innovation agree with the analytical values when population size is small ($(2 \leq N \leq 4)$). Hence, the assumptions on which Eq. (2.2) is based would appear to hold, for best-of-2 as well as for random oblique SL, when the saturation effect can be neglected.

The situation for the complex cultural traits that we see in Fig. 2.2d is quite different. Here, we find that the analytical values of $\hat{C}_{pop}$ are approximately linear in N, and the IBS values coincide almost exactly with the analytical values. In particular, the saturation effect is not observed, at least in the range of population sizes examined. Comparison of Fig. 2.2b, c suggests that the approximate linearity of the analytical values follows from the assumption of β small rather than μ small.

Let us now confirm that Eq. (2.6) gives a good approximation when $\beta = 0.1$, as in Fig. 2.2b, d. In Fig. 2.2b where $\mu = 1$, the analytical value at $N = 250$ obtained from Eqs. (2.1) and (2.2) with Eq. (2.4) is $\hat{C}_{pop} = 278.0$, while Eq. (2.6) gives $249 \times 1 = 249$. Similarly, in Fig. 2.2d where $\mu = 0.04$, the analytical value is $\hat{C}_{pop} = 11.1$, and Eq. (2.6) gives $249 \times 0.04 = 9.96$. In each case the discrepancy is about 10 %.

With regard to the proposal by Henrich (2004) that complex skills are more easily lost than simple skills, Fig. 2.2a shows that it is a valid inference if the imagination space is limited. That is, the IBS values plateau at $\hat{C}_{pop} \approx 500$ for $N \geq 5$, so that a decrease of population size should not result in the reduction of the equilibrium number of simple cultural traits. On the other hand, a drastic reduction is predicted if the imagination space is unlimited.

2.7 Results for Success Bias Social Learning with $K = 2$

In Fig. 2.3a–d, we plot $\hat{C}_{pop}$ against N for success bias SL. To enable comparison with direct bias, which we have modeled as best-of-2, we assume that the newborn samples just two adults and chooses the one who is more "successful"—i.e. the one carrying more 1-variants of the cultural traits—to be his/her exemplar. The short-dash broken line gives the IBS values with infinite-sites innovation, and the long-dash-and-dot broken line gives the IBS values with recurrent innovation. Analytical values are not available for success bias SL.

Figure 2.3a shows the results with $\beta = 0.9$ and $\mu = 1$ for simple cultural traits. Compared to best-of-2, the values of $\hat{C}_{pop}$ do not rise as quickly as N increases; hence the saturation effect is not as pronounced. In fact, these values do not appreciably exceed the corresponding values for random oblique.

Figure 2.3d shows the results for complex cultural traits, where $\beta = 0.1$ and $\mu = 0.04$. The values of $\hat{C}_{pop}$ for infinite-sites and recurrent innovation are both approximately linear in N and nearly indistinguishable. Moreover, these plots apparently have the same slope as the corresponding plots for best-of-2.

2.8 Results for One-to-Many Social Learning

In Fig. 2.4a–d, we plot $\hat{C}_{pop}$ against N for one-to-many SL. The short-dash broken line gives the IBS values with infinite-sites innovation, and the long-dash-and-dot broken line gives the IBS values with recurrent innovation. Again, analytical values are not available.

In the case of simple cultural traits, comparison of Fig. 2.4a with Fig. 2.1a shows that $\hat{C}_{pop}$ is larger with one-to-many SL than with random oblique SL when N is small, with a major reversal as N increases. For example, with recurrent innovation, the respective values when $N = 250$ are $\hat{C}_{pop} = 220.4$ and $\hat{C}_{pop} = 365.6$ (see Fig. 2.5). This last result accords with intuition, since the effective

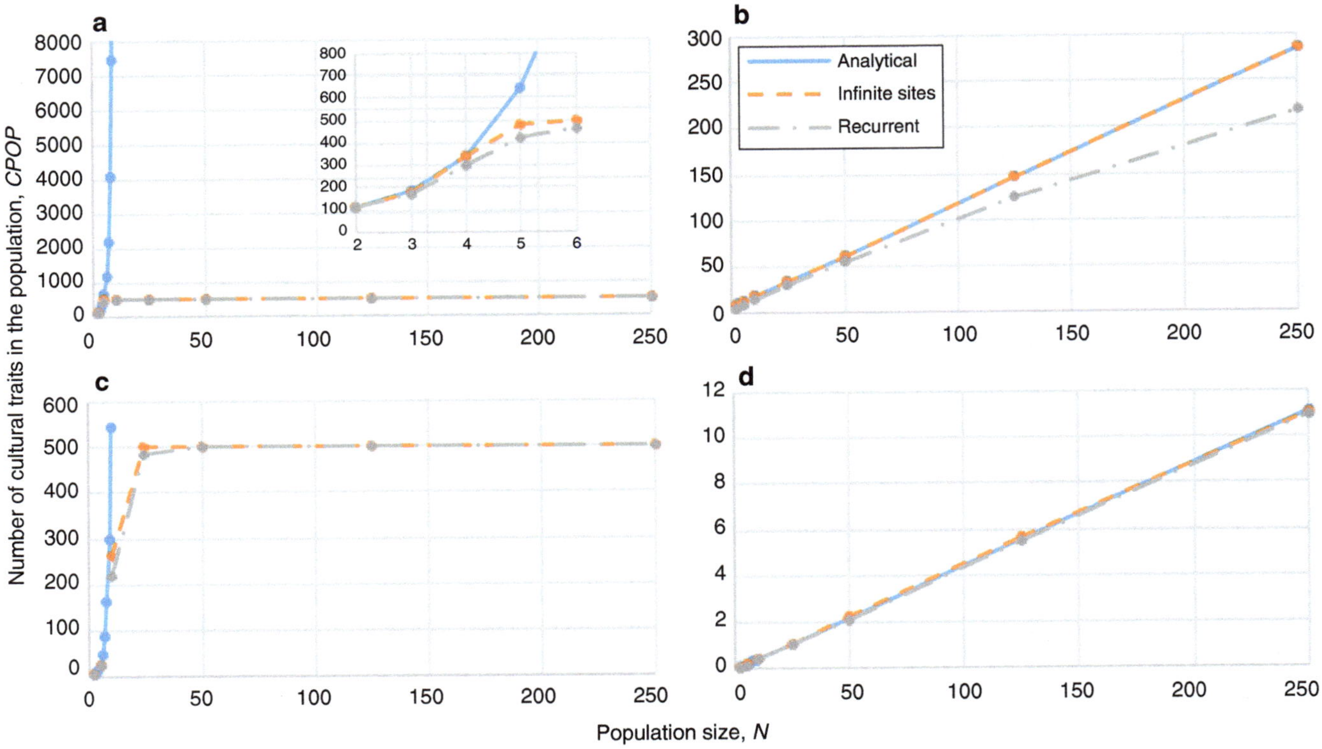

Fig. 2.2 Expected number of cultural traits at equilibrium, $\hat{\bar{C}}_{pop}$, is plotted against the population size, N, for the case of best-of-2 (direct bias) social learning. Solid (*blue*) line: analytical values. Short-dash broken (*orange*) line: individual-based simulation values with infinite-sites innovation. Long-dash-and-dot broken (*grey*) line: individual-based simulation values with recurrent innovation. (**a**) $\beta = 0.9$, $\mu = 1$. (**b**) $\beta = 0.1$, $\mu = 1$. (**c**) $\beta = 0.9$, $\mu = 0.04$. (**d**) $\beta = 0.1$, $\mu = 0.04$

population size for one-to-many SL is significantly smaller than for random oblique SL (Aoki et al. 2011).

Finally, we note from Fig. 2.4d that the plots of the IBS values are approximately linear in N and almost identical to what we observe in Figs. 2.1d, 2.2d, and 2.3d. That is, mode of SL appears to make little difference in the number of complex cultural traits that can be maintained. Hence, we suggest that mode of SL should be invoked with caution in any attempt to explain an observed variation among societies in the number of complex cultural traits.

2.9 Discussion

We have obtained theoretical predictions for the expected number of independent cultural traits that can be maintained in a population at equilibrium, $\hat{\bar{C}}_{pop}$, by a combination of analytical method and individual-based simulations (IBS). We have done this for random oblique SL, best-of-2 SL, success bias SL, and one-to-many SL, in order to compare the effects of these four different modes of SL. We have investigated how $\hat{\bar{C}}_{pop}$ depends on the efficiency of social learning (SL) by the newborn, β, the rate of innovation by older individuals, μ, and the population size, N. In addition to these three parameters, we have introduced in the IBS an upper bound on the number of imaginable cultural traits, M, which we call the imagination space. For the analytically-derived values of $\hat{\bar{C}}_{pop}$, we have assumed no such limit. It is an interesting and open question whether imagination space is indeed finite, and if so what its magnitude might be.

$\hat{\bar{C}}_{pop}$ is monotone non-decreasing in N for all modes of SL considered in this paper, for both the infinite-sites and recurrent innovation processes, and all combinations of the parameters β and μ. We say "non-decreasing" rather than "increasing," because when M is finite the values of $\hat{\bar{C}}_{pop}$ eventually saturate as N increases and plateau at M. We have set $M = 500$ in the IBS.

The class of cultural traits for which β and μ are both large may be called simple cultural traits, as they can be more easily acquired by social learning or by innovation (Henrich 2004). The number of simple cultural traits in the population saturate at smaller values of the population size than the number of complex cultural traits (see below), as can be seen from a comparison of Fig. 2.2a, d, for example (saturation has not occurred in the latter figure). Comparison of Figs. 2.1a, 2.2a, 2.3a, and 2.4a shows that saturation occurs first for best-of-2 SL, second for success bias SL, third for random oblique SL, and last for one-to-

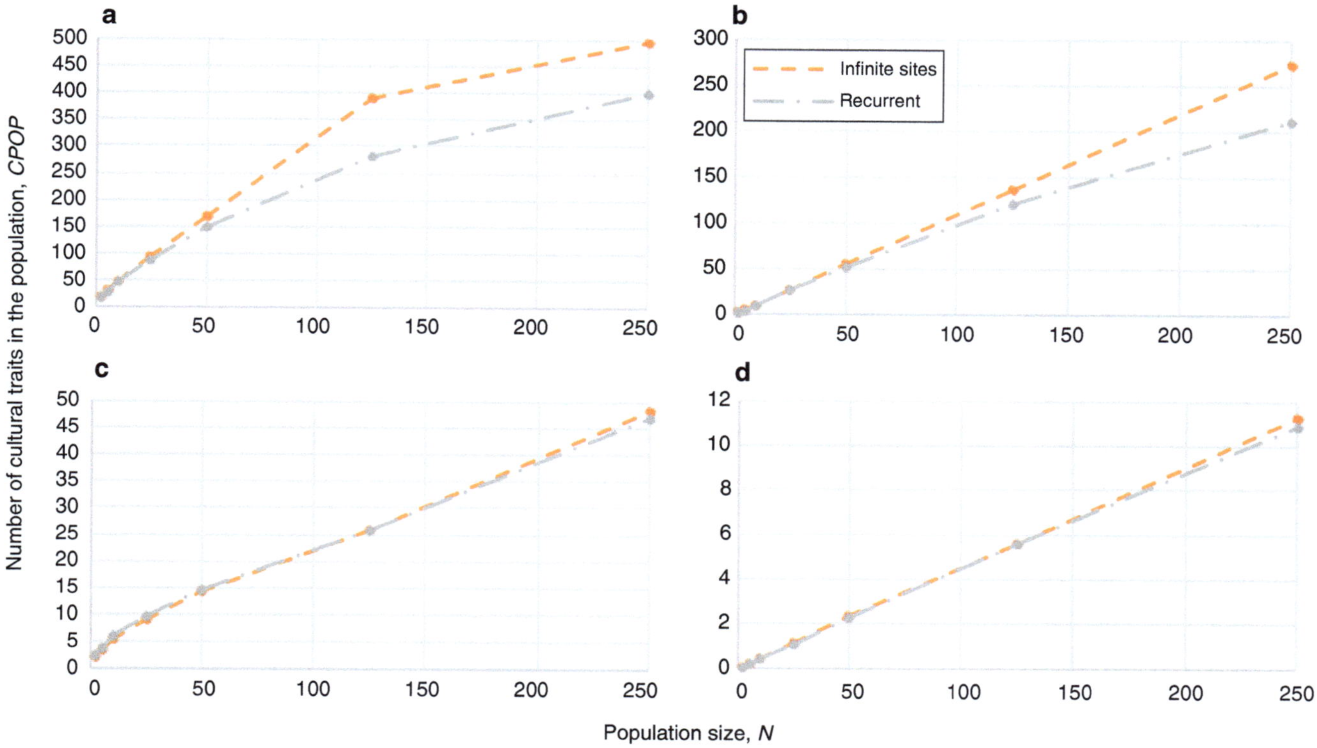

Fig. 2.3 Expected number of cultural traits at equilibrium, $\hat{\overline{C}}_{pop}$, is plotted against the population size, *N*, for the case of success bias social learning. Short-dash broken (*orange*) line: individual-based simulation values with infinite-sites innovation. Long-dash-and-dot broken (*grey*) line: individual-based simulation values with recurrent innovation. (**a**) $\beta = 0.9$, $\mu = 1$. (**b**) $\beta = 0.1$, $\mu = 1$. (**c**) $\beta = 0.9$, $\mu = 0.04$. (**d**) $\beta = 0.1$, $\mu = 0.04$

many SL. Figure 2.5 compares the IBS values of $\hat{\overline{C}}_{pop}$ at the five population sizes $N = 5,\ 25,\ 50,\ 125,$ and 250; saturation has occurred only for best-of-2 SL. We see that within this range of population sizes, the mode of SL has a significant effect on $\hat{\overline{C}}_{pop}$.

The saturation effect is most pronounced for best-of-2 SL, which is our model of direct bias. In particular, Fig. 2.2a shows that the IBS values of $\hat{\overline{C}}_{pop}$ for simple cultural traits do not depend on population size when $N \geq 5$. This means that a hunter-gatherer family of size 5, say, can support the same number of simple cultural traits as a tribe (ethnolinguistic group) numbering 500 individuals, say. Of course, the lower limit for the population size will vary with the specific values of β, μ, and *M*, but the principle remains the same. To reiterate, population size above a certain, relatively low, threshold does not predict the number of simple cultural traits to be found in a society, if SL is directly biased and imagination space is limited.

This theoretical result may help to explain why Collard et al. (2005, 2013a, b) and Read (2006) find no statistically significant association between number of tool types—"subsistants" and "technounits," as defined by Oswalt (1973, 1976)—and population size among hunter-gatherer societies. However, the tool types counted in these statistical studies apparently include the digging stick, which likely qualifies as a simple cultural trait, but also the boomerang, crossbow, and harpoon, which do not strike us as being so simple. We suggest that statistical analyses should in future be done separately for cultural traits that can be classified as simple or complex.

Our reason for making this point is that the situation is quite different for complex cultural traits, which are the class of cultural traits defined by small values of β and μ. Here, $\hat{\overline{C}}_{pop}$ is approximately linear in *N* for all modes of SL, so we expect to see a correlation between the number of complex cultural traits and population size. Moreover, comparison of Figs. 2.1d, 2.2d, 2.3d, and 2.4d shows that the plots are almost identical; thus mode of SL appears not to affect the outcome. The plots of $\hat{\overline{C}}_{pop}$ in these four panels are all close to $(N-1)\,\mu$ (see Eq. (2.6)), which is the expected number of 1-variants in a population comprising $N-1$ adults and one newborn, where the former have on average spontaneously acquired μ infinite-sites (i.e. non-overlapping) innovations each over their lifetimes, and the latter has not yet innovated. (Recall that μ is the innovation rate per adult per generation.) Social learning apparently contributes little to the number of

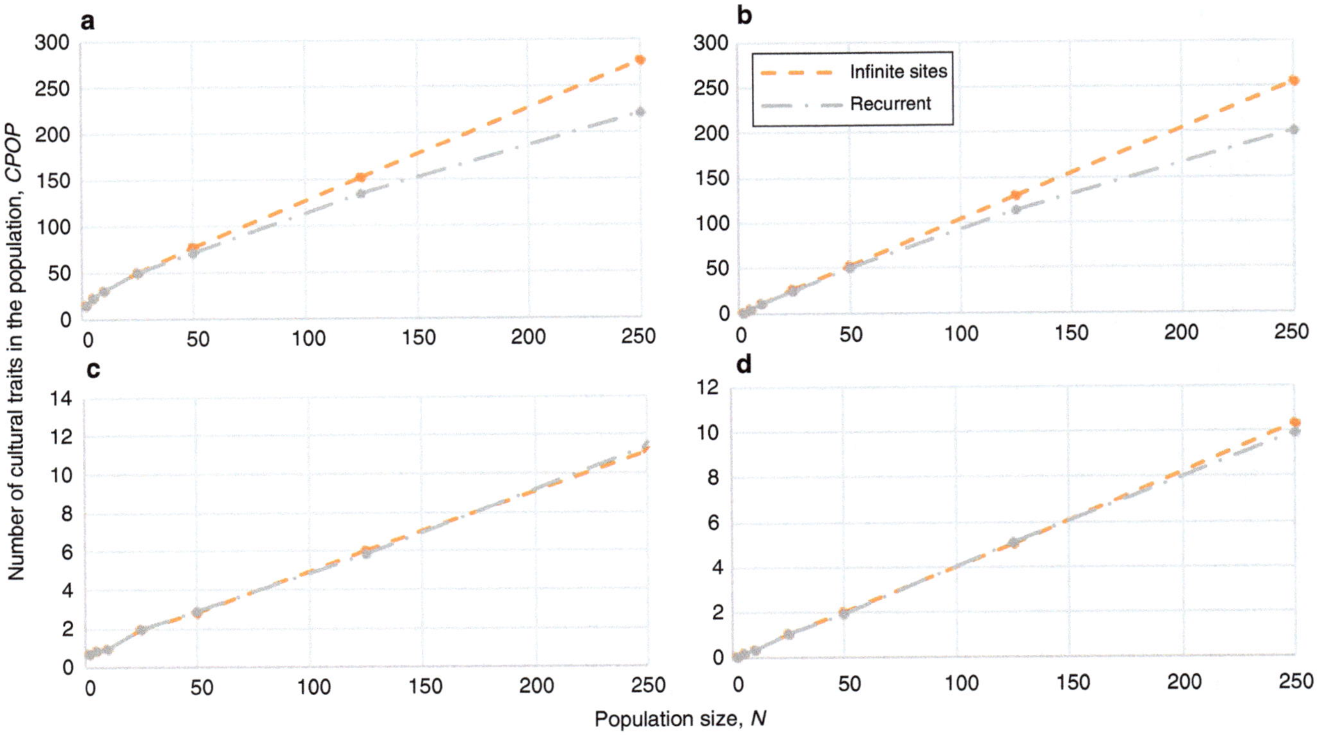

Fig. 2.4 Expected number of cultural traits at equilibrium, $\hat{C}_{pop}$, is plotted against the population size, N, for the case of one-to-many bias social learning. Short-dash broken (*orange*) line: individual-based simulation values with infinite-sites innovation. Long-dash-and-dot broken (*grey*) line: individual-based simulation values with recurrent innovation. (**a**) $\beta = 0.9, \ \mu = 1$. (**b**) $\beta = 0.1, \ \mu = 1$. (**c**) $\beta = 0.9, \ \mu = 0.04$. (**d**) $\beta = 0.1, \ \mu = 0.04$

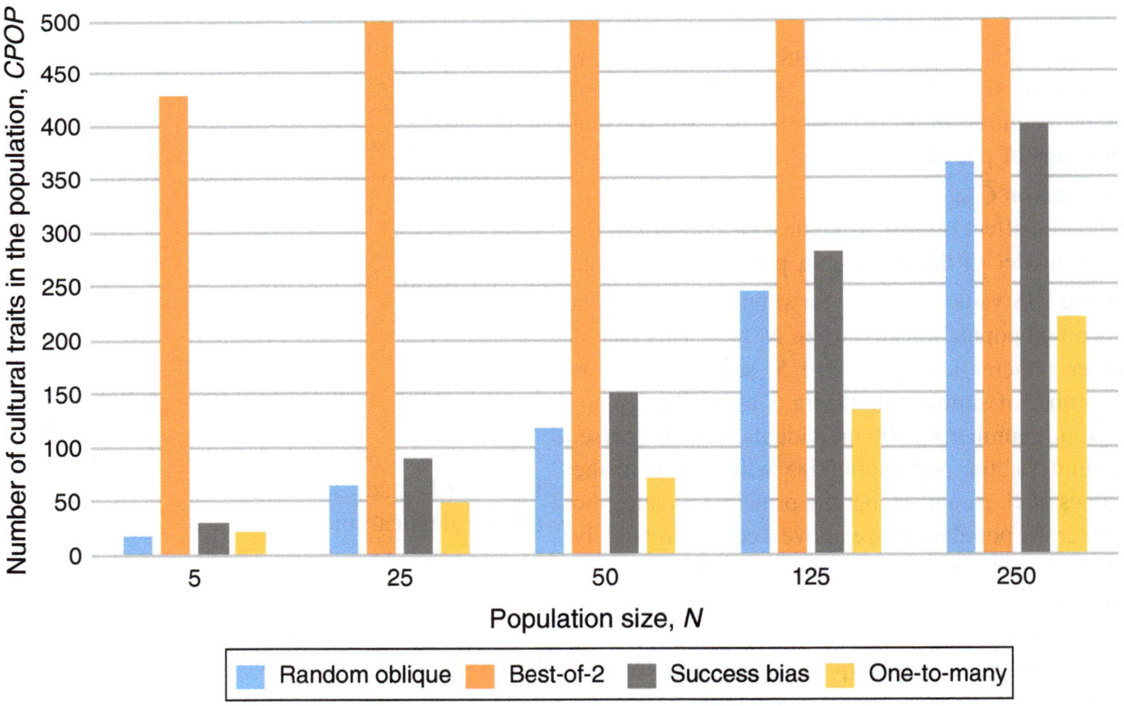

Fig. 2.5 Individual-based simulation values of $\hat{C}_{pop}$ for the four modes of social learning with recurrent innovation are compared at five population sizes, $N = 5, \ 25, \ 50, \ 125$, and 250. *Blue bars*: random oblique. *Brown bars*: best-of-2. *Grey bars*: success bias. *Orange bars*: one-to-many. $\beta = 0.9, \ \mu = 1$

complex cultural traits carried by an individual, which is why the mode of SL has little effect.

We now return to the unanswered question of what we should include among the cultural traits. Field workers (e.g. Whiten et al. 1999) have suggested that unique occurrences should be excluded, because evidence of social learning—a necessary condition for culture—is lacking in such cases. Theoretically, there is no reason to discriminate against those that are represented only once in the population. However, we have seen above that most of the complex cultural traits that are maintained in the population can be accounted for by innovation alone. We do not wish to argue either way, but unique occurrences can make a large contribution to our count of the number of cultural traits, especially when they are complex (data not shown).

We have been using the term complex cultural trait to denote a cultural trait that is difficult to acquire either by SL (small β) or by innovation (small μ). On the other hand, this term more often refers to a cultural trait that comprises many component parts—e.g. technounits (Oswalt 1973, 1976), procedural units (Perreault et al. 2013)—each of which can in itself be regarded as a cultural trait. Thus, there is an ambiguity in the concept of a cultural trait. However, the two descriptions of a complex cultural trait are not inconsistent, if it is assumed that a complex cultural trait in the latter sense is functional/serviceable only if all of its component parts have been successfully acquired. It may be difficult for one individual by him/herself to acquire, either by SL or by innovation, a complex cultural trait in its entirety. If each member of society specializes in the acquisition of the knowledge and skill to manufacture a different but complementary component part and shares the product with others (Nakahashi and Feldman 2014), then a greater number of complex cultural traits may be maintained, especially if these component parts can be used interchangeably in various complex cultural traits.

Saturation or plateauing has also been observed in a theoretical study of cultural evolution by Mesoudi (2011, see his Fig. 5B). Mesoudi (2011) proposes a model in which "each individual learns a set of cultural traits that are functionally sequential, such that earlier traits must be learned before later traits . . . ." This model is clearly an improvement over models such as ours that assume independent cultural traits. We note in passing that the research program outlined by Enquist et al. (2011), which takes into account the many types of interdependency among cultural traits, should be useful in advancing our understanding of how culture evolves toward greater complexity. Mesoudi (2011) compares three modes of SL, random oblique (unbiased), direct bias, and success bias (indirect bias), and shows that, with direct bias in particular, the mean "cultural complexity" at equilibrium plateaus when population size exceeds a certain threshold. Thus, here again theory does not necessarily predict a statistically significant relationship between number of tool types and population size.

We wish to apply our theoretical results to the archaeological record of Palaeolithic technologies and thereby further our understanding of why Neanderthals went extinct to be replaced by modern humans. Several theories, including the "learning hypothesis" (e.g. Aoki 2013), argue for a difference in cultural level or technological complexity between the two species, which is in turn attributed to a difference in learning strategies. Mode of SL is an important element of a learning strategy, which we would like to infer from the archaeological record (e.g. O'Brien et al. this volume, Chap. 9). Perhaps counterintuitively, it is the simple cultural traits that may be diagnostic.

Suppose we are able to sample Neanderthal and modern human archaeological sites that can be regarded as reflecting the activities of populations of similar size, or population size can be statistically controlled. Then, provided this population size lies within a suitable range, the number of simple cultural traits is predicted to be largest when SL is directly-biased, smallest for one-to-many SL, and somewhere in between with random oblique or success bias SL (Fig. 2.5). Hence, an observed difference among these sampled sites in the number of simple cultural traits would be consistent with a difference in mode of SL. It may also reflect a difference in the size of the imagination space. For complex cultural traits, on the other hand, we expect to see an effect of population size, but not of mode of SL, on their numbers.

However, before we can do this type of analysis, an objective measure must be devised for estimating the number of cultural traits in archaeological assemblages (Perreault et al. 2013).

Acknowledgments We thank Seiji Kadowaki for information on archaeological issues. We also thank Alex Mesoudi and the reviewers for insightful comments. This research was supported in part by Monbukagakusho grant 22101004.

Appendix 1: Partial Proof of Eq. (2.2)

To illustrate, we show the derivation of the first term on the right hand side of Eq. (2.2), which represents the contribution of up-transitions—i.e. the expected number of cultural traits of popularity $i-1$ immediately after innovation that have popularity i at the end of the time step. Let X_{i-1} be an integer-valued random variable giving the number of cultural traits of popularity $i-1$ immediately after innovation. Note P^*_{i-1} is the expected value of X_{i-1}. Similarly, let Y_i be an integer-valued random variable giving the number of cultural traits of popularity i after SL by the newborn—in the temporarily extended population comprising the N adults and the one newborn—that had popularity $i-1$ immediately after innovation. Finally, let Z_i be an integer-valued random variable giving the number of cultural traits of popularity i after the death of a random adult.

Then,

$$\Pr(Y_i \mid X_{i-1}) \sim binomial\,(X_{i-1}, b_{i-1})$$

and

$$\Pr(Z_i \mid Y_i, X_{i-1}) \sim binomial\,(Y_i, 1-d_{i-1})$$

That is, the conditional probability of Y_i given X_{i-1} is binomial with number of trials X_{i-1} and probability of success b_{i-1}. Similarly, the conditional probability of Z_i given Y_i and X_{i-1} is binomial with number of trials Y_i and probability of success $1-d_{i-1}$. Hence, using $E(\cdot)$ to denote expected value, we obtain

$$E(Z_i \mid Y_i, X_{i-1}) = Y_i\,(1-d_{i-1})\,.$$

Thus

$$E(Z_i \mid X_{i-1}) = X_{i-1}\;\; b_{i-1}\,(1-d_{i-1})\,,$$

and finally,

$$E(Z_i) = P^*_{i-1}\;\; b_{i-1}\,(1-d_{i-1})\,,$$

which is exactly the first term on the right hand side of Eq. (2.2).

Appendix 2: Proof of Eq. (2.6)

Since $\beta = 0$ entails $b_i = 0$ for $1 \le i \le N$, Eq. (2.5) reduces to

$$\begin{aligned}
-\hat{P}_1 d_1 + \hat{P}_2 d_2 &= -\mu\,(1-d_1)\\
-\hat{P}_2 d_2 + \hat{P}_3 d_3 &= 0\\
&\vdots\\
-\hat{P}_{N-1} d_{N-1} + \hat{P}_N d_N &= 0\\
-\hat{P}_N d_N &= 0
\end{aligned}$$

where $d_i = \frac{i}{N}$ for $1 \le i \le N$ from either Eqs. (2.3) or (2.4). Hence, $\hat{P}_N = 0$, $\hat{P}_{N-1} = 0, \cdots, \hat{P}_2 = 0$ and $\hat{P}_1 = \frac{1-d_1}{d_1}\mu = (N-1)\,\mu$.

References

Aoki K (2013) Determinants of cultural evolutionary rates. In: Akazawa T, Nishiaki Y, Aoki K (eds) Dynamics of learning in Neanderthals and modern humans volume 1: cultural perspectives. Springer, Tokyo, pp 199–210

Aoki K, Lehmann L, Feldman MW (2011) Rates of cultural change and patterns of cultural accumulation in stochastic models of social transmission. Theor Popul Biol 79:192–202

Atkinson QD (2011) Phonemic diversity supports a serial founder effect model of language expansion from Africa. Science 332:346–349

Brown MJ, Feldman MW (2009) Sociocultural epistasis and cultural exaptation in footbinding, marriage form, and religious practices in early 20th century Taiwan. Proc Natl Acad Sci U S A 106:22139–22144

Cavalli-Sforza LL, Feldman MW (1973a) Models for cultural inheritance, I: group mean and within-group variation. Theor Popul Biol 4:42–55

Cavalli-Sforza LL, Feldman MW (1973b) Cultural versus biological inheritance: phenotype transmission from parent to children (a theory of the effect of parental phenotypes on children's phenotype). Am J Hum Genet 25:618–637

Cavalli-Sforza LL, Feldman MW (1981) Cultural transmission and evolution: a quantitative approach. Princeton University Press, Princeton

Collard M, Kemery M, Banks S (2005) Causes of toolkit variation among hunter-gatherers: a test of four competing hypotheses. Can J Archaeol 29:1–19

Collard M, Buchanan B, Morin J, Costopoulos A (2011) What drives the evolution of hunter-gatherer subsistence technology? A reanalysis of the risk hypothesis with data from the Pacific northwest. Philos Trans R Soc B 366:1129–1138

Collard M, Ruttle A, Buchanan B, O'Brien MJ (2012) Risk of resource failure and toolkit variation in small-scale farmers and herders. PLoS One 7:e40975

Collard M, Buchanan B, O'Brien MJ (2013a) Population size as an explanation for patterns in the Paleolithic archaeological record. More caution is needed. Curr Anthropol 54:S388–S396

Collard M, Buchanan B, O'Brien MJ, Scholnick J (2013b) Risk, mobility, or population size? Drivers of technological diversity among recent western North American hunter-gatherers. Philos Trans R Soc B 368:20120412

Collard M, Ruttle A, Buchanan B, O'Brien MJ (2013c) Population size and cultural evolution in nonindustrial food-producing societies. PLoS One 8:e72628

Enquist M, Ghirlanda S, Eriksson K (2011) Modelling the evolution and diversity of cumulative culture. Philos Trans R Soc B 366:412–423

Gale JS (1990) Theoretical population genetics. Unwin Hyman, London

Guglielmino CR, Viganotti C, Hewlett B, Cavalli-Sforza LL (1995) Cultural variation in Africa: role of mechanisms of transmission and adaptation. Proc Natl Acad Sci U S A 92:7585–7589

Henrich J (2004) Demography and cultural evolution: how adaptive cultural processes can produce maladaptive losses—the Tasmanian case. Am Antiq 69:197–214

Hewlett B (2013) *"Ekeloko"* the spirit to create: innovation and social learning among Aka adolescents of the central African rainforest. In: Akazawa T, Nishiaki Y, Aoki K (eds) Dynamics of learning in Neanderthals and modern humans volume 1: cultural perspectives. Springer, Tokyo, pp 187–195

Ihara Y, Feldman MW (2004) Cultural niche construction and the evolution of small family size. Theor Popul Biol 65: 105–111

Jordan P, O'Neill S (2010) Untangling cultural inheritance: language diversity and long-house architecture on the Pacific northwest coast. Philos Trans R Soc B 365:3875–3888

Jordan P, Shennan S (2009) Diversity in hunter-gatherer technological traditions: mapping trajectories of cultural 'descent with modification' in northeast California. J Anthropol Archaeol 28:342–365

Kobayashi Y, Aoki K (2012) Innovativeness, population size and cumulative cultural evolution. Theor Popul Biol 82:38–47

Lehmann L, Aoki K, Feldman MW (2011) On the number of independent cultural traits carried by individuals and populations. Philos Trans R Soc B 366:424–435

Lycett SJ, Gowlett AJ (2008) On questions surrounding the Acheulean tradition. World Archaeol 40:295–315

Mesoudi A (2011) Variable cultural acquisition costs constrain cumulative cultural evolution. PLoS One 6:e18239

Morand-Ferron J, Cole EF, Rawles JEC, Quinn JL (2011) Who are the innovators? A field experiment with 2 passerine species. Behav Ecol 22:1241–1248

Nakahashi W, Feldman MW (2014) Evolution of division of labor: emergence of different activities among group members. J Theor Biol 348:65–79

Odling-Smee FJ, Laland KN, Feldman MW (2003) Niche construction. Princeton University Press, Princeton

Oswalt WH (1973) Habitat and technology: the evolution of hunting. Holt, Rinehart and Winston, New York

Oswalt WH (1976) An anthropological analysis of food-getting technology. Wiley, New York

Perreault C, Brantingham PJ, Kuhn SL, Wurz S, Gao X (2013) Measuring the complexity of lithic technology. Curr Anthropol 54:S397–S406

Powell A, Shennan S, Thomas MG (2009) Late Pleistocene demography and the appearance of modern human behavior. Science 324:1298–1301

Read D (2006) Tasmanian knowledge and skill: maladaptive imitation or adequate technology? Am Antiq 71:164–184

Reader SM, Laland KN (2001) Primate innovation: sex, age, and social rank differences. Int J Primatol 22:787–805

Rogers DS, Ehrlich PR (2008) Natural selection and cultural rates of change. Proc Natl Acad Sci U S A 105:3416–3420

Rogers DS, Feldman MW, Ehrlich PR (2009) Inferring population histories using cultural data. Proc R Soc Lond Ser B 276:3835–3843

Shennan S (2001) Demography and cultural innovation: a model and its implications for the emergence of modern human culture. Camb Archaeol J 11:5–16

Strimling P, Sjöstrand J, Enquist M, Errikson K (2009) Accumulation of independent cultural traits. Theor Popul Biol 76:77–83

Whiten A, Goodall J, McGrew WC, Nishida T, Reynolds V, Sugiyama Y, Tutin CEG, Wrangham RW, Boesch C (1999) Culture in chimpanzees. Nature 399:682–685

The Evolution of Culturally Transmitted Teaching Behavior

3

Wataru Nakahashi

Abstract

The replacement of Neanderthals by modern humans may possibly have been influenced by the different cultural transmission mechanisms of the two hominins. Since teaching is widespread in modern human societies, but extremely rare in animals, it may have played an important role in human cultural evolution. In modern humans, how and whom to teach may, in part, be transmitted culturally. Therefore, in this paper, I develop a cultural transmission model of teaching. I show that even when costly, teaching can evolve provided that teachers transmit their cultural traits more actively than non-teachers. Teaching is more likely to evolve when the cost of social learning is low relative to individual learning, social learning is accurate, the environment is stable, and the effect of teaching is extensive. Under certain conditions, two states, existence and non-existence of teaching in the population, are evolutionarily stable (bistable). When this happens, social learning is sometimes maintained by teaching under unstable environments where social learning cannot exist without teaching. Differences in subsistence strategy and group structure between Neanderthals and modern humans may have affected the evolution of the teaching behaviors of the two hominins.

Keywords

Evolutionarily stable strategy (ESS) • Bistability • Cultural hitchhiking • Cumulative culture

3.1 Introduction

The replacement of Neanderthals by modern humans is one of the greatest mysteries in hominin evolution. Many anthropologists and archeologists have investigated the differences between the two hominins in an attempt to clarify the causes of the replacement. It has often been argued that Neanderthals lacked so-called "modern behavior", or at least modern behavior appeared later in Neanderthals than in modern humans, which led to demographic disadvantage and subsequent extinction. Although this simple view has recently been questioned (Zilhão 2013), various useful cultural skills may have appeared earlier in African modern humans than in Eurasian Neanderthals (McBrearty and Brooks 2000; McBrearty 2012). For example, the backed pieces, an innovative way of hafting spears for composite weapon armature, appeared in Howiesons Poort, South Africa, around 59,000–65,000 years ago, but there were no comparable industries in the Neanderthal hunting equipment, al least until the Châtelperronian, some 20,000 years later (Villa and Soriano 2010).

Why did African modern human culture evolve faster than Eurasian Neanderthal culture? Theoretical research on cultural evolution has shown that accurate social learning and exploratory individual learning are critical for rapid

W. Nakahashi (✉)
School of Advanced Sciences, SOKENDAI (The Graduate University for Advanced Studies), Shonan Village, Hayama, Kanagawa 240-0193, Japan
e-mail: w.nakahashi@gmail.com

A. Mesoudi and K. Aoki (eds.) *Learning Strategies and Cultural Evolution during the Palaeolithic*, Replacement of Neanderthals by Modern Humans Series, DOI 10.1007/978-4-431-55363-2_3,

cultural evolution (Henrich 2004; Kobayashi and Aoki 2012; Nakahashi 2013c, 2014), suggesting that cultural traits may possibly have been transmitted more accurately and explored more extensively in modern humans than in Neanderthals.

In present-day humans, accurate transmission is often supported by teaching. Here, teaching is defined as a costly (or a neutral) change in behavior on the part of the teacher that increases the rate or probability of learning in the pupil (Caro and Hauser 1992). Both experimental and theoretical studies suggest that teaching is essential in cumulative cultural evolution, one of the most significant characteristics of human culture (Dean et al. 2012; Castro and Toro 2014). Although cultural transmission between non-relatives is essential for human cultural evolution, teaching to non-relatives is less common in animals, suggesting that our strong dependence on teaching may have evolved after splitting from the chimpanzee lineage; i.e., the evolution of teaching to non-relatives may cause the difference with regard to cultural cumulativeness between humans and non-human animals. Despite the uncertainty as to whether Neanderthals taught cultural traits to non-relatives, the difference in cultural evolutionary speeds between Neanderthals and modern humans may possibly be explained by the different teaching behaviors of the two hominins.

Recently, there has been an increase in research on teaching in animals that discusses the selection pressures that may favor the evolution of teaching (reviewed in Caro and Hauser 1992; Thornton and Raihani 2008). Since teaching is, to a certain degree, similar to cooperation, because teachers pay a cost and pupils receive a benefit through teaching events, theoretical research on the evolution of teaching has focused primarily on the cost of teaching, the benefits of being taught, and the genetic relatedness between teacher and pupil (Fogarty et al. 2011; Aoki et al. 2013). In these studies, teaching behavior was assumed to be genetically transmitted. However, many human behaviors are in reality not genetically, but culturally transmitted. Since teaching behavior is very different among different societies (Hewlett et al. 2011; Terashima 2013), it may, in part, be culturally transmitted, even though our ability (and preference) for teaching is genetically transmitted. This situation is similar to our language; i.e., our language capacity (e.g., throat structure, universal grammar) is genetically transmitted, but our language itself (e.g., vocabulary, pronunciation) is culturally transmitted. Since the transmission mechanism differs greatly between genetic traits and cultural traits, previous studies that have assumed the genetic transmission of teaching are insufficient for understanding the evolution of human teaching behavior.

One of the most important properties of cultural transmission is that the fitness (transmissibility) of a cultural trait is not always the same as the fitness (reproductive success) of its carriers, while the fitness of a genetic trait is essentially the same as that of its carriers; i.e., even if a cultural trait is harmful for its carriers, its frequency may increase, provided it is more frequently transmitted than other cultural traits. This mechanism is often discussed in memetics (Blackmore 2000). For example, costly prestige-seeking behavior can spread by prestige-biased social learning (Ihara 2008). In other words, cultural traits easily "hitchhike" on other characteristics of carriers. The evolution of teaching behavior is likely to have been influenced by this property of cultural transmission because teaching directly affects the cultural transmission mechanism. Therefore, the study of the evolution of human teaching behavior requires consideration of cultural transmission.

In this paper, I develop a new model that considers the evolution of culturally transmitted teaching behavior by extending the learning capacity model proposed by Nakahashi (2010, 2013a, b). I obtain the conditions required for culturally transmitted teaching behavior to evolve. First, I explain the learning capacity model, and then include teaching behavior in the model.

3.2 Model

3.2.1 Basic Model of Nakahashi (2010)

Consider a very large population with infinitely many kinds of cultural traits (know-how, information, principle, etc.) generated by individual learning and the mistakes of social learning. Cultural traits can be classified into two categories, beneficial and neutral. Organisms cannot distinguish between beneficial and neutral cultural traits, and acquire them by individual learning and social learning. The number of cultural traits they learn (learning capacity) is determined by their strategy gene, i.e., an organism with strategy (n, m) acquires n cultural traits by individual learning and m by social learning. That is, individual learning capacity is n and social learning capacity is m. Although organisms may or may not share the same cultural traits, the possibility that an organism learns the same cultural trait more than once can be disregarded because, by assumption, there are infinitely many kinds of cultural traits and organisms acquire a finite number of cultural traits.

When an organism learns individually, it acquires new (previously non-existent) cultural traits by itself, for example, by trial-and-error or insight. Let the probability of acquiring a beneficial cultural trait (success rate of individual learning) be r $(0 < r \leq 1)$. When an organism learns socially, it copies cultural traits from random, possibly different, members of its parental generation (oblique transmission), where each cultural trait is randomly picked from the repertory of a target's cultural traits pool. The probability of copying a target's cultural trait accurately (accuracy of social learning) is a ($0 < a \leq 1$) for both beneficial and neutral cultural traits. That is, a fraction a of socially learned cultural traits remain unchanged but the remaining fraction mutates into different cultural traits. It may be reasonable to assume that a mutant cultural trait often loses the essential function of the wild-type cultural trait but seldom gains a new function, as is the common assumption with regard to genetic mutation. Therefore, we can assume that beneficial cultural traits become neutral with probability $1 - a$ because of a loss of function, but all neutral cultural traits remain neutral because they function as "pseudogenes".

The life cycle of organisms consists of two stages: learning/viability-selection stage and fertility-selection stage. In the learning/viability-selection stage, each organism learns cultural traits piece by piece and pays a small viability cost for each learning activity. It costs $1 - e^{-c}$ to acquire a cultural trait by individual learning and $1 - e^{-d}$ by social learning ($c > 0$, $d > 0$), regardless of whether that cultural trait is beneficial or neutral. Therefore, the probability that an organism with strategy (n, m) can survive and proceed to the next stage is $[1 - (1 - e^{-c})]^n \left[1 - \left(1 - e^{-d}\right)\right]^m = e^{-cn-dm}$. For example, if we assume that individual and social learning requires c and d units of time, respectively, and there exists a constant exogenous mortality rate in this stage, the survival probability after learning one cultural trait individually and socially becomes e^{-c} and e^{-d}, respectively, and the survivability of an organism with strategy (n, m) is e^{-cn-dm}.

In the fertility-selection stage, the fitness of an organism is determined by the number of beneficial cultural traits acquired during the learning/viability-selection stage. Neutral cultural traits do not affect the fertility of organisms. The fitness of an organism that has k beneficial cultural traits is expressed as $w + k$, where w is the baseline fitness of organisms ($w > 0$). We assume fertility selection: the number of their offspring is proportional to $w + k$. All organisms die soon after their cultural traits are passed on to the next generation.

In each generation before the learning/viability-selection stage, (part of) the environment changes and some of the beneficial cultural traits lose their essential role (become neutral) while the remaining parts remain beneficial and all neutral cultural traits remain neutral. For example, cultural traits specialized for mammoth hunting may have become neutral after the extinction. The probability that a beneficial cultural trait becomes neutral in the next generation is $1 - s$ ($0 < s < 1$). That is, parameter s represents the stability of the environment. The possibility that a neutral cultural trait reverts to being beneficial in the next generation can be neglected when we assume that (part of) the environment always changes to an inexperienced state, like the situation of the infinite environmental state model (Feldman et al. 1996). We can consider a different type of environmental change. That is, a beneficial cultural trait does not become (completely) neutral by environmental change, but its effect on fitness decreases to s times its previous value. For example, organisms may get only eight head of game by a hunting method that previously brought ten head of game to the parental generation because the method becomes outdated. Nakahashi (2013a) shows that even if we consider the latter type of environmental change, the following result is unchanged. Therefore, we can apply the stability of the environment, s, to both types of environmental change.

Baseline fitness, w, is the contribution of genetic (innate) traits and corresponds to the fitness of organisms with no learned cultural traits. The baseline fitness may be greater when genetic traits are sufficiently adaptive in a particular environment. Such an environment can be considered to be mild for organisms, so baseline fitness w reflects the mildness of the environment.

Let us consider the evolutionarily stable learning strategy $(\hat{n}, \hat{m})$. Here, $\hat{n}$ and $\hat{m}$ are the cultural capacities, i.e. the numbers of cultural traits acquired by individual learning and social learning. An evolutionarily stable strategy (ESS) is a strategy that is stable to the invasion of rare mutants of small effect (mutants cannot increase their frequency) (Maynard Smith 1982). Assume that a mutant strategy (n, m) (strategy B) is introduced at low frequency into an equilibrium population whose members all use the strategy (n^*, m^*) (strategy A). Then, the fitness of the mutant is

$$W(B, A) = (w + rn + a\hat{x}m)\, e^{-cn-dm} \tag{3.1}$$

where

$$\hat{x} = \frac{rsn^*}{n^* + m^* - asm^*}. \tag{3.2}$$

is the fraction of cultural traits that are beneficial at this equilibrium. The derivation of $\hat{x}$ is shown in Appendix1. Equation (3.1) implies that the survivability of a mutant over the learning/viability-selection stage is e^{-cn-dm}, during which it acquires rn beneficial cultural traits by individual learning and $a\hat{x}m$ beneficial traits by social learning, so that its number of offspring in the fertility-selection stage is proportional to $w + rn + a\hat{x}m$. Since $W(A, A)$ is the fitness of the wild type, if $W(B, A) < W(A, A)$ is satisfied for all B ($B \neq A$), strategy A is the ESS.

As shown in Appendix 2, the ESS $(\hat{n}, \hat{m})$ can be obtained analytically (Table 3.1). Since Nakahashi (2013a) showed

Table 3.1 The ESS $(\hat{n}, \hat{m})$

	$c \leq d$	$c > d$
$cw < r$	$(1/c - w/r, 0)$	$(1/c - w/r, 0)$ if $acs \leq d$; $\left(\frac{(1-as)(r-cw)}{r(c-d)}, \frac{(acs-d)(r-cw)}{rd(c-d)}\right)$ if $acs > d$
$cw \geq r$	$(0, 0)$	$(0, 0)$

by numerical simulations that the strategy of the population converges to the ESS by natural selection in this model, the ESS can be considered to be the evolutionary outcome of competition among learning strategies under each condition.

3.2.2 Including Teaching Behavior

Next, let us include teaching behavior in the basic model. Assume that teaching behavior is transmitted culturally. If an organism has the cultural trait for teaching (teaching trait), it actively transmits cultural traits to learners. We assume for the moment that teaching behavior is a neutral cultural trait so that it does not affect the fertility of its carriers (I will consider the fertility cost of teaching later). When an organism learns from teachers (organisms with the teaching trait), it can acquire t times ($t > 1$) more cultural traits than when learning from non-teachers (organisms without the teaching trait) per same viability cost (unit of time). That is, teaching decreases the cost of social learners (social learning). Since the probability of copying another's cultural trait accurately (accuracy of social learning) is a, we may assume that every cultural trait including teaching trait is accurately transmitted to the next generation with probability a. We also assume that environmental change does not affect the effect of teaching behavior.

Assume as before that every organism has the learning strategy (n, m). Then, if the population size remains constant, an organism of the parental generation transmits, on average, m cultural traits, accurately or inaccurately, to the offspring generation, because the number of cultural traits that the offspring generation acquires from the parental generation by social learning should be the same as the number that the parental generation transmits to the offspring generation. Let the proportion of teachers in the parental generation be p. Then, on average, a teacher and a non-teacher transmit $mt/(1-p+pt)$ and $m/(1-p+pt)$ cultural traits to the offspring generation, respectively, because the former transmits t times as many cultural traits as the latter. Note that $pmt/(1-p+pt)+(1-p)m/(1-p+pt) = m$. Since every organism including a teacher has $n+m$ cultural traits and the accuracy of social learning is a, the expected number of (individuals with) a *particular* cultural trait carried by a teacher that is accurately transmitted to the next generation, j, is

$$j = \frac{amt}{(n+m)\ (1-p+pt)}. \tag{3.3}$$

Since the teaching trait is necessarily transmitted from teachers, when j is smaller than one, the expected number of teaching trait in the population decreases in the next generation so that the proportion of teachers, p, also decreases, and when j is larger than one, both increase. Therefore, when the population is fixed for the learning strategy (n, m), the proportion of teachers in the population converges to where j equals unity, i.e.,

$$\tilde{p}\,(n,m) = \begin{cases} \frac{(at-1)m-n}{(t-1)(n+m)} & \text{if } (at-1)\,m > n \\ 0 & \text{otherwise} \end{cases}. \tag{3.4}$$

Setting $p = 0$ in Eq. (3.3), we find that the sufficient condition for the teaching trait to invade the population is

$$\frac{at\hat{m}}{\hat{n}+\hat{m}} > 1 \tag{3.5}$$

where $(\hat{n}, \hat{m})$ is the same as in Table 3.1 because learning strategy evolves to $(\hat{n}, \hat{m})$ provided teachers are absent. This condition can be satisfied only when $\hat{m} > 0$, i.e., $c > d$, $cw < r$ and $acs > d$. Then, Ineq. (3.5) can be rewritten as

$$\frac{t\,(acs-d)}{s\,(c-d)} > 1. \tag{3.6}$$

Next, let us obtain the stable equilibrium with teaching. Since we have already obtained the equilibrium proportion of teachers when the population is fixed for a learning strategy (see Eq. (3.4)), let us consider how learning strategy evolves when there is a fixed proportion of teachers in the population. When a proportion p of the population is teachers, learning strategy evolves to be different from $(\hat{n}, \hat{m})$ in Table 3.1. We should change d to $d/(1-p+pt)$ in Table 3.1, because each organism can socially learn $1-p+pt$ times more cultural traits per same viability cost (unit of time) when teachers are present. That is, with probability $1-p$ it acquires one cultural trait from a non-teacher and with probability p it acquires t cultural traits from a teacher, per d units of time.

Write the evolutionarily stable learning strategy with proportion p of teachers as $(\tilde{n}(p), \tilde{m}(p))$. Both the proportion of teachers and learning strategy can be (evolutionarily) stable at the intersection points of $p = \tilde{p}\,(n,m)$ and $(n,m) = (\tilde{n}(p), \tilde{m}(p))$, which we write $(\hat{n}, \hat{m}, \hat{p})$, i.e., $\hat{p} = \tilde{p}\,(\hat{n}, \hat{m})$ and $(\hat{n}, \hat{m}) = (\tilde{n}\,(\hat{p})\,, \tilde{m}\,(\hat{p}))$. Writing $y = 1-\hat{p}+\hat{p}t$ $(1 \le y \le t)$, we can infer from Table 3.1 that

$$(\hat{n}, \hat{m}) = \left(\frac{y\,(1-as)\,(r-cw)}{r\,(cy-d)}, \frac{y\,(acsy-d)\,(r-cw)}{rd\,(cy-d)}\right), \tag{3.7}$$

when $cy > d$, $cw < r$ and $acsy > d$. Hence, from Eq. (3.4), non-zero $\hat{p}$ must satisfy

$$\hat{p} = \frac{(at-1)\,\hat{m}-\hat{n}}{(t-1)\,(\hat{n}+\hat{m})} = \frac{1}{t-1}\left[\frac{t\,(acsy-d)}{s\,(cy-d)} - 1\right]. \tag{3.8}$$

But $\hat{p} = (y-1)/(t-1)$, so Eq. (3.8) can be rewritten as

$$y = \frac{t\,(acsy - d)}{s\,(cy - d)}. \tag{3.9}$$

To obtain the condition for Eq. (3.9) to be satisfied, let us write

$$f(y) = csy^2 - s\,(d + act)\,y + td. \tag{3.10}$$

When $f(y) = 0$ (condition Eq. (3.9)) has the roots between $1 < y \le t$, there can be a stable equilibrium (evolutionarily stable strategy) with teaching. Since $f(y) = 0$ is a quadratic equation of y and $f(t) > 0$ and $f'(t) > 0$ are satisfied, $f(y) = 0$ has the larger root between $1 < y < t$ when $f(1) < 0$ is satisfied, and has two roots between $1 < y < t$ when $f(1) > 0$, $f'(1) < 0$, and

$$s(d + act)^2 - 4ctd > 0 \tag{3.11}$$

(the discriminant is positive) are satisfied. As shown in Appendix 3, the larger root is stable and the smaller root is unstable. Condition $f(1) < 0$ is the same as condition Ineq. (3.6); i.e., provided teaching trait can invade the population, an evolutionarily stable strategy where teachers are present always exists. When conditions $f(1) > 0$, $f'(1) < 0$ ($act > 2c - d$), and Ineq. (3.11) are satisfied, the situation is "bistable". Note that condition Ineq. (3.6) is more stringent than condition Ineq. (3.11) provided $act > 2c - d$. The bistable situation implies that teaching behavior cannot invade when there are no teachers, but once there are many teachers, teaching trait stably exists in the population. This is because, when there are many teachers, the cost of social learning decreases so that each organism learns more cultural traits socially ($\tilde{m}(p)/\,[\tilde{n}(p) + \tilde{m}(p)]$ increases as p increases), which in turn entails that the teaching trait is more easily maintained in the population.

Figure 3.1 shows the conditions for each type of situation to be achieved. When $acs \le d$, $act > 2c - d$ and Ineq. (3.11) are satisfied (region IV), two states, only individual learning exists, $(\hat{n}, \hat{m}, \hat{p}) = (+, 0, 0)$, and both individual learning and social learning with teaching coexist, $(\hat{n}, \hat{m}, \hat{p}) = (+, +, +)$, are bistable. That is, teaching maintains social learning.

Teachers may have lower fitness (fertility) than non-teachers because teaching entails additional costs to teachers. This may affect the evolution of learning strategy $(\tilde{n}(p), \tilde{m}(p))$, because an organism who learns more cultural traits socially is more likely to be a teacher and suffers teaching cost so that the dependence on social learning decreases. From Eq. (3.4), when the proportion of social learning, $\tilde{m}(p)/\,[\tilde{n}(p) + \tilde{m}(p)]$, decreases, the proportion of teachers, p, decreases. However, teaching cost does not affect the condition for the invasion of teaching (Ineq. (3.5)) because $(\tilde{n}(0), \tilde{m}(0))$ is the same as the case of costless teaching. Note that when teachers are absent, social learners do not suffer teaching cost. That is, culturally transmitted teaching behavior can evolve even when teaching is costly. This is because, even if teachers have lower fertility than non-teachers, teaching trait still has higher fitness (transmissibility) than other cultural traits. Teaching cost may decrease the region where the stable equilibrium with teaching exists (bistable region), but general tendency of the above results may still hold.

In the above model, teachers are assumed to facilitate social learning by pupils, so that the number of cultural traits each pupil learns increases to t times the value without teaching. Teaching is sometimes assumed to increase the accuracy of social learning in pupils (e.g., Castro and Toro 2014). If we assume that the accuracy of social learning increases to t times ($1 < t \le 1/a$) when an organism learns a cultural trait from a teacher, instead of the number of cultural traits increasing to t times, the results are different. In this case, since the number of cultural traits each teacher transmits is the same as non-teachers, when the population is fixed for a learning strategy (n, m) and population size is constant, each teacher transmits (on average) m cultural traits to the next generation, of which the fraction at is accurately transmitted. Since every organism has $n + m$ cultural traits, the expected frequency of teaching trait increases when

$$\frac{atm}{n + m} > 1 \tag{3.12}$$

However, since $at \le 1$, Ineq. (3.12) is never satisfied. That is, teaching behavior never evolves when it affects the accuracy of social learning, not the number of cultural traits. This is because each organism always acquires a part of its cultural traits by individual learning, so that even when the teaching trait is transmitted with complete accuracy ($at = 1$), its expected frequency must decrease from the previous generation. How teaching behavior affects pupils' social learning is important for its evolution.

3.3 Discussion

I have analyzed a mathematical model of cultural transmission in which teaching behavior is also culturally transmitted. I have shown from Ineqs. (3.6) and (3.11) that teaching is more likely to evolve when the cost of social learning is low relative to individual learning, social learning is accurate, the environment is stable, and the effect of teaching is extensive, provided teaching affects the number of cultural traits each pupil learns from teachers (see also Fig. 3.1). The first three conditions are those required for the dependence on social

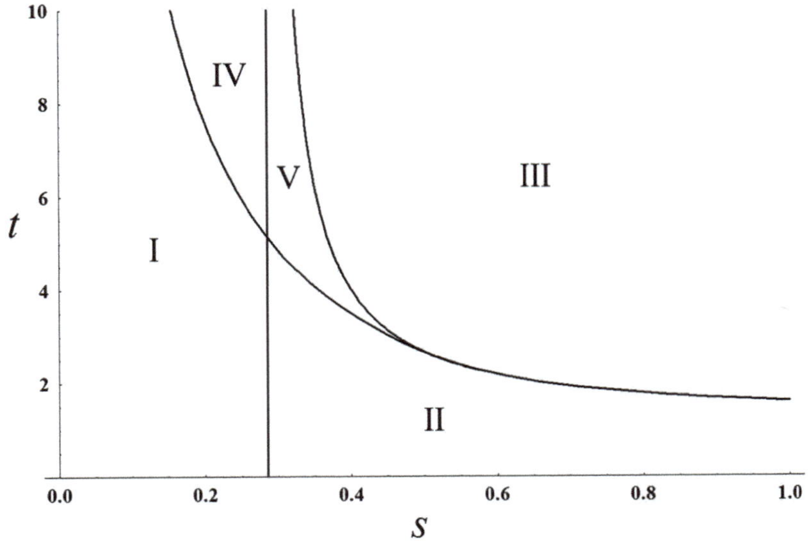

Fig. 3.1 Evolutionarily stable learning and teaching strategy, $(\hat{n}, \hat{m}, \hat{p})$, is shown in the (s, t)-parameter space. The strategies in each region are (I) individual learning only: $(\hat{n}, \hat{m}, \hat{p}) = (+, 0, 0)$, (II) coexistence of individual learning and social learning without teaching: $(\hat{n}, \hat{m}, \hat{p}) = (+, +, 0)$, (III) coexistence of individual learning and social learning with teaching: $(\hat{n}, \hat{m}, \hat{p}) = (+, +, +)$, (IV) bistability of $(\hat{n}, \hat{m}, \hat{p}) = (+, 0, 0)$ and $(\hat{n}, \hat{m}, \hat{p}) = (+, +, +)$, and (V) bistability of $(\hat{n}, \hat{m}, \hat{p}) = (+, +, 0)$ and $(\hat{n}, \hat{m}, \hat{p}) = (+, +, +)$. Parameters are $c = 0.005$, $d = 0.001$, and $a = 0.7$. Here, we assume $cw < r$

learning to increase, implying that teaching evolves when we strongly depend on social learning. This is because teaching behavior is a socially transmitted cultural trait in the model so that it is more likely to be maintained in the population when the dependence on social learning is significant. Such dependence may have increased during human (hominin) evolution because the cost of social learning decreased and the accuracy of social learning increased as human social learning ability (e.g., communication ability, language ability) improved. Therefore, it is reasonable to assume that teaching may have evolved at a late stage in human evolution. The importance of social learning on the evolution of teaching may also explain why teaching is less common in animals, whose dependence on social learning is minimal.

Teaching can have two properties: teachers may more actively transmit their cultural traits than non-teachers, and pupils may more accurately learn cultural traits from teachers than from non-teachers. The model shows that the former promotes the evolution of teaching, while the latter does not, i.e., teaching evolves because pupils learn teaching behavior more frequently, not because they learn it more accurately. Although accurate social learning through teaching may be important for cumulative cultural evolution, it does not contribute to the maintenance of teaching in the population. Since humans actually learn from teachers more often than from non-teachers, the assumption of an active teacher may be realistic and the model results are applicable to the evolution of human teaching, although accurate learning from teachers is not simultaneously included in the model. It may be that we prefer to learn from good teachers, which also increases the number of cultural traits learned from teachers and enhances the spread of teaching behavior in the population.

The model shows that teaching is more likely to evolve when the effect of teaching (number of cultural traits each teacher transmits per single transmission event) is significant. The effect of teaching may be considered to indicate teaching ability; however we do not consider its evolution in this model. Teaching to offspring may have been an essential selection pressure for the evolution of our high genetic ability for teaching (e.g., theory of mind, empathic sensitivity) because genetic relatedness between teacher and pupil is important for the evolution of genetically transmitted teaching behavior (Fogarty et al. 2011; Aoki et al. 2013). The ability to teach offspring applies also to teaching to non-relatives. In other words, teaching to offspring is a preadaptation for teaching to non-relatives. Therefore, once (genetically transmitted) teaching to offspring evolved, (culturally transmitted) teaching to non-relatives would have followed. Since an increase in the dependence on a behavior may trigger an increase in the ability to perform the behavior (Nakahashi 2010), teaching ability may improve to a greater extent when culturally transmitted teaching behavior spreads in the population.

The model also shows that there is a bistable region where teaching both exists and does not exist, i.e., even when teaching can exist in a stable form in the population, it cannot evolve if there are relatively few teachers initially. This implies that the existence of teaching in a population sometimes depends on the past experience of the population.

For example, once a population experiences a sufficiently stable environment for teaching to be able to spread, teaching is maintained, even when the environment becomes unstable and becomes a bistable condition. If the population does not experience a stable environment, teaching does not evolve under the bistable condition. Moreover, teaching and social learning may exist in a stable form under an unstable environment where social learning cannot exist without teaching. Note that teaching never spreads in the population under the condition where social learning does not exist. The experience of a stable environment and the evolution of teaching are important for the maintenance of social learning under unstable environments.

The model only considers individual learning and unbiased oblique social learning, which may not be sufficiently sophisticated to describe human learning strategies. For example, modification of socially learned cultural traits (social improvement) is important for human cumulative cultural evolution (Nakahashi 2013a, b). Nevertheless, cultural traits are always acquired by individual learning or (various types of) social learning. Since teaching is important in social learning, but meaningless in individual learning, the result that teaching is more likely to evolve when the dependence on social leaning is significant should be robust. That is, the simple model may be sufficient for discussing the evolution of teaching, although using the model results to discuss more complex factors may have no justification.

The model shows that, when teaching is culturally transmitted, it evolves because it is more likely to be transmitted than ordinary cultural traits. That is, teachers apply the principle that we should actively teach many cultural traits to non-relatives, which are then more likely to be transmitted than with the principle that we do not need to teach cultural traits to others. In other words, cultural traits of influential persons are more likely to be transmitted than those of ordinary persons. This mechanism is important when we consider the evolution of human behavior because our behavior is often culturally transmitted. For example, we tend to rely on honest and kind persons, who may follow the principle that we should be honest and kind to others, which may promote a more cooperative human society. It is important to consider this cultural mechanism when we investigate human behaviors that may relate to cultural transmission.

Although social learning and teaching are less common in non-human animals, such a cultural transmission mechanism may also be important in animal behavior. For example, consolation is observed in some species, which may reflect their empathy for others' distress (reviewed in de Waal 2008). Although the machinery of empathy for recognizing distress may be genetically transmitted, the behavior of consoling others may possibly be culturally transmitted. For example, maternal behavior is often influenced by the early experience of being cared for by the mother and observing the maternal behaviors of others (Gonzalez et al. 2001; Champagne and Meaney 2007), although in mammals mothers may also take care of their offspring through instinct. In other words, consolation behavior may remain in the population even when consolation is costly because the behaviors of empathetic consolers are more likely to be transmitted culturally.

Although archeological evidence suggests that the Upper Paleolithic modern humans may have taught each other how to make stone tools (Pigeot 1990; Takakura 2013), the timing of the initiation of teaching to non-relatives in hominins is uncertain. Since stone tools became more complex and difficult to make as hominins evolved, transmission of the skills for making them may have become essential at a certain stage in hominin evolution. Of course, the information transmitted by teaching was never restricted to stone tools, but the difficulty of stone tool making would reflect the accuracy of cultural transmission, which may have been significant if teaching existed. That is, the existence of complex stone tools suggests the existence of teaching in the society, although we cannot distinguish between teaching to offspring and to non-relatives.

When there is one-to-many transmission, cultural evolution may proceed more rapidly, provided that each individual can recognize the most skilled person or that the teachers tend to have higher skills (Henrich 2004; Kobayashi and Aoki 2012; but see Aoki et al. 2011; Aoki 2013; Nakahashi 2013c, 2014). One-to-many transmission may imply cultural transmission between non-relatives. Therefore, rapid progress of complex technologies suggests the existence of teaching to non-relatives in the society.

Archeological evidence suggests that cultural evolutionary rates were extremely low before the emergence of modern humans. For example, the Acheulean tradition of *Homo erectus* remained much the same in the archaeological record for over a million years (Jelinek 1977), and the Mousterian tradition of Neanderthals was almost static for hundreds of thousands of years (Akazawa et al. 1998). On the other hand, modern human culture changed rapidly during and after the Upper Paleolithic in Europe (Bar-Yosef 2002) and the Middle Stone Age in Africa (McBrearty and Brooks 2000), although the exact timing of the increase in the rate of cultural evolution is controversial. Therefore, we can assume that teaching to non-relatives began at the latest during this period.

Although it is uncertain whether Neanderthals actually engaged in teaching, if a difference in teaching behavior between Neanderthals and modern humans existed, we could explain the different cultural evolutionary speeds between the two hominins. We can speculate about two reasons for the evolution of different teaching behaviors. One possible explanation is that the African environment was more stable than the Eurasian at that time, so that teaching evolved only in African modern humans. Even if the environmental stability was similar between the two regions, the difference in subsistence strategies may have affected the evolution of teaching. For example, animals and plants may have different sensitivities to environmental changes, so that a hominin who strongly depends on hunting and one that depends on gathering may also respond differently to environmental changes. Since modern humans may have engaged in a wider range of economies than Neanderthals (Kuhn and Stiner 2006; Richards and Trinkaus 2009), the cultural skills of modern humans may possibly have been less sensitive to environmental changes, so that modern humans may have depended more strongly on social learning and, therefore, teaching evolved more easily.

The other explanation is that teaching ability evolved differently between the two hominins because of their different group structures. For example, teaching to non-relatives may barely evolve if the group is formed only by relatives and there is little chance to communicate with non-relatives. Ancient genome sequences suggest that Neanderthal groups may have been formed only of relatives (Prüfer et al. 2014; Castellano et al. 2014), which may possibly have inhibited the evolution of teaching to non-relatives. Neanderthals may have possessed different communication and cognitive abilities, as compared to modern humans (Green et al. 2010), which suggests that the two hominins may have had different teaching abilities. Given that numerous factors may have affected the evolution of Neanderthals and modern humans in a complex fashion, it is difficult to ascertain the most critical cause of the differences in teaching behavior between the two hominins. However, we can speculate that the different subsistence strategies and group structures may have played an important role in the evolution of teaching behaviors.

Although the model is simple, it allows us to discuss how teaching affects human cultural evolution. The model shows that teaching may cause the evolution of social learning under an unstable environment where social learning cannot exist without teaching. The frequency of change in the utility of a cultural trait may depend on the property of the cultural trait. For example, information on the location of an animal may change often, but the strategy of how to trace the animal may seldom change. We may individually learn cultural traits whose utility often changes and socially those whose utility seldom changes. The model suggests that teaching entails social learning of cultural traits that need to be learned individually if teachers are absent. In other words, teaching broadens the variety of cultural traits learned socially. Compared with other animals, humans depend strongly on social learning, which may, in part, be supported by teaching. Moreover, since teaching increases the dependence on social learning, it may strengthen social ties between individuals, which would lead to closer cooperation within a group. Indirect reciprocity is one of the most important mechanisms for the evolution of human cooperation, and is possibly supported by information on the reputation of others (Nowak and Sigmund 1998; Ohtsuki et al. 2009). The existence of teaching may be important for the spread of information on reputation. We like to gossip. Furthermore, mutual teaching may give rise to the emergence of discussion. That is, when two persons have different information and attempt to teach one another, discussion may arise. Animals never discuss, and the emergence of discussion may promote the evolution of logical thinking, which is one of the most important factors in human intelligence.

In conclusion, when teaching behavior is culturally transmitted and affects the number of cultural traits each teacher transmits, teaching to non-relatives evolves when the dependence on social learning is strong. The cost of teaching does not inhibit the spread of teaching in the population. Teaching may have affected human cultural evolution in numerous situations. Despite the uncertainty as to how teaching influenced the replacement of Neanderthals by modern humans, it is important to consider the evolution of behaviors that relate to cultural transmission mechanisms when investigating cultural evolution.

Acknowledgements This research was supported in part by Grant-in-Aid for Scientific Research on Innovative Areas No. 25118006 "The Evolutionary Origin and Neural Basis of the Empathetic Systems" to Hisashi Ohtsuki.

Appendix 1

Let x be the fraction of cultural traits that are beneficial for offspring in the parental generation. Then, when all organisms use the strategy (n^*, m^*), they learn n^* cultural traits individually where proportion r is beneficial so that, on average, rn^* beneficial cultural traits are acquired through individual learning. Organisms also learn m^* cultural traits socially, where proportion x is beneficial, and those beneficial cultural traits are learned accurately with probability a, so that on average, axm^* beneficial cultural traits are acquired by social learning. Thus, the proportion of beneficial cultural traits in the offspring generation is

$$\frac{rn^* + axm^*}{n^* + m^*}. \tag{3.13}$$

Since a fraction s of cultural traits that are beneficial for offspring remain beneficial in the grand-offspring generation, the fraction of cultural traits that are beneficial for grand-offspring in the offspring generation, x', is

$$x' = \frac{rn^* + axm^*}{n^* + m^*}s. \tag{3.14}$$

Since $x' = x = \hat{x}$ at equilibrium, we have

$$\hat{x} = \frac{rsn^*}{n^* + m^* - asm^*} \tag{3.15}$$

and the fraction of beneficial cultural traits always converges to $\hat{x}$ because

$$x' - \hat{x} = \frac{asm^*}{n^* + m^*} (x - \hat{x}). \tag{3.16}$$

Appendix 2

Ignoring double mutants and pleiotropy among learning capacities, the necessary and sufficient condition for the strategy $(\hat{n}, \hat{m})$ to be evolutionarily stable for the invasion of rare mutants of small effect is

$$\begin{cases} \left.\frac{\partial W(B,A)}{\partial n}\right|_{(n,m)=(\hat{n},\hat{m})} = 0 \text{ and } \left.\frac{\partial^2 W(B,A)}{\partial n^2}\right|_{(n,m)=(\hat{n},\hat{m})} < 0 \\ \text{or} \\ \hat{n} = 0 \text{ and } \left.\frac{\partial W(B,A)}{\partial n}\right|_{(n,m)=(\hat{n},\hat{m})} < 0 \end{cases} \tag{3.17}$$

$$\begin{cases} \left.\frac{\partial W(B,A)}{\partial m}\right|_{(n,m)=(\hat{n},\hat{m})} = 0 \text{ and } \left.\frac{\partial^2 W(B,A)}{\partial m^2}\right|_{(n,m)=(\hat{n},\hat{m})} < 0 \\ \text{or} \\ \hat{m} = 0 \text{ and } \left.\frac{\partial W(B,A)}{\partial m}\right|_{(n,m)=(\hat{n},\hat{m})} < 0 \end{cases} \tag{3.18}$$

Since $W(B, A) = (w + rn + a\hat{x}m) e^{-cn-dm}$, we have

$$\begin{cases} r = c(w + r\hat{n} + a\hat{x}\hat{m}) \\ \text{or} \\ \hat{n} = 0 \text{ and } r < c(w + r\hat{n} + a\hat{x}\hat{m}) \end{cases} \tag{3.19}$$

$$\begin{cases} a\hat{x} = d(w + r\hat{n} + a\hat{x}\hat{m}) \\ \text{or} \\ \hat{m} = 0 \text{ and } a\hat{x} < d(w + r\hat{n} + a\hat{x}\hat{m}) \end{cases} \tag{3.20}$$

where

$$\hat{x} = \frac{rs\hat{n}}{\hat{n} + \hat{m} - as\hat{m}}. \tag{3.21}$$

Solving them, we have the ESS $(\hat{n}, \hat{m})$ shown in Table 3.1.

Appendix 3

Let us consider the stability of the equilibrium$(\hat{n}, \hat{m}, \hat{p})$ against a perturbation. As shown above, $f(y) = csy^2 - s(d + act)y + td = 0$ has the larger root between $1 < y < t$ when $cy > d$, $cw < r$, $acsy > d$, and $t(acs - d) > s(c - d)$ are satisfied, and has two roots between $1 < y < t$ when $cy > d$, $cw < r$, $acsy > d$, $t(acs - d) < s(c - d)$, $act > 2c - d$, and $s(d + act)^2 - 4ctd > 0$ are satisfied. Consider a perturbation, for example, a slight change of environmental stability resulted in slight changes of learning strategy and the proportion of teachers in the population from the equilibrium $(\hat{n}, \hat{m}, \hat{p})$. Then, assuming that the environment reverts to the previous stability, let us consider whether learning strategy and the proportion of teachers in the population also revert to the previous state (equilibrium). Assume that the present state is $(n, m, p) = (\tilde{n}(\hat{p} + \Delta), \tilde{m}(\hat{p} + \Delta), \hat{p} + \Delta)$ where $|\Delta| << 1$. From Eq. (3.4), provided the population is fixed for the learning strategy, the proportion of teachers converges to

$$\begin{aligned} \tilde{p}(\tilde{n}(\hat{p}+\Delta), \tilde{m}(\hat{p}+\Delta)) &= \frac{(at-1)\tilde{m}(\hat{p}+\Delta) - \tilde{n}(\hat{p}+\Delta)}{(t-1)[\tilde{n}(\hat{p}+\Delta) + \tilde{m}(\hat{p}+\Delta)]} \\ &= \frac{1}{t-1}\left[\frac{t\{acs[y+(t-1)\Delta]-d\}}{s\{c[y+(t-1)\Delta]-d\}} - 1\right]. \end{aligned} \tag{3.22}$$

Writing that

$$\tilde{p}(\tilde{n}(\hat{p} + \Delta), \tilde{m}(\hat{p} + \Delta)) = \hat{p} + \lambda\Delta, \tag{3.23}$$

the equilibrium is unstable if $|\lambda| > 1$ and stable if $|\lambda| < 1$. Then, Eq. (3.22) can be rewritten as

$$\frac{y-1}{t-1} + \lambda\Delta = \frac{1}{t-1}\left[\frac{t\{acs[y + (t-1)\Delta] - d\}}{s\{c[y + (t-1)\Delta] - d\}} - 1\right], \tag{3.24}$$

i.e.,

$$y + (t-1)\lambda\Delta = \frac{t[acsy - d + acs(t-1)\Delta]}{s[cy - d + c(t-1)\Delta]}. \tag{3.25}$$

From Eq. (3.9), $y = 1 - \hat{p} + \hat{p}t$ satisfies $sy(cy - d) = t(acsy - d)$, so that Eq. (3.25) can be rewritten as

$$\begin{aligned} s(t-1)\Delta[\lambda(cy - d) + cy] + c\lambda(t-1)^2\Delta^2 \\ = acst(t-1)\Delta. \end{aligned} \tag{3.26}$$

Neglecting the term of Δ^2, we have

$$\lambda\left(cy-d\right)+cy=act, \tag{3.27}$$

i.e.,

$$\lambda=-1+\frac{act-d}{cy-d}. \tag{3.28}$$

Since $cy-d>0$ and $act-d>0$ provided $f(y)=0$ has a root between $1<y<t$, $|\lambda|>1$ is satisfied when

$$act-d>2\left(cy-d\right), \tag{3.29}$$

i.e.,

$$y<\frac{d+act}{2c}. \tag{3.30}$$

That is, the equilibrium is unstable if Ineq. (3.30) is satisfied and stable if Ineq. (3.30) is reversed. Since the axis of symmetry of $f(y)$ is $(d+act)/2c$, the smaller root of $f(y)=0$ is unstable and the larger root is stable.

References

Akazawa T, Aoki K, Bar-Yosef O (1998) Neandertals and modern humans in western Asia. Plenum Press, New York

Aoki K (2013) Determinants of cultural evolutionary rates. In: Akazawa T, Nishiaki Y, Aoki K (eds) Dynamics of learning in Neanderthals and modern humans, volume 1: cultural perspectives, Replacement of Neanderthals by modern humans series. Springer, Tokyo, pp 199–210

Aoki K, Lehmann L, Feldman MW (2011) Rates of cultural change and patterns of cultural accumulation in stochastic models of social transmission. Theor Popul Biol 79:192–202

Aoki K, Wakano JY, Feldman M (2013) Modeling the evolution of parental altruistic teaching. In: Terashima H (ed) Replacement of neanderthals by modern humans: testing evolutionary models of learning. Proceedings of the 8th conference on RNMH, pp 70–71

Bar-Yosef O (2002) The Upper Paleolithic revolution. Annu Rev Anthropol 31:363–393

Blackmore S (2000) The meme machine. Oxford University Press, Oxford

Caro TM, Hauser MD (1992) Is there teaching in nonhuman animals? Q Rev Biol 67:151–174

Castellano S, Parra G, Sánchez-Quinto FA, Racimo F, Kuhlwilm M, Kircher M, ... Pääbo S (2014) Patterns of coding variation in the complete exomes of three Neandertals. Proc Natl Acad Sci 111:6666–6671

Castro L, Toro MA (2014) Cumulative cultural evolution: the role of teaching. J Theor Biol 347:74–83

Champagne FA, Meaney MJ (2007) Transgenerational effects of social environment on variations in maternal care and behavioral response to novelty. Behav Neurosci 121:1353–1363

Dean LG, Kendal RL, Schapiro SJ, Thierry B, Laland KN (2012) Identification of the social and cognitive processes underlying human cumulative culture. Science 335:1114–1118

de Waal FB (2008) Putting the altruism back into altruism: the evolution of empathy. Annu Rev Psychol 59:279–300

Feldman MW, Aoki K, Kumm J (1996) Individual versus social learning: evolutionary analysis in a fluctuating environment. Anthropol Sci 104:209–232

Fogarty L, Strimling P, Laland KN (2011) The evolution of teaching. Evolution 65:2760–2770

Gonzalez A, Lovic V, Ward GR, Wainwright PE, Fleming AS (2001) Intergenerational effects of complete maternal deprivation and replacement stimulation on maternal behavior and emotionality in female rats. Dev Psychobiol 38:11–32

Green RE, Krause J, Briggs AW et al (2010) A draft sequence of the Neandertal genome. Science 328:710–722

Henrich J (2004) Demography and cultural evolution: how adaptive cultural processes can produce maladaptive losses-the Tasmanian case. Am Antiq 69:197–214

Hewlett BS, Fouts HN, Boyette AH, Hewlett BL (2011) Social learning among Congo Basin hunter-gatherers. Philos Trans R Soc B 366:1168–1178

Ihara Y (2008) Spread of costly prestige-seeking behavior by social learning. Theor Popul Biol 73:148–157

Jelinek AJ (1977) The Lower Paleolithic: current evidence and interpretations. Annu Rev Anthropol 6:11–32

Kobayashi Y, Aoki K (2012) Innovativeness, population size and cumulative cultural evolution. Theor Popul Biol 82: 38–47

Kuhn SL, Stiner MC (2006) What's a mother to do? The division of labor among Neandertals and modern humans in Eurasia. Curr Anthropol 47:953–980

Maynard Smith J (1982) Evolution and the theory of games. Cambridge University Press, Cambridge

McBrearty S (2012) Sharpening the mind. Nature 491:531–532

McBrearty S, Brooks A (2000) The revolution that wasn't: a new interpretation of the origin of modern human behavior. J Hum Evol 39:453–563

Nakahashi W (2010) Evolution of learning capacities and learning levels. Theor Popul Biol 78:211–224

Nakahashi W (2013a) Evolution of improvement and cumulative culture. Theor Popul Biol 83:30–38

Nakahashi W (2013b) Cultural evolution and learning strategies in hominids. In: Akazawa T, Nishiaki Y, Aoki K (eds) Dynamics of learning in Neanderthals and modern humans, volume 1: cultural perspectives, Replacement of Neanderthals by modern humans series. Springer, Tokyo, pp 245–254

Nakahashi W (2013c) A mathematical model of cultural interactions between modern and archaic humans. In: Akazawa T, Nishiaki Y, Aoki K (eds) Dynamics of learning in Neanderthals and modern humans, volume 1: cultural perspectives, Replacement of Neanderthals by Modern humans series. Springer, Tokyo, pp 255–263

Nakahashi W (2014) The effect of cultural interaction on cumulative cultural evolution. J Theor Biol 352:6–15

Nowak MA, Sigmund K (1998) Evolution of indirect reciprocity by image scoring. Nature 393:573–577

Ohtsuki H, Iwasa Y, Nowak MA (2009) Indirect reciprocity provides only a narrow margin of efficiency for costly punishment. Nature 457:79–82

Pigeot N (1990) Technical and social actors flintknapping specialists and apprentices at Magdalenian Etiolles. Archaeol Rev Camb 9: 126–141

Prüfer K, Racimo F, Patterson N, Jay F, Sankararaman S, Sawyer S, ... Pääbo S (2014) The complete genome sequence of a Neanderthal from the Altai Mountains. Nature 505:43–49

Richards MP, Trinkaus E (2009) Isotopic evidence for the diets of European Neanderthals and early modern humans. Proc Natl Acad Sci 106:16034–16039

Takakura J (2013) Using lithic refitting to investigate the skill learning process: lessons from Upper Paleolithic assemblages at the Shirataki sites in Hokkaido, Northern Japan. In: Akazawa T, Nishiaki Y, Aoki K (eds) Dynamics of learning in Neanderthals and modern humans, volume 1: cultural perspectives, Replacement of Neanderthals by modern humans series. Springer, Tokyo, pp 151–172

Terashima H (2013) The evolutionary development of learning and teaching strategies in human societies. In: Akazawa T, Nishiaki Y, Aoki K (eds) Dynamics of learning in Neanderthals and modern humans, volume 1: cultural perspectives, Replacement of Neanderthals by modern humans series. Springer, Tokyo, pp 141–150

Thornton A, Raihani NJ (2008) The evolution of teaching. Anim Behav 75:1823–1836

Villa P, Soriano S (2010) Hunting weapons of Neanderthals and early modern humans in South Africa: similarities and differences. J Anthropol Res 66:5–38

Zilhão J (2013) Neandertal-modern human contact in Western Eurasia: issues of dating, taxonomy, and cultural associations. In: Akazawa T, Nishiaki Y, Aoki K (eds) Dynamics of learning in Neanderthals and modern humans, volume 1: cultural perspectives, Replacement of Neanderthals by modern humans series. Springer, Tokyo, pp 21–58

A Population-Genetics Based Model for Explaining Apparent Cultural Continuity from the Middle to Upper Palaeolithic in Eurasia

4

Yutaka Kobayashi, Seiji Kadowaki, and Masaki Naganuma

Abstract

There has been heated debate over the interpretation of the Châtelperronian industry as a result of Neanderthals' acculturation through direct or distant influence from *Homo sapiens*. In contrast, relatively little attention has been paid to the possibility that *Homo sapiens* populations may also have undergone acculturation through the influence from indigenous archaic hominins during the expansion of their habitat range. In this article, we argue that the acculturation of the invading *Homo sapiens* population may provide a consistent explanation for varying degrees of cultural continuity across the MP-UP transition widely observed in Eurasia. We simulate the spread of a *Homo sapiens* population in a region preoccupied by an archaic hominin population by means of mathematical modeling to investigate the possible effects of the acculturation of the *Homo sapiens* population on the pattern of the cultural transition associated with biological replacement. The results suggest that cultural continuity is likely to be observed when the *Homo sapiens* population is initially small and/or spreads slowly. This in turn implies that whether the local physical environment was relatively favorable to *Homo sapiens* or to the archaic hominin is an important determinant of the degree of the continuity.

Keywords

Simulation • MP-UP transition • Cultural transmission • Acculturation • Demography

4.1 Introduction

The Late Pleistocene Out-of-Africa models of *Homo sapiens*, as variously delineated by genetic and palaeoanthropological studies (e.g., Smith and Ahern 2013), have greatly influenced interpretations of behavioral and cultural records from the Middle and early Upper Palaeolithic periods in Eurasia. For example, several archaeologists have proposed that lithic techno-typological similarity among some of the Initial Upper Palaeolithic or "transitional" industries in the Levant (i.e., the Emiran), central Europe (i.e., the Bohunician), and the Altai (i.e., Kara Bom) can be regarded as "road-signs" of *Homo sapiens* groups expanding their geographic distributions from the Levant to other parts of Eurasia (Bar-Yosef and Belfer-Cohen 2013; Škrdla 2003; Svoboda 2007; Tostevin 2007). Similarly, recently increasing Middle Palaeolithic (MP) records in the Arabian Peninsula have greatly contributed to discussions on the geographic expansion of *Homo sapiens* along the "southern route" (e.g., Armitage et al. 2011; Delagnes et al. 2013; Rose et al. 2011).

On the other hand, given possible encounters between dispersing *Homo sapiens* and indigenous archaic hominin

Y. Kobayashi (✉)
Department of Management, Kochi University of Technology, Tosayamada, Kami-city, Kochi 782-8502, Japan
e-mail: kobayashi.yutaka@kochi-tech.ac.jp

S. Kadowaki
University Museum, Nagoya University, Nagoya, Japan

M. Naganuma
Center for Ainu & Indigenous Studies, Hokkaido University, Sapporo, Japan

A. Mesoudi and K. Aoki (eds.) *Learning Strategies and Cultural Evolution during the Palaeolithic*, Replacement of Neanderthals by Modern Humans Series, DOI 10.1007/978-4-431-55363-2_4,

populations, as suggested by genetic studies (Green et al. 2010; Prüfer et al. 2014), interactions between the two groups can also be represented in archaeological records. This viewpoint is illustrated by previous and recent interpretations of the Châtelperronian industry as a result of Neanderthals' acculturation through direct or distant influence from *Homo sapiens* groups (Mellars 1999, 2005; Hublin 2013). On the basis of renewed chronological data of European "transitional" industries, some researchers propose that the Châtelperronian was created by Neanderthals, receiving indirect, distant cultural influence from *Homo sapiens* groups who had already colonized parts of central/eastern Europe by this time (Hublin 2013; Hublin et al. 2012). In central Europe, Tostevin (2007) examined a similar acculturation scenario for the Szeletian industry that hypothetically emerged from the indigenous Micoquian through acculturation by the invasive Bohunician Behavioral Package. Because the Bohunician lithic technology is regarded to have originated in the Levant, its influence over the Micoquian and the Szeletian, i.e., indigenous cultures in central Europe, can provide implications for cultural interactions between local Neanderthals and *Homo sapiens* groups that are supposed to have dispersed from the Levant.

While these previous studies focus on a scenario of indigenous populations that are more or less acculturated by invasive groups, there have been few discussions on the process of invasive groups receiving cultural influences from local groups. The latter case might initially sound counter-intuitive but not unlikely for various reasons. For example, indigenous populations must have acquired behaviors and technology adapted to the surrounding environments, with which invasive groups are going to cope. It is also possible that modern humans found archaic cultural traits somehow attractive. A cultural diffusion from Neanderthals to modern humans has been recently proposed as one of possible explanations for the discovery of formal bone tools, *lissoirs*, from two Mousterian sites in southwest France (Soressi et al. 2013). This special type of bone tools continues to occur in subsequent "transitional" and Upper Palaeolithic cultures, including the Châtelperronian, Protoaurignacian, and Aurignacian.

We propose this scenario of acculturation on the side of invasive groups as a potential interpretive framework for archaeological records that apparently show varying degrees of continuity from the MP to UP, during which local, archaic hominin groups are supposed to have been replaced or assimilated by incoming *Homo sapiens* populations. For example, in the Levant, some researchers point out similarity or gradual differences in core reduction technology between the late Levantine Mousterian industry, associated with Neanderthal remains at Dederiyeh, Amud, and Kebara Caves, and the Initial Upper Palaeolithic or the Emiran (Belfer-Cohen and Goring-Morris 2007; Meignen 2012; Olszewski 2009; but see Tostevin 2003 for a different view). Both industries are similarly characterized by prepared core technologies designated as Levallois or adapted Levallois for producing triangular blanks, i.e., morphologically Levallois points. Although we have little evidence regarding the makers of the IUP/Emiran, fragmentary human fossils from Ksal Akil Layer XXV (Ethelruda) and Üçağızlı are reported to include features of *Homo sapiens* (Douka et al. 2013; Kuhn et al. 2009).

In the Zagros region, the Middle Palaeolithic industry, i.e., the Zagros Mousterian, is followed by the Early Baradostian which marks the beginning of the Upper Palaeolithic. The former industry is associated with Neanderthal fossil remains at Shanidar and Bisitun Caves. Although the two lithic industries clearly differ from each other in core reduction technology, the Early Baradostian assemblages at Warwasi rockshelter include some tool types (e.g., retouched points and side scrapers) and truncated-faceted cores that characterize the Zagros Mousterian (Olszewski 2009; Olszewski and Dibble 1994). This observation is supported by recent reanalyses of the formation processes of stratified deposits at this site, suggesting the integrity of the stratified assemblages (Tsanova 2013).

In this study, we assume cultural contact between indigenous and invasive groups as one of the possible factors for the above mentioned techno-typological similarity between the late MP and early UP in west Asia. This assumption is supported by chronological overlap between some late MP and early UP sites in west Asia (Kadowaki 2013) although it needs to be verified further with additional, reliable radiometric dates like the case recently made for Europe (Higham et al. 2014). Interactions between Neanderthals and *Homo sapiens* in west Asia are also recently suggested by their genetic analyses (Green et al. 2010; Gibbons 2014). However, considering different views on the timing and nature of the MP-UP transition in west Asia (Shea 2008; Rebollo et al. 2011; Douka et al. 2013), we do not exclude other possible reasons, such as independent innovation or behavioral convergence, for some of the apparent cultural similarity between the late MP and early UP.

On the basis of lithic techno-typological continuity between the UP and earlier assemblages or industries in eastern Eurasia, several archaeologists have recently reappraised the multi-regional model for the origin of *Homo sapiens* (e.g., Derevianko 2011; Derevianko and Shunkov 2012). However, in light of the hypothesis that *Homo sapiens* could have experienced cultural influences from indigenous archaic hominin populations while they expanded their habitat range, the observed techno-typological continuity does not necessarily imply biological continuity. This is theoretically conceivable as we argue later (see Discussion); the theory predicts that the pattern of the cultural transition associated with a biological transition is largely dependent on the mode of the underlying demographic process. Importantly, cultural continuity is rather likely to occur for a broad range of parameter values.

In central Asia and southern Siberia, it has been reported that lithic assemblages changed continuously from MP to EUP based on the archaeological records from several multilayered sites. The excavators interpreted the long-term preservation of the Levallois technique and MP type tools and coexistence with UP type tools in the toolkits during the MP-UP transition as autochthonous formation of Upper Paleolithic without any influence from invasive groups (Derevianko ibid.). Several local MP assemblages in these regions (i.e., Sibiryachikha industries) are associated with Neanderthals fossil remains at Teshk-Tash, Okladnikov and Chagyrskaya caves (e.g., Okladnikov 1940; Derevianko et al. 2013), while the local EUP assemblage similar to preceding MP layers has been uncovered in association with personal ornaments, bone tools and archaic human bone ("Denisovian") from Denisova cave stratum 11 (Krause et al. 2010).

In China, elaborated bone tools and personal ornaments are considered as principal markers of Upper Paleolithic or modern human behaviors and several cases, such as Shuidonggou and Shiyu, suggest the emergence of blade technique and artifacts of symbolic behaviors (e.g., Qu 2012; Hou et al. 2013; Qu et al. 2013; Guan et al. 2012). On the other hand, there is a controversy over the definition of the MP industries or period (Gao 1999; Norton et al. 2009), and taxonomic attribution of archaic indigenous population earlier than *Homo sapiens* groups is still an open question (e.g., Wu 2004; Stringer 2012). Moreover, traditional local core-flake industries (mode I) continued for a long time period from the Lower to Upper Paleolithic, and some of them are possibly associated with bone tools, pendants (at Xiaogushan cave) and *Homo sapiens* fossils (at Zhoukoudian Upper cave) (e.g., Qu et al. ibid.; Zhang et al. 2010).

As these examples suggest, the MP-UP transition is not always a discontinuous shift but rather shows diverse degrees and modes of continuity depending on regions. Thus, if *Homo sapiens* was the maker of EUP industries as usually postulated, we need to reconsider the validity of our implicit assumption that biological replacement generally causes discontinuous cultural shifts. In this paper, we construct a mathematical model to simulate the invasion of a population of *Homo sapiens* into a region preoccupied by an archaic hominin population and the associated acculturation on the side of *Homo sapiens*. The purpose is to identify the major determinants of the mode of cultural continuity associated with biological replacement. We analyze the model using population-genetic approximation techniques and individual-based simulations. Based on the results, we discuss the possible causes of spatial variation in the pattern of cultural shifts during the MP-UP transition.

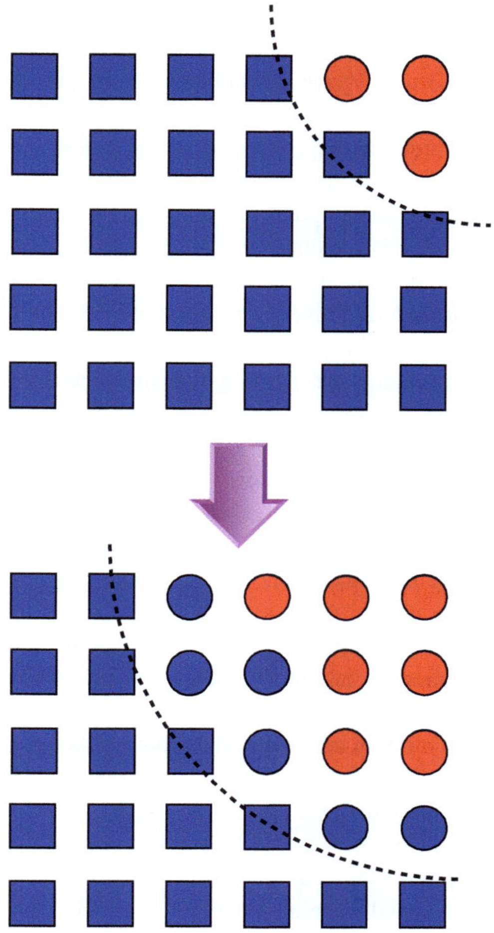

Fig. 4.1 Schematic diagram representing the situation considered in the model. *Squares* and *circles* represent archaic and modern humans, respectively. Cultural phenotypes are represented by colors, where *blue* and *red* represent C_A and C_M, respectively. *Dotted lines* represent the boundary between the two species. Modern humans are initially rare and all bear C_M (*upper panel*). As time goes, modern humans gradually replace archaic humans and at the same time the indigenous phenotype C_A spreads into the population of modern humans due to interspecific oblique transmission (*lower panel*)

4.2 Mathematical Model

4.2.1 Description

We note that the following model is too simplistic in many respects as it is intended to extract the essence of the problem and also to maintain analytical tractability. Imagine that a small population of *Homo sapiens* is invading a region preoccupied by a population of an archaic hominin species (see Fig. 4.1). Let us for simplicity call the former and the latter modern and archaic humans, respectively. Suppose that this region accommodates a total of N individuals of the archaic or modern humans, where N is constant. Suppose, again for simplicity, that both populations reproduce asexually, so that each individual has only one parent. Generations are overlapping insofar as cultural transmission occurs. Modern humans are initially rare in frequency, given that they have just arrived in this region. Let p_0 denote this initial frequency, and p the frequency of modern humans in any subsequent generation. Because the anatomical and genetic features of present-day human populations are predominantly derived from the ancestral modern humans, the model assumes that the modern humans increase in frequency to eventually replace the archaic humans. To simulate this increase, we formally assume that the relative fitnesses of modern and the archaic humans are 1 and 1-s, respectively. We do not specify the cause of this fitness difference; it may be cultural and/or genetic and may even depend on regions. Since the population size is finite, modern humans can go extinct with a positive probability due to random sampling drift despite selective superiority. However, we set Ns and Np_0 sufficiently large, so that the probability of extinction for modern humans is negligibly small.

We assume that cultural transmission may occur from the archaic to modern humans, while we neglect transmission in the opposite direction for simplicity. We assume that the archaic and modern humans initially have distinctive phenotypes, referred to as C_A and C_M, respectively. A phenotype in this case may for example be the use of a particular type of tools or the use of a particular stone-processing technique such as the Levallois technique. The phenotypes are transmitted through social learning from one generation to the next. We assume that each modern human learns vertically and obliquely with probabilities 1-γ and γ, respectively. When a modern human learns obliquely, he/she may copy one of the archaic humans with a certain probability. As a result, the phenotype C_A can gradually spread into the population of modern humans. We assume that the probability of copying an archaic human when transmission is oblique, denoted by $X(u,p)$, is given by

$$X(u, p) = 1 - Y(u, p) = \frac{u(1-p)}{(1-u)p + u(1-p)}, \quad (4.1)$$

where u is a parameter indicating modern humans' preference for archaic role models as opposed to modern role models and may range from 0 to 1; $u = 0$ ($u = 1$) implies that modern humans always choose modern (archaic) humans as their role models when they learn obliquely. In the special case of $u = 1/2$, there is no bias in choice. In the above equation, $Y(u,p) = 1\text{-}X(u,p)$ gives the probability that a modern human copies a modern human. It must be noted that the probability of learning C_A by oblique transmission is in general higher than given by Eq. (4.1) because modern humans may also bear C_A. We assume that transmission of a phenotype is always faithful and perfect, so that social learning of a phenotype never fails when attempted.

We formally assume that the two phenotypes C_A and C_M are selectively neutral. In reality, different phenotypes would contribute to fitness differently. However, we assume that the influences of the focal traits on fitness are so small compared to the overall fitness difference between the two species that we can regard the traits approximately as selectively neutral. We will investigate below how the pattern of cultural dynamics depends on selection coefficient s, initial frequency p_0, oblique transmission rate γ, and preference for archaic role models u. We are particularly interested in the probability that C_A goes extinct or the conditional expected time until the extinction of C_A.

4.2.2 Population Dynamics

Prior to the analysis of cultural dynamics, let us consider the population dynamics of modern humans. Given that $Ns >> 1$ and $Np_0 >> 1$, we may neglect the effect of random genetic drift as far as population dynamics is concerned. Thus, the dynamics of the frequency of modern humans, p, is described by the following difference equation:

$$p' = \frac{p}{p + (1-p)(1-s)} = \frac{p}{1 - s(1-p)}, \quad (4.2)$$

where p' represents the frequency in the next generation. The difference in the frequencies between succeeding generations, denoted by Δp, is given by subtracting p from both sides of Eq. (4.2):

$$\Delta p = \frac{sp(1-p)}{1 - s(1-p)}. \quad (4.3)$$

Furthermore, if as we assume $s \ll 1$, the denominator (the average fitness) is approximately one and the difference Δp can be replaced by the differential dp/dt. Thus, we obtain the following approximate differential equation:

$$\frac{dp}{dt} \approx sp(1-p). \quad (4.4)$$

This equation can easily be solved by means of logit transformation (i.e. let $z = p/(1-p)$) to yield the frequency of modern humans as a function of generation t:

$$z = \frac{p}{1-p} \approx \frac{p_0}{1-p_0} e^{st} \quad (4.5)$$

or

$$p \approx \frac{p_0 e^{st}}{1 - p_0 + p_0 e^{st}}. \quad (4.6)$$

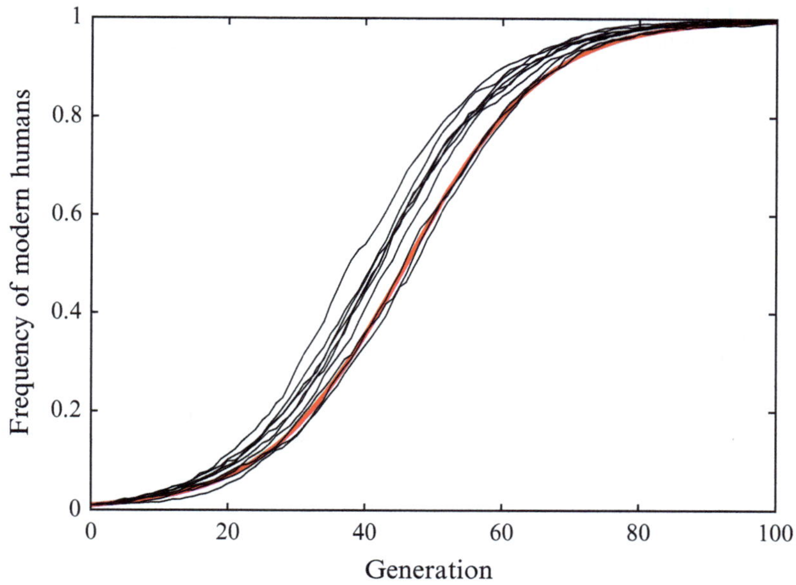

Fig. 4.2 The population dynamics of modern humans in the model. *Black lines* represent 10 sample paths obtained from individual-based simulations and the *red line* represents the prediction of the analytical approximation. Parameter values are $N = 10{,}000$, $s = 0.1$, and $p_0 = 0.01$

Figure 4.2 compares the results of individual-based simulations and Eq. (4.6). As the figure shows, the results of individual-based simulations vary between runs even under fixed parameter values due to random genetic drift. There are analytical techniques to take this stochastic variation into account (e.g., Otto and Barton 1997). However, we refrain from using those techniques here because they are so complicated that our main conclusions would rather be obscured. Our deterministic approximation Eq. (4.6) is justified if the initial *number* of modern humans is sufficiently large (say, >50), so that the effect of genetic drift is negligible from the start.

4.2.3 Patterns of Cultural Dynamics

Here we loosely categorize the patterns of cultural dynamics based on results of individual-based simulations. Detailed mathematical analysis for each pattern is provided in later subsections. We found the following four distinguishable patterns.

Pattern A (Rapid extinction of C_A): C_A goes extinct quickly during biological replacement (Fig. 4.3a). This pattern occurs when selection is strong (s is large), the initial frequency of modern humans is high (p_0 is large), the preference for archaic role models is low (u is small), and/or the oblique transmission rate is low (γ is small).

Pattern B (Persistence of C_A): C_A remains at a high frequency during biological replacement and persists indefinitely (Fig. 4.3b). This pattern occurs when selection is weak (s is small), the initial frequency of modern humans is low (p_0 is small), the preference for archaic role models is high (u is large), and/or the oblique transmission rate is high (γ is large).

Pattern C (Slow extinction of C_A): C_A decreases to some extent in frequency during biological replacement. It persists for a while after the replacement but eventually goes extinct (Fig. 4.3c). This pattern occurs under conditions intermediate between the patterns A and B.

Pattern D (Temporary decrease of C_A): C_A decreases transiently in frequency during biological replacement. However, it eventually reaches fixation and persists indefinitely (Fig. 4.3d). This pattern also occurs under conditions intermediate between A and B.

4.2.4 Fixation Probability of C_A

Here we compute the eventual fixation probability of the archaic phenotype C_A in the population of modern humans. In terms of the four patterns described in the previous subsection, this is the probability that pattern B or D occurs. Note that the change in the frequency of C_A is stochastic due to random cultural drift. Thus, we need to cope with a complex stochastic process, in which biological replacement and random cultural drift occur simultaneously. We analyze this process using the following approximation technique. Given that selection favoring modern humans is strong ($Ns >> 1$), we may consider that the entire process consists of two stages. The first stage constitutes a fast process, in which modern humans replace the archaic ones. In this stage, the frequency of C_A in the population of modern humans increases due to interspecific transmission in an approximately deterministic manner. This deterministic process can be analyzed by the method used in the studies of

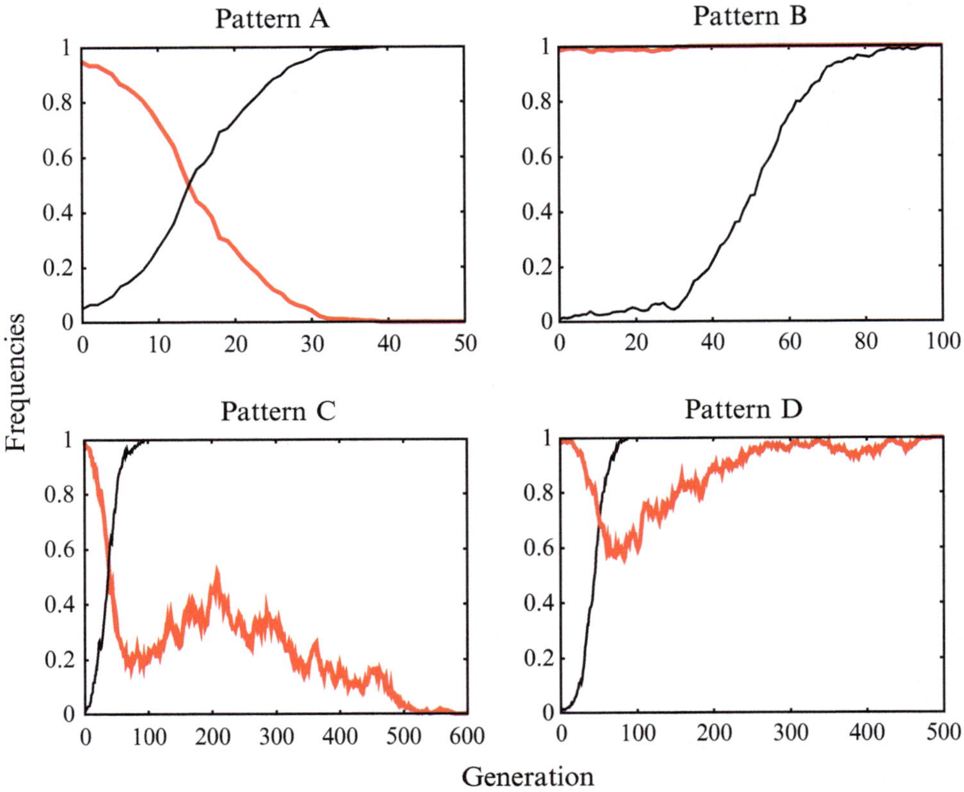

Fig. 4.3 Four distinct patterns of cultural transition associated with biological replacement. The *red* and *black* lines represent the frequencies of C_A and modern humans, respectively, in the entire population. Population size is $N = 1{,}000$ in all panels. Other parameter values are (*A*) $u = 0.01$, $q = 0.01$, $s = 0.2$, and $p_0 = 0.05$, (*B*) $u = 0.1$, $q = 0.1$, $s = 0.1$, and $p_0 = 0.01$, (*C*) $u = 0.05$, $q = 0.05$, $s = 0.1$, and $p_0 = 0.01$, (*D*) $u = 0.02$, $q = 0.05$, $s = 0.1$, and $p_0 = 0.01$

genetic hitchhiking (Maynard Smith and Haigh 1974). The second stage constitutes a relatively slow process, where the archaic humans are already extinct and the frequency of C_A in the population of modern humans fluctuates stochastically due to random cultural drift. This stage can be analyzed using diffusion-approximation techniques.

Let us first consider the first stage. Let q denote the frequency of C_A *in the population of modern humans*, so that the total frequency of C_A is given by $1-p+pq$. Provided that we may neglect the effect of sampling drift, the frequency dynamics of q is described by the following difference equation:

$$q' = (1-\gamma)\,q + \gamma\,[Y(u,p)\,q + X(u,p)], \qquad (4.7)$$

where q' represents the value of q in the next generation, given the current frequency q. The first term on the right hand side represents the contribution of vertical transmission, while the second term represents the contribution of oblique transmission. The first and second terms in the brackets represent the contributions of oblique learning from modern and archaic humans, respectively. The former term is multiplied by q because only a fraction q of modern humans bears C_A.

Subtracting q from both sides of Eq. (4.7) and assuming γ is sufficiently small, we can replace the difference q'-q by dq/dt to obtain the following approximate differential equation:

$$\frac{dq}{dt} \approx \gamma X(u,p)\,(1-q)\,. \qquad (4.8)$$

Substituting Eq. (4.6) into Eq. (4.8) yields

$$\frac{dq}{dt} \approx \frac{u\gamma\,(1-p_0)}{(1-u)\,p_0 e^{st} + u\,(1-p_0)}\,(1-q)\,. \qquad (4.9)$$

Solving this equation with the initial condition $q = 0$ yields

$$q \approx 1 - \left[Y(u,p_0) + X(u,p_0)\,e^{-st}\right]^{\frac{\gamma}{s}}. \qquad (4.10)$$

From this equation, we find that after biological replacement the frequency of C_A in the population of modern humans—and hence in the entire population—is given by

$$\lim_{t\to\infty} q \approx 1 - Y(u,p_0)^{\frac{\gamma}{s}}. \qquad (4.11)$$

In the second stage, the archaic humans are extinct and the frequency of C_A fluctuates due to sampling drift. Since C_A and C_M are selectively neutral, the fixation probability of C_A, denoted by π_A, is given by the initial frequency in the second process, which is the final frequency in the first process.

$$\pi_A \approx 1 - Y(u,p_0)^{\frac{\gamma}{s}}. \qquad (4.12)$$

Figure 4.4 compares the results of individual-based simulations and Eq. (4.12). As the figure shows, the probability of fixation of C_A is high when γ/s is large, p_0 is small, and u is large. In particular, these results imply that the archaic phenotype C_A is likely to persist when selection is weak and the initial frequency of modern humans is low; i.e. when modern humans slowly replace the archaic humans. It should be noticed that qualitative effects of parameters on the fixation probability of C_A would be unchanged even if one of the two phenotypes is weakly favored over the other. This is because the frequency of C_A quickly approaches to Eq. (4.11) as long as the selection pressure on the two phenotypes is sufficiently small compared to the overall fitness difference s. In this case, the selection pressure only affects the fixation probability of C_A in the second slow process, but this does not alter the effects of u, γ, s, and p_0. This justifies the use of the neutral model as an approximation of weakly selected cases.

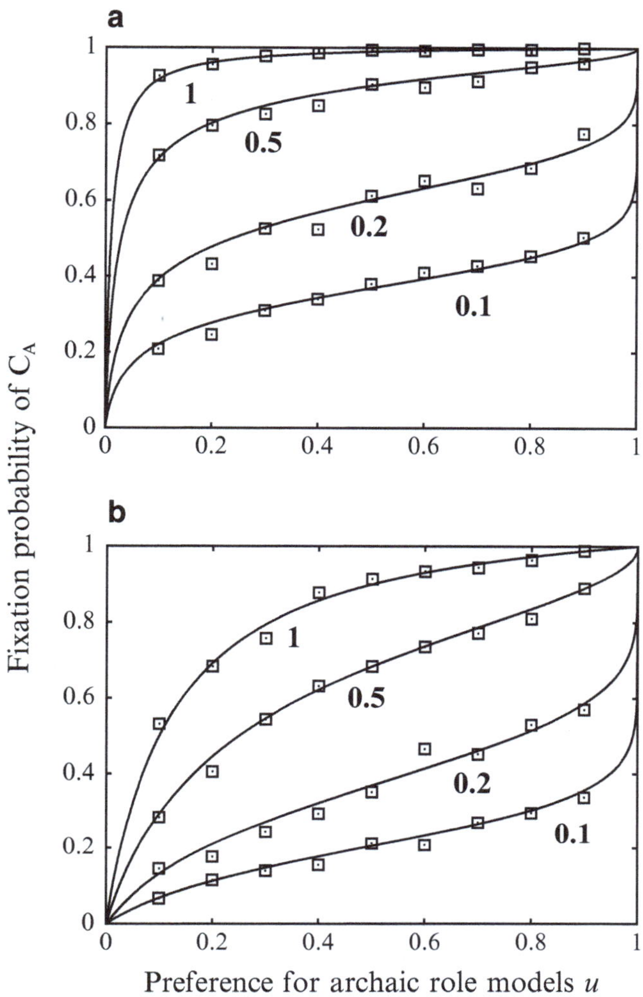

Fig. 4.4 Effect of parameters on the fixation probability of C_A. *Dots* and *lines* represent the results of individual-based simulations (computed using data of 500 replications) and analytical results, respectively. Population size is 5,000 in both panels. The initial frequency of modern humans is (**a**) $p_0 = 0.01$ and (**b**) $p_0 = 0.1$. The *number* next to each line represents the value of the ratio γ/s for the corresponding result

4.2.5 Expected Time Until C_A Is Lost

In Sect. 2.3, we showed that, even when C_A is eventually lost, it may persist for a long time after the biological replacement (Pattern C). Here, we compute the *conditional* expected time t_0 elapsed after the completion of biological replacement until the extinction of C_A. Here "conditional" means that we only consider the cases where extinction eventually occurs. For this purpose, we can use a formula derived by Kimura and Ohta (1969):

$$t_0 \approx -2N\left(\frac{q}{1-q}\right)\ln q, \tag{4.13}$$

where q is the initial frequency of C_A in the second process, which is the final frequency in the first process Eq. (4.11). Note that the coefficient $4N$ in the original article is replaced by $2N$ here because cultural phenotypes behave in the same way as alleles on haploid loci. Substituting Eq. (4.11) for q in Eq. (4.13) yields the following approximate formula:

$$t_0 \approx -2N\left(Y(u,p_0)^{-\frac{\gamma}{s}} - 1\right)\ln\left(1 - Y(u,p_0)^{\frac{\gamma}{s}}\right), \tag{4.14}$$

Figure 4.5 compares the results of individual-based simulations and Eq. (4.14). Comparison of Figs. 4.4 and 4.5 reveals that the time until the extinction of C_A tends to be long when the fixation probability of C_A is high. As the figure shows, the time until the extinction can be very long (several hundred generations) depending on parameter values. In such cases, we should consider that C_A effectively persisted. In this sense, the formula Eq. (4.12) and Fig. 4.4 may underestimate the probability that cultural continuity is observed.

4.2.6 Expected Time Until C_M Is Lost

In Sect. 2.3, we showed that phenotype C_A may temporarily decrease in frequency even when it eventually replaces C_M (Pattern D). Here, we compute the (conditional) expected time t_1 until phenotype C_A reaches fixation after the completion of biological replacement under the condition that the fixation of C_A occurs. For this purpose, we again use a formula derived by Kimura and Ohta (1969):

$$t_1 \approx -2N\left(\frac{1-q}{q}\right)\ln(1-q), \tag{4.15}$$

where q is again the initial frequency of C_A in the second process, which is the final frequency in the first process Eq.

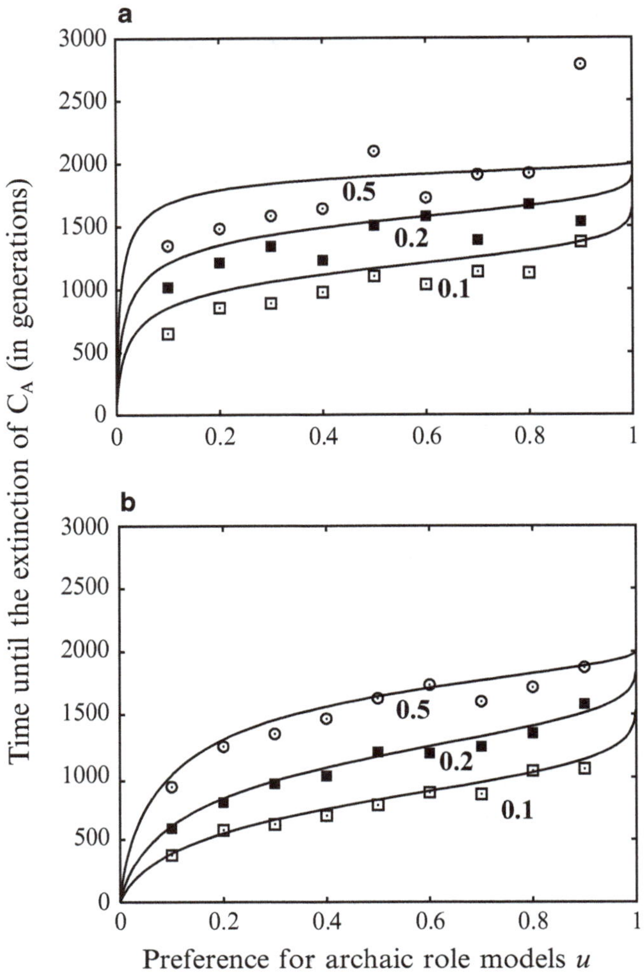

Fig. 4.5 Average number of generations until C_A is lost. *Dots* are results of individual-based simulations (averages of 500 replications) and *lines* are analytical results. Open squares, filled squares, and open circles are results for $\gamma/s = 0.1$, $\gamma/s = 0.2$, and $\gamma/s = 0.5$, respectively. The population size is 1,000 in both panels. The initial frequency of modern humans is (**a**) $p_0 = 0.01$ and (**b**) $p_0 = 0.1$. The number beside each line represents the value of γ/s for the corresponding result

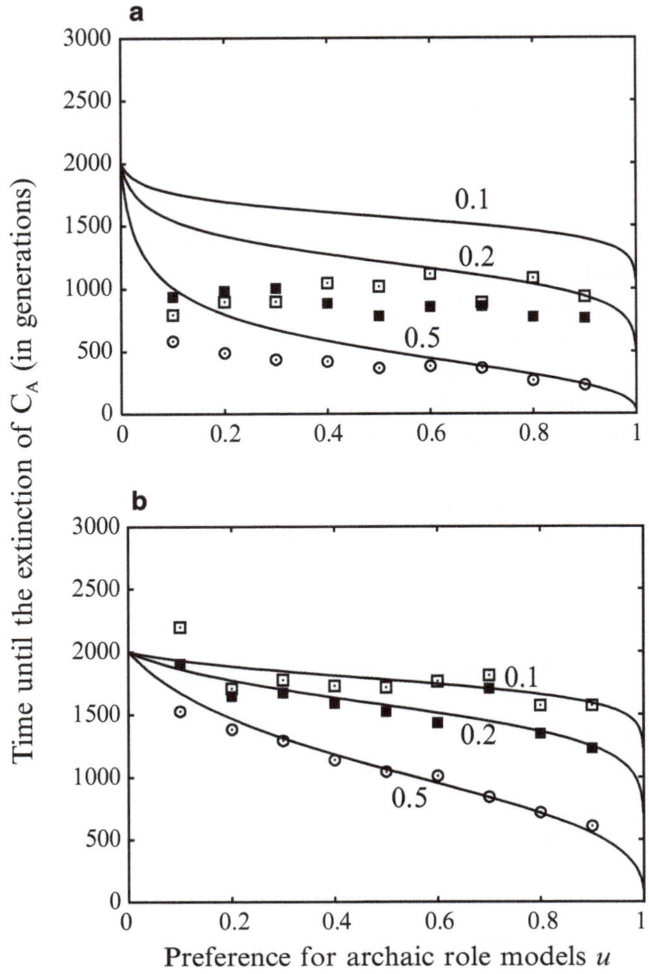

Fig. 4.6 Average number of generations until C_M is lost. *Dots* are results of individual-based simulations (averages of 500 replications) and *lines* are analytical results. Open squares, filled squares, and open circles are results for $\gamma/s = 0.1$, $\gamma/s = 0.2$, and $\gamma/s = 0.5$, respectively. The population size is 1,000 in both panels. The initial frequency of modern humans is (**a**) $p_0 = 0.01$ and (**b**) $p_0 = 0.1$. The number beside each line represents the value of γ/s for the corresponding result

(4.11). Substituting Eq. (4.11) for q in Eq. (4.15) yields the following approximate formula:

$$t_1 \approx -\frac{2N\gamma}{s}\left(\frac{Y(u,p_0)^{\frac{\gamma}{s}}}{1-Y(u,p_0)^{\frac{\gamma}{s}}}\right)\ln Y(u,p_0), \qquad (4.16)$$

Figure 4.6 compares the results of individual-based simulations and Eq. (4.16). Comparison between Figs. 4.4 and 4.6 reveals that in general the time until the fixation of C_A is long when the fixation probability of C_A is low. In Fig. 4.6a, where $Np_0 = 10$, the simulation results largely deviate from analytical results. Thus, in this case, the analytical approximation formula is not reliable. This is because the approximation assumes that the initial number of modern humans Np_0 is not very small. To check the mathematical soundness of our approximation method, we have confirmed that this deviation disappears for sufficiently large population size (results not shown).

4.3 Discussion

The four patterns found in our study have important implications for interpretation of archaeological records. That is, they show that, in the presence of acculturation on the side of the invader species, the degree of cultural continuity depends on underlying demographic dynamics in a non-trivial manner. Specifically, cultural continuity is likely to be observed when the invader species is initially rare and is only weakly favored over the indigenous species. This suggests that in regions where local environmental conditions were favorable to the indigenous population rather than to *Homo sapiens* culture tends to be continuous beyond the biological replacement. For example, the fact that archaeological records from eastern Eurasia show on average less clear transitions than those from western parts might imply that the spread of modern humans decelerated for some reasons in those regions. It would be worth investigating the relationship between the local environmental conditions during the MP-UP transition and the mode of cultural continuity over different regions to test the conclusion of the model.

Our model may also help understand the case in which only some tool types show continuity beyond the MP-UP transition while most of the old toolkits are replaced, as in the case of the transition from the Zagros Mousterian to the Early Baradostian. Namely, in our model, whether a cultural element is inherited to the new industry is determined largely by chance under intermediate conditions (Pattern C or D). Of course, in reality, modern humans may have had an ability to distinguish useful tools from less useful ones and hence they might have copied only useful tools while ignoring the rest. However, it is unlikely that humans can perfectly evaluate the usefulness of a tool, and hence we believe that our conclusion is qualitatively unchanged even if adaptive social learning is taken into consideration.

Perhaps one of the most important claims of the present study is that cultural continuity is not an obvious indicator of biological continuity, but rather provides mixed information about biological continuity and the way biological replacement might have occurred; for example, pattern B is observed either when extremely slow biological replacement occurred or when biological replacement did not occur at all. Although our study is aimed at understanding the MP-UP transition, the model itself applies to any other invasion events, including those within a single species (invasion of one population into another of the same species). Thus in a sense, the present study reveals a general methodological limitation of using artifact remains alone as clues for inferring underlying biological processes and emphasizes the importance of using biological remains in combination.

There are several restrictive assumptions in our model, which need to be relaxed in future work. For example, we did not take acculturation on the side of archaic hominin populations into account. If the Châtelperronian was created by Neanderthals under the influence of the invading culture of *Homo sapiens*, we need to incorporate bidirectional interspecific social learning to explain this case. Then, the behavior of the model may change, but we believe that the change would be only quantitative. More important, it is still premature to exclude the possibility that the Châtelperronian was made by modern humans. Given that there is no reliable evidence of cultural transmission from modern humans to Neanderthals, it is possible that interspecific cultural transmission was asymmetric. However, it would be worth generalizing the model to incorporate bidirectional cultural transmission.

A possible extension of the model would be to incorporate sexual reproduction, which may be important given the genetic evidence of interbreeding between archaic and modern humans (Green et al. 2010; Prüfer et al. 2014). Under interbreeding, the indigenous culture can be transmitted to *Homo sapiens* even through vertical transmission. If interbreeding is the major cause of acculturation, we expect that the direction of cultural influence coincides with the direction of gene flow between the two species. Specifically, if F1 hybrids of Neanderthals and modern humans were not able to mate and/or reproduce with Neanderthals, then the culture of *Homo sapiens* would have never been transmitted to Neanderthal populations. It would be interesting in future to estimate the rate of interspecific cultural transmission from genetic data and examine whether the model explains observed cultural continuity under the estimated rate.

Acknowledgments This research derives from a joint investigation of two research realms, entitled "Archaeological Research of the Learning Behaviors of the Neanderthals and Early Modern Humans" and "Research on Evolutionary Models of Human Learning Abilities", as part of the "Replacement of Neanderthals by Modern Humans" project that is funded by the Grant-in-Aid for Scientific Research on Innovative Areas (Grant No. 22101001) from the Ministry of Education, Culture, Sports, Science and Technology, Japan.

References

Armitage S, Jasim S, Marks A, Parker A, Usik V, Uerpmann H-P (2011) The southern route "Out of Africa": evidence for an early expansion of modern humans into Arabia. Science 331:453–456

Bar-Yosef O, Belfer-Cohen A (2013) Following Pleistocene road signs of human dispersals across Eurasia. Quat Int 285:30–43

Belfer-Cohen A, Goring-Morris N (2007) From the beginning: Levantine Upper Palaeolithic cultural change and continuity. In: Mellars P, Boyle K, Bar-Yosef O, Stringer C (eds) Rethinking the human revolution. University of Cambridge, Cambridge, pp 199–206

Delagnes A, Crassard R, Bertran P, Sitzia L (2013) Cultural and human dynamics in southern Arabia at the end of the Middle Paleolithic. Quat Int 300:234–243

Derevianko AP (2011) The Upper Paleolithic in Africa and Eurasia and the origin of anatomically modern humans, vol 1. Institute of Archaeology and Ethnography Press, Novosibirsk, p 560

Derevianko AP, Shunkov MV (2012) A new model of formation of the anatomically modern human. Her Russ Acad Sci 82(3):202–213

Derevianko AP, Markin SV, Shunkov MV (2013) The Sibiryachikha facies of the Middle Paleolithic of the Altai. Archaeol Ethnol Anthropol Eurasia 41(1):89–103

Douka K, Bergman C, Hedges R, Wesselingh F, Higham T (2013) Chronology of Ksar Akil (Lebanon) and implications for the colonization of Europe by anatomically modern humans. PLoS One 8(9):e72931. doi:10.1371/journal.pone.0072931

Gao X (1999) A discussion "Chinese Middle Paleolithic". Acta Anthripol Sin 18(1):1–15 (in Chinese)

Gibbons A (2014) Oldest Homo sapiens genome pinpoints Neandertal input. Science 343:1417

Green R, Krause J, Briggs A, Maricic T, Stenzel U, Kircher M, Patterson N, Li H, Zhai W, Fritz M, Hansen N, Durand E, Malaspinas A-S, Jensen J, Marques-Bonet T, Alkan C, Prüfer K, Meyer M, Burbano H, Good J, Schultz R, Aximu-Petri A, Butthof A, Höber B, Siegemund M, Weihmann A, Nusbaum C, Lander E, Russ C, Novod N, Affourtit J, Egholm M, Verna C, Rudan P, Brajkovic D, Kucan Ž, Gušic I, Doronichev V, Golovanova L, Lalueza-Fox C, de la Rasilla M, Fortea J, Rosas A, Schmitz R, Johnson P, Eichler E, Falush D, Birney E, Mullikin J, Slatkin M, Nielsen R, Kelso J, Lachmann M, Reich D, Pääbo S (2010) A draft sequence of the Neandertal genome. Science 328:710–722

Guan Y, Gao X, Li F, Pei S, Chen F, Zhou Z (2012) Modern human behaviors during the late stage of the MIS3 and the broad spectrum revolution: evidence from a Shuidonggou Late Paleolithic site. Chin Sci Bull 57(4):379–386

Higham T, Douka K, Wood R, Bronk Ramsey C, Brock F, Basell L, Camps M, Arrizabalaga A, Baena J, Barroso-Ruíz C, Bergman C, Boitard C, Boscato P, Caparrós M, Conard N, Draily C, Froment A, Galván B, Gambassini P, Garcia-Moreno A, Grimaldi S, Haesaerts P, Holt B, Iriarte-Chiapusso M-J, Jelinek A, Jordá Pardo JF, Maíllo-Fernández J-M, Marom A, Maroto J, Menéndez M, Metz L, Morin E, Moroni A, Negrino F, Panagopoulou E, Peresani M, Pirson S, de la Rasilla M, Riel-Salvatore J, Ronchitelli A, Santamaria D, Semal P, Slimak L, Soler J, Soler N, Villaluenga A, Pinhasi R, Jacobi R (2014) The timing and spatiotemporal patterning of Neanderthal disappearance. Nature 512:306–309

Hou Y, Yang S, Dong W, Zhang J, Liu Y (2013) Late Pleistocene representative sites in north China and their indication of evolutionary human behavior. Quat Int 295:183–190

Hublin J-J (2013) The makers of the early Upper Paleolithic in western Eurasia. In: Smith F, Ahern J (eds) The origins of modern humans: biology reconsidered. John Wiley & Sons, Inc., New Jersey, pp 223–252

Hublin J-J, Talamo S, Julien M, David F, Connet N, Bodu P, Vandermeersch B, Richards MP (2012) Radiocarbon dates from the Grotte du Renne and Saint-Césaire support a Neandertal origin for the Châtelperronian. Proc Natl Acad Sci U S A 109(46):18743–18748

Kadowaki S (2013) Issues of chronological and geographical distributions of Middle and Upper Palaeolithic cultural variability in the Levant and implications for the learning behavior of Neanderthals and *Homo sapiens*. In: Akazawa T, Nishiaki Y, Aoki K (eds) Dynamics of learning in Neanderthals and modern humans vol. 1: cultural perspectives. Springer, New York, pp 59–91

Kimura M, Ohta T (1969) The average number of generations until fixation of a mutant gene in a finite population. Genetics 61:763–771

Krause J, Qiaomei F, Good JM, Viola B, Shunkov MV, Derevianko AP, Paabo S (2010) The complate mitochondrial DNA genome of an unknown hominin from southern Siberia. Nature 464(8):894–897

Kuhn S, Stiner M, Güleç E, Özer I, Yılmaz H, Baykara I, Ayşen A, Goldberg P, Martínez Molina K, Ünay E, Suata-Alpaslan F (2009) The early Upper Paleolithic occupations at Üçağızlı Cave (Hatay, Turkey). J Hum Evol 56:87–113

Maynard Smith J, Haigh J (1974) The hitch-hiking effect of a favorable gene. Genet Res 23:23–35

Meignen L (2012) Levantine perspectives on the Middle to Upper Paleolithic "transition". Archaeol Ethnol Anthropol Eurasia 40(3):12–21

Mellars P (1999) The Neanderthal problem continued. Curr Anthropol 40(3):341–350

Mellars P (2005) The impossible coincidence. A single-species model for the origins of modern human behavior in Europe. Evol Anthropol 14:12–27

Norton C, Gao X, Feng X (2009) The east Asian Middle Paleolithic reexamined. In: Camps M, Chauhan P (eds) Sourcebook of Paleolithic transitions. Springer, New York, pp 245–254

Okladnikov AP (1940) Neanderthal people and their cultural footsteps in central Asia (as a preliminary data and conclusion about the excavation of Teshik-Tash grotto). Sov Arkheol 6:5–19 (in Russian)

Olszewski D (2009) Comparing Middle to Upper Paleolithic transitions in the Middle East and Egypt. In: Camps M, Chauhan P (eds) Sourcebook of Paleolithic transitions. Springer, New York, pp 315–332

Olszewski D, Dibble H (1994) The Zagros Aurignacian. Curr Anthropol 35(1):68–75

Otto SP, Barton NH (1997) The evolution of recombination: removing the limits to natural selection. Genetics 147:879–906

Prüfer K, Racimo F, Patterson N, Jay F, Sankararaman S, Sawyer S, Heinze A, Renaud G, Sudmant P, de Filippo C, Li H, Mallick S, Dannemann M, Fu Q, Kircher M, Kuhlwilm M, Lachmann M, Meyer M, Ongyerth M, Siebauer M, Theunert C, Tandon A, Moorjani P, Pickrell J, Mullikin J, Vohr S, Green R, Hellmann I, Johnson P, Blanche H, Cann H, Kitzman J, Shendure J, Eichler E, Lein E, Bakken T, Golovanova L, Doronichev V, Shunkov M, Derevianko A, Viola B, Slatkin M, Reich D, Kelso J, Pääbo S (2014) The complete genome sequence of a Neanderthal from the Altai Mountains. Nature 505:43–49

Qu T (2012) The emergence and the regional variability of modern humans and their behavior. Acta Anthripol Sin 31(3):269–278 (in Chinese)

Qu T, Bar-Yosef O, Wang Y, Wu X (2013) The Chinese Upper Paleolithic: geography, chronology, and techno-typology. J Archaeol Res 21:1–73

Rebollo NR, Weiner S, Brock F, Meignen L, Goldberg P, Belfer-Cohen A, Bar-Yosef O, Boaretto E (2011) New radiocarbon dating of the transition from the Middle to the Upper Paleolithic in Kebara Cave, Israel. J Archaeol Sci 38:2424–2433

Rose J, Usik V, Marks A, Hilbert Y, Galletti C, Parton A, Geiling J, Černý V, Morley M, Roberts R (2011) The Nubian Complex of Dhofar, Oman: an African Middle Stone Age industry in southern Arabia. PLoS One 6(11):e28239. doi:10.1371/journal.pone.0028239

Shea J (2008) Transitions or turnovers? Climatically-forced extinctions of Homo sapiens and Neanderthals in the east Mediterranean Levant. Quat Sci Rev 27:2253–2270

Škrdla P (2003) Comparison of Boker Tachtit and Stránská skála MP/UP transitional industries. J Israel Prehist Soc 33:37–73

Smith F, Ahern J (2013) The origins of modern humans: biology reconsidered. John Wiley & Sons, Inc., New Jersey

Soressi M, McPherronc S, Lenoire M, Dogandžić T, Goldberg P, Jacobs Z, Maigrot Y, Martisius N, Miller C, Rendu W, Richards M, Skinner M, Steele TE, Talamo S, Texier J-P (2013) Neandertals made the first specialized bone tools in Europe. Proc Natl Acad Sci USA 110(35):14186–14190

Stringer C (2012) The status of Homo heidelbergensis (Schoetensack 1908). Evol Anthropol 21:101–107

Svoboda J (2007) On modern human penetration into northern Eurasia: the multiple advances hypothesis. In: Mellars P, Boyle K, Bar-Yosef O, Stringer C (eds) Rethinking the human revolution. University of Cambridge, Cambridge, pp 329–339

Tostevin G (2003) A quest for antecedents: a comparison of the terminal Middle Palaeolithic and early Upper Palaeolithic of the Levant. In: Goring-Morris AN, Belfer-Cohen A (eds) More than meets the eye: studies on Upper Palaeolithic diversity in the Near East. Oxbow Books, Oxford, pp 54–67

Tostevin G (2007) Social intimacy, artefact visibility and acculturation models of Neanderthal-modern human interaction. In: Mellars P, Boyle K, Bar-Yosef O, Stringer C (eds) Rethinking the human revolution. University of Cambridge, Cambridge, pp 341–357

Tsanova T (2013) The beginning of the Upper Paleolithic in the Iranian Zagros. A taphonomic approach and techno-economic comparison of early Baradostian assemblages from Warwasi and Yafteh (Iran). J Hum Evol 65:39–64

Wu X (2004) On the origin of modern humans in China. Quat Int 117:131–140

Zhang J, Huang W, Yuan B, Fu R, Zhou L (2010) Optically stimulated luminescence dating of cave deposits at the Xiaogushan prehistoric site, northeastern China. J Hum Eval 59:514–524

Mobility and Cultural Diversity in Central-Place Foragers: Implications for the Emergence of Modern Human Behavior

5

L.S. Premo

Abstract

Although anthropologists have long recognized the importance of mobility to hunter-gatherers, much work remains to be done on the issue of how mobility impacts levels of cultural diversity in central-place foraging populations. Many archaeologists identify signs of increased diversity in culture material and of increased differentiation between regions as indicators of modern human behavior. A better understanding of how mobility affects these variables may provide us with an additional line of evidence for explaining the appearance of archaeological indicators of modernity. Here, I introduce a spatially explicit agent-based model, based on Kelly's (The foraging spectrum: diversity in hunter-gatherer lifeways. Smithsonian Institution Press, Washington, 1995) central-place foraging model, to address the following research question: how does length of the effective foraging radius (r_e) affect the effective size of a metapopulation composed of central-place foraging groups? The results show that mobility strategies that emphasize logistical mobility can inhibit intergroup interaction and, in turn, increase the effective size of a subdivided population. Considered within the larger context of Sewall Wright's work on the effects of isolation by distance, the findings have interesting implications not only for neutral cultural diversity at the level of the metapopulation but also for cultural differentiation between groups. To the extent that we can identify shifts in hominin mobility strategies in the Paleolithic archaeological record, these theoretical findings may help us better understand the appearance of modern behavior.

Keywords

Human evolution • Hunter-gatherers • Metapopulation model • Mobility • Modern human behavior

5.1 Introduction

> . . . how important is mobility, as opposed to other variables such as population density, in explaining the rates of spread of cultural behaviours in a population of foraging groups? (Perreault and Brantingham 2011, p. 67)
>
> . . . there are two ways in which a population can be regarded as increasing. One is the obvious one of demographic growth in a local population. The other arises when an isolated population establishes contact with another population . . . (Shennan 2001, p. 12)

L.S. Premo (✉)
Department of Anthropology, Washington State University, Pullman, WA 99164-4910, USA

Department of Human Evolution, Max Planck Institute for Evolutionary Anthropology, 04103 Leipzig, Germany
e-mail: luke.premo@wsu.edu

A. Mesoudi and K. Aoki (eds.) *Learning Strategies and Cultural Evolution during the Palaeolithic*, Replacement of Neanderthals by Modern Humans Series, DOI 10.1007/978-4-431-55363-2_5,

5.1.1 The Emergence of Modern Human Behavior: Individual- vs. Population-Level Explanations

Explaining the emergence of so-called modern human behavior presents one of anthropology's "grand challenges" (Kintigh et al. 2014). Most researchers agree that modern behavior includes most if not all of the following: symbolic thought/art, personal ornamentation, ritual burials, relatively high rates of change in material culture, increased regional differentiation in material culture, and increased complexity in tool technology (e.g., Ambrose 2001; d'Errico 2003; Klein 1995; McBrearty and Brooks 2000; Mellars 1996; for a useful review, see Nowell 2010). Many modern behaviors are represented in the Paleolithic archaeological record in the form of implements made of stone, bone, shell, or mineral pigments that have survived tens of thousands to hundreds of thousands of years. Abstract engravings on ochre (Henshilwood et al. 2002, 2009) and ostrich eggshell (Ambrose 1998; Texier et al. 2010) as well as carved figurines, cave paintings, and musical instruments are thought to serve as evidence of symbolic thought, if not full-blown artistic traditions. Hominin-modified shell beads serve as evidence of personal ornamentation (Henshilwood et al. 2004; Kuhn et al. 2001; Vanhaeren et al. 2006), just as the presence of bone tools (Henshilwood et al. 2001; Yellen et al. 1995), composite tools (Ambrose 2001, 2010), and special stone tool production techniques, such as heat treatment (Brown et al. 2009), serve as evidence for increased technological complexity.

Even though some modern behaviors differ in degree rather than in kind from non-modern behavior, the material culture associated with the former is generally thought to signal a substantial shift to a lifestyle in which symbols and technological know-how played more significant roles in hominin societies than ever before. Yet, it remains unclear whether the explanation for this shift resides at the level of the individual or at the level of the population. Was this shift caused by a change that endowed individuals with more sophisticated cognitive abilities, including high fidelity social learning mechanisms and a knack for innovation? Or was it caused by changes to the size and/or connectedness of populations composed of hominins who had already possessed such cognitive capabilities for hundreds, if not thousands, of generations?

At the risk of oversimplifying a nuanced debate, two families of potential explanations have been proposed for the emergence of modern behaviors; one views modernity as an individual-level property and the other views it as an emergent property of a population. As recently as two decades ago, it seemed as if the Paleolithic record clearly showed a large temporal gap between the earliest appearance of *anatomically* modern humans, dating to approximately 190,000 years ago (MacDougall et al. 2005), and *behaviorally* modern humans, dating to no more than 50,000 years ago (Ambrose 1998; Klein 1992, 1995). What is more, it was thought that this lag might hold the key to understanding the emergence of modern human behavior. Given this context, it is not surprising that one family of potential explanations views the appearance of modern behavior as an abrupt "cognitive revolution," which occurred relatively late in human evolution and resulted in a more-or-less fully formed behavioral package (Ambrose 1998; Ambrose and Lorenz 1990; Klein 1992, 1995).

Perhaps the best-known example of this school of thought is Richard Klein's (1992, 1995, 2000, 2008) hypothesis that the apparently abrupt behavioral shift at the boundary of the MP/UP in Europe and the MSA/LSA in Africa may have been caused by a genetic mutation that affected the neural architecture of the brains of anatomically modern humans. Such a mutation is proposed to have endowed those who carried it with the kind of cognitive abilities that archaeologists recognize as "modern." These abilities presumably include the full range of high fidelity social learning mechanisms humans rely on today. Although this is undoubtedly an interesting and bold idea, Klein has consistently stated that it suffers from a major obstacle: "the main problem with a neural explanation has long been that it cannot be tested with fossils" (2008, p. 272). But on this point it is worth noting that recent paleogenomic work by Svante Paabo and colleagues comparing the ancient DNA recovered from Neandertal and so-called Denisovan fossils to modern human DNA is moving us closer to the possibility of testing the neural hypothesis more directly and systematically than most would have thought feasible even 15 short years ago (e.g., Burbano et al. 2010; Green et al. 2010; Maricic et al. 2013; Prufer et al. 2014; Reich et al. 2010).

By contrast, a second family of potential explanations views the appearance of modern behavior more as a gradual evolutionary process than a sudden revolution (e.g., Foley and Lahr 1997; McBrearty and Brooks 2000). McBrearty and Brooks (2000) question the interpretation there was a long gap between the earliest appearance of anatomically modern humans and the earliest signs of modern behavior in the archaeological record. McBrearty and Brooks suggest that the gap more likely reflects a sampling bias of the archaeological record, and they might be right. The more closely and systematically archaeologists study the Middle Stone Age record, the more it appears that many of the archaeological proxies of modern behavior, once thought to appear rather suddenly and only after approximately 50,000 years ago, may in fact have a much deeper and more complicated history in Africa (Brown et al. 2009; d'Errico and Henshilwood 2007; d'Errico et al. 2005; d'Errico and Stringer 2011; Foley and Lahr 1997; McBrearty and Brooks 2000; Marean et al. 2007; Tryon and Faith 2013; Wadley 2013).

Contrary to the predictions of the neural change hypothesis, the Middle Stone Age record of Africa shows that artifact classes indicative of modern behaviors—e.g., bone tools, composite tools, personal ornaments, possibly even symbolic art—appear in different regions in Africa at different times prior to 50,000 years ago (and, in some cases, then disappear again) in a piecemeal fashion rather than all at once as a fully formed package of behaviors (d'Errico and Stringer 2011; Foley and Lahr 1997; McBrearty and Brooks 2000; Wurz 2013).

The decidedly more gradualist and nonlinear perspective of change espoused in Foley and Lahr's (1997) so-called "Mode 3 hypothesis," for example, suggests that the emergence of modern behavior may be better understood as a mosaic evolutionary process in a subdivided population, marked by fits and starts in different regions at different times, rather than as a sudden revolution. This perspective does not necessarily restrict modern behavior to anatomically modern humans. To wit, formal bone tools, once thought to be associated exclusively with behaviorally modern humans, are now known to exist outside of Africa in stratigraphic contexts that predate the earliest known incursions of anatomically modern humans (Soressi et al. 2013). Also note that the gradualist perspective dovetails with the view of modern behavior as an emergent property of a *population* of social learners rather than a property of every *individual* in a population. Thus, unlike the neural change hypothesis, there is no a priori reason why explanations that treat modern behavior as an emergent property of a population cannot be applied to precocious archaeological evidence from stratigraphic contexts that predate 50 kya as well as to ethnographically documented cases of MSA- or MP-like technology among recent and undeniably behaviorally modern human hunter-gatherers.

A fundamental concept underlying most of the population-level explanations of modern behavior is *effective population size*, represented by N_e (Wright 1931). Although those who study cultural evolution have described the effective size of a population as the number of "teachers" who pass their cultural variants to the next generation of social learners (e.g., Premo and Scholnick 2011)[1] or as the number of distinct cultural variants (such as pottery motifs) available for copying by "naïve" individuals (e.g., Shennan and Wilkinson 2001), neither characterization is entirely correct. I find Hartl and Clark's (2007, p. 121) general definition useful: "the effective population size of an actual population is the number of individuals in a theoretically ideal population having the same magnitude of random genetic drift as the actual population." As Hartl and Clark (2007) further explain, a different measure of effective population size corresponds to each of the three ways that the magnitude of random drift can be measured. The details concerning how and why the values provided by the inbreeding effective size (based on the change in probability of identity by descent), the variance effective size (based on the change in variance in variant frequency) and the eigenvalue effective size (based on the rate of loss of heterozygosity) can differ are beyond the scope of the present study (see Crow and Kimura 1970, p. 362).

Here are the four general points concerning effective population size that are important in the context of the present study. First, populations with larger effective sizes are marked by higher levels of neutral diversity and by slower rates of loss in diversity than populations with smaller effective sizes. Second, the "theoretically ideal" populations to which Hartl and Clark refer are those that meet the assumptions of the Wright-Fisher model of reproduction (Fisher 1930; Wright 1931). Those assumptions are worth recounting here. In its most basic form, the Wright-Fisher model of reproduction assumes a constant and finite census population size (*N*), a freely mixing population (i.e., panmixia, or random mating), discrete non-overlapping generations, discrete variants, and the absence of selection. Population genetic research that employs the Wright-Fisher model often assumes an infinite-allele model of mutation, which holds that each copying error results in an entirely new allele, one that has never existed in the population before. Despite the fact that empirical populations violate one or more of these assumptions, N_e is still a useful concept. Third, the effective size of an actual population can be less than or greater than its census size. For example, if an empirical population of census size *N* shows less diversity and faster rates of loss in diversity than would a "theoretically ideal" population of the same census size, the effective size of the empirical population is less than its census size, or $N_e < N$. This can occur if traits are passed via conformist biased transmission (Premo and Scholnick 2011). On the other hand, if an empirical population of census size *N* shows greater diversity and slower rates of loss in diversity than would a "theoretically ideal" population of the same census size, the effective size of the empirical population exceeds its census size, or $N_e > N$. This can occur if traits are passed via vertical cultural transmission such that each member of the "skilled" generation passes its variant to a different member of the "naïve" generation. Local-scale unbiased cultural transmission (Premo and Scholnick 2011) and forms of frequency-dependent biased cultural transmission that favor rare variants can both also result in $N_e > N$. Fourth,

[1]Under Wright-Fisher model conditions, only approximately 63 % of the members of the "skilled" generation serve as teachers for members of the "naïve" generation. Thus, even under the "theoretically ideal" conditions, the number of individuals who actually serve as "teachers" to the next generation is lower than the effective size of the population. Thus, in the context of cultural evolution, it is more accurate to describe the effective population size as the number of individuals who could *potentially* serve as teachers for the naïve generation rather than the number of individuals who actually serve in that role.

the clear-cut relationship between effective population size and copy error rate on one side and cultural diversity on the other is not expected to hold if variants are subject to natural selection or cultural selection. I turn now to cultural evolutionary models that include some form of selection.

A separate but related line of inquiry to the one introduced above investigates the effect of population size on the mean fitness of a population of social learners. Interesting theoretical work by Shennan (2001) and others (Henrich 2004; Powell et al. 2009) on *adaptive* culturally transmitted variants, shows that larger populations are less likely than smaller populations to lose beneficial, fitness-enhancing innovations due to drift. These results are consistent with the understanding that the strength of drift decreases relative to the strength of selection as population size increases. Under conditions in which there is at least a small chance per transmission event that the fitness of the cultural variant adopted by the naïve individual will be higher than the target variant and where each individual's reproductive fitness is a function of culturally transmitted variation only, simulations show that the equilibrium mean fitness of a population increases nonlinearly with population size (see Shennan 2001). Of course this holds true when each offspring inherits cultural variants only from its parent (i.e., vertical cultural transmission of haploid traits), in which case the relative frequency of an adaptive cultural variant increases in a population solely due to the fact that individuals who display it have a greater number of offspring on average than those who do not (see Shennan's (2001) "simple model" in which the cultural variants are analogous to haploid genes). But it also holds true under some conditions in the presence of oblique, direct-biased cultural transmission (Boyd and Richerson 1985), which has been modeled in a number of ways (see Henrich 2004; Powell et al. 2009; Shennan 2001) (but see Vaesen (2012)). The results of these models show that population size can affect equilibrium mean fitness even if the offspring of the most fit parents occasionally adopt lower-fitness variants as a result of oblique cultural transmission.

Note that while neutral theory (Kimura 1968, 1983; Kimura and Crow 1964; see also Neiman 1995) provides the means to predict levels of diversity as a function of effective population size and copy error rate, the work reviewed in the paragraph above focuses instead on how population size affects mean fitness and thus does not directly address the effect of population size on the *diversity* of adaptive traits. When cultural variants are marked by differential fitness (regardless of whether this difference is imbued by natural selection or cultural selection), population size and copy error rate alone cannot be used to derive clear-cut expectations of cultural diversity in the same way as under Wright-Fisher model conditions. In short, the relationship between population size and the diversity of adaptive variants is more complicated than the relationship between effective population size and the diversity of selectively neutral variants. On one hand, holding copy error rate constant, one would expect larger populations to generate greater diversity if for no other reason than where there are more social learners there are more transmission events and, thus, more copy errors. But on the other hand, because natural selection is more efficient in larger populations due to the fact that the strength of drift decreases relative to the strength of selection as population size increases, many of the additional variants produced will not persist for long if their fitness effects are not positive. In general, unless the underlying fitness landscape is very "rugged" indeed, one can expect to see less cultural diversity in the presence of selection than in the absence of selection. In sum, unlike under Wright-Fisher model conditions, population size and copy error rate alone do not show a clear relationship to the diversity of culturally transmitted traits in the presence of either natural or cultural selection.

Questions concerning the origins of modern behavior remain unresolved due in part to the fact that we are still in the process of learning more about factors that affect the level of cultural diversity, technological complexity, and rates of cumulative cultural change. Some of the factors that have been investigated to date include the rate of migration between groups (Powell et al. 2009), local group extinction (Premo and Kuhn 2010), and intergroup social network topology (Premo 2012a; White 2013). Other important factors of interest, which include mobility, risk avoidance, and resource structure, are more closely related to hunter-gatherer behavioral ecology and optimal foraging theory than they are to cultural evolutionary theory.

Although I certainly do not mean to imply that any single factor was the only, or even the most important, variable for the emergence of modern behavior, here, I use a spatially explicit model to isolate the effects of hunter-gatherer mobility on effective population size. In the context of selectively neutral cultural variants in a subdivided population, effective population size is directly related to cultural diversity. The goal of this modeling exercise is not to reproduce all of the conditions of the Pleistocene, nor is it to investigate all of the many nuances of hunter-gatherer mobility. Rather, I employ modeling here as a relatively coarse-grained heuristic tool with the hope that what it can teach us about how residential and logistical mobility affect effective population size (and, by extension, cultural evolution of selectively neutral traits) will improve the research questions we ask of the Paleolithic record. Before describing the model and discussing my results, I use the rest of the introduction to review previous research on the effects of mobility on hunter-gatherer technology.

5.1.2 Hunter-Gatherer Mobility and Cultural Diversity

Anthropologists have long recognized how important mobility is to hunter-gatherers. But it was not until the 1960s and 1970s that there was sustained effort to build a theory of hunter-gatherer mobility. Binford's (1980) "Willow smoke and dogs' tails" is surely among the most influential papers on the topic in part because it makes a useful distinction between residential and logistical mobility. Residential mobility refers to the movement of all of the members of the group from one residential base location to another. Logistical mobility refers to the movement of some of the group's members on task-specific forays between a resource and the residential base. While all hunter-gatherer societies include a mix of logistical mobility and residential mobility, one may be emphasized over the other if doing so improves net foraging returns. Indeed, Binford viewed the particular mix of residential and logistical mobility displayed by a hunter-gatherer group as a useful way to place it along a continuum of adaptive foraging strategies. According to Binford's terminology, hunter-gatherers who emphasize logistical over residential mobility are located closer to the "collector" end of the spectrum, and those who emphasize residential over logistical mobility are located closer to the "forager" end of the spectrum. Roughly speaking, "collectors" move resources to people while "foragers" move people to resources. Binford (1980) argues that residential mobility and logistical mobility are interdependent such that a reduction in residential mobility *causes* an increase in logistical mobility, just as a reduction in logistical mobility *causes* an increase in residential mobility.

The goal of Kelly's (1983, p. 277) exploratory cross-cultural study of hunter-gatherer mobility is to "initiate development of a theory of hunter-gatherer mobility strategies." Kelly defines measures of residential mobility (number of residential moves per year and average distance per residential move) and logistical mobility (one-way distance covered during logistical forays). He uses ethnographic data to explore relationships between mobility and environmental factors, such as effective temperature (ET), primary production, and primary biomass. He uncovers a number of interesting relationships between resource structure and hunter-gatherer mobility, some of which are pertinent to the present study. For instance, in the absence of storage, the number of residential moves per year increases and the average distance per residential move decreases as resource accessibility decreases. Hunter-gatherers who rely heavily on fauna can afford to pay the higher energetic costs associated with "commuting" to and from logistical locations farther from the residential base than can hunter-gatherers who rely more heavily on plants. Finally, hunter-gatherers who rely heavily on fauna and live in areas where resource accessibility is low (i.e., habitats marked by high primary biomass and low ET) employ very high residential mobility during the winter unless they have access to resources that can be stored in bulk, like salmon.

These empirical relationships inform Kelly's (1990, 1991, 1995) "central-place foraging model," which employs the concept of the *effective foraging radius*, r_e. Effective foraging radius is the distance at which the daily net return rate of a logistical foray (the gross caloric return minus the costs of commuting round-trip between the residential base and the logistical camp) equals the minimum daily caloric requirement of the forager, which includes the caloric needs of the individuals he or she is supporting. Because r_e varies as a function of the return rate of available resources and the forager's daily caloric requirement (see Kelly 1991, p. 142, 1995, p. 134), effective foraging radius is not the same for all hunter-gatherer societies. In fact, the effective foraging radius is likely to vary through time and over space even within a single society. According to the assumptions of Kelly's simple central-place foraging model, it is optimal for hunter-gatherers to move the residential base a distance of $2r_e$ rather than to conduct logistical forays targeting resources that are located at a distance greater than r_e from the residential base. While it may be possible for hunter-gatherers to increase logistical and residential mobility simultaneously under special conditions—Kelly (1995) submits equestrian bison hunters of North America as a possible example of this—residential and logistical mobility are interdependent in the central-place foraging model. Echoing Binford (1980), any decrease to a forager's r_e (decreased logistical mobility) causes an increase in the frequency of residential moves, just as any increase to r_e (increased logistical mobility) causes a decrease in the frequency of residential moves. Kelly's central-place foraging model turns out to be as useful as it is elegant. Because it incorporates both residential and logistical mobility, one can represent the full forager-collector continuum, at least in an idealized form where logistical and residential mobility are interdependent, by varying a single parameter: r_e. Lower values of r_e represent more forager-like strategies, and higher values of r_e represent more collector-like strategies.

Although Binford (1980) and Kelly (1983) did not directly address the issue of how hunter-gatherer mobility might affect technological complexity or levels of cultural diversity—variables of special interest for the study of the emergence of modern behavior—other studies soon would. In an interesting paper on how time budgeting might affect hunter-gatherer toolkits, Torrence (1983, p. 13) speculates that the "high mobility" of hunter-gatherer groups would place a limit on the "gross number of artefacts which can be carried between residences." Holding constant the number of tasks that a group needs to perform, it stands to reason that switching to a smaller and more generalized toolkit provides one adaptive response to the transport limits imposed by increased residential mobility.

Clearly inspired by Torrence's interesting speculation, Shott (1986) investigates the effects of mobility on hunter-gatherer technology. In particular, he examines relationships between Oswalt's (1976) measures of toolkit diversity (the number of different subsistant[2] classes per toolkit, which is actually a measure of tool class richness) and toolkit versatility (or mean complexity, calculated as total number of technounits[3] divided by the number of subsistant classes) and Kelly's (1983) measures of residential and logistical mobility in 14 ethnographically documented hunter-gatherer groups. Shott (1986) finds a statistically significant negative relationship between the frequency of residential moves and the number of subsistants per toolkit as well as a significant positive relationship between the length of stays in winter/wet-season camps and the number of subsistants per toolkit. Both findings are consistent with Torrence's (1983) hypothesis that an adaptive strategy for dealing with increased residential mobility entails decreasing the number of tools (presumably, in this case, by decreasing the number of subsistant classes) per toolkit. However, Shott does not find a significant relationship between the total distance covered by residential moves per year and the number of subsistants per toolkit. Nor is there a significant relationship between toolkit versatility, as measured by mean complexity, and either of Kelly's measures of residential mobility. Thus, Shott does not find evidence that increased residential mobility corresponds with more generalized toolkits. Territory size, effective temperature, and net primary productivity each show no significant relationship to either the number of subsistants per toolkit or to mean complexity. In light of the results, Shott (1986, p. 34) is justified in stating: "Relationships such as those involving mobility and technology are likely to be complex and to be complicated by other factors (Hitchcock 1982, p. 372). Under these circumstances, simple and clear relationships should be the exception rather than the rule."

A string of more recent studies use data culled largely from Oswalt (1976) and Binford (2001) to test a series of alternative hypotheses for explaining levels of diversity and complexity in hunter-gatherer toolkits. Collard et al. (2005) make use of Oswalt's (1976) technological data on the number of subsistants per toolkit (as mentioned above, a measure of richness), the average number of technounits per tool class (a measure of complexity), the total number of subsistant technounits per toolkit (another measure of complexity), and proxy data from Binford (2001) to investigate four factors possibly responsible for variation in subsistence-related toolkits used by recent hunter-gatherers. The four factors include the structure of food resources, risk of failure to procure resources (effective temperature and net above-ground productivity), residential mobility (number of residential moves per year and distance traveled annually during residential moves), and population size. Forward stepwise linear multiple regressions show that both of the proxies for risk of resource failure were the only significant predictors of the measures of toolkit diversity and complexity. Although Collard et al. (2005) caution against interpreting the results as unqualified support for the risk-buffering hypothesis (Torrence 1989, 2000), their study suggests that residential mobility, resource structure, and population size do not serve as useful predictors of subsistant richness and complexity in Oswalt's sample of ethnographically known hunter-gatherers.

Read (2008) conducts a similar set of tests on a slightly different data set, one that includes three additional technological variables (total number of complex subsistant types, total number of complex subsistants, and total number of complex technounits) and an additional proxy for risk of resource failure (length of growing season). Risk of resource failure is thought to decrease as length of growing season increases. In contrast to Collard et al. (2005), Read's results suggest that the structure of subsistence-related toolkits is best explained by an interaction between length of growing season and number of residential moves per year. Although the results of both studies suggest that risk-reduction plays an important role in explaining variation in toolkit structure, Read's regressions find an additional significant relationship between toolkit structure and residential mobility. It is worth noting that subsequent studies by Collard et al. (2011, 2013a, b), which apply similar techniques to different hunter-gatherer samples—and in the case of Collard et al. (2013a), to different technological data as well—fail to find a similar effect of residential mobility on toolkit richness.[4]

A recent study by Perreault and Brantingham (2011) is more closely aligned with the goals of the present paper, although their model and mine are very different. Perreault and

[2]"Subsistant" is Oswalt's (1976) term for any tool used to procure food. An interesting empirical question is to what extent tools used to procure food are subject to natural selection and/or cultural selection.

[3]"Technounit" is Oswalt's (1976) term for a distinct technological component of a tool (e.g., the technounits of a projectile would include the shaft, hafting, point, and so on).

[4]Collard et al. (2013a) do not find an effect of distance moved per year during residential moves on their more general measure of "technological richness." But, unlike the other studies, Collard et al. (2013a) does not include proxy data for number of residential moves per year. Collard et al. (2013b) find a significant negative effect of number of residential moves on the average number of technounits per tool class in their North American sample and a significant negative effect of total distance moved per year during residential moves on the average number of technounits per tool class in their global sample. They find no effect of mobility on number of subsistants per toolkit in any of their samples.

Brantingham employ a spatially explicit model to investigate how the number of "steps" in a Lévy walk affects the average time elapsed before two foragers happen upon each other in a featureless 2-dimensional space. In their study, each simulation begins with two foragers located at their respective residential bases, which are separated by 10 arbitrary units of space. Both foragers are characterized by the same "number of foraging moves," n, a parameter that designates how many "steps" each will conduct on its random walk before returning to its starting location (i.e., residential base). The random walk follows a Lévy flight, where the length of each step is defined by the Lévy parameter, $\mu = 2$. If after n steps the foragers do not "meet" (i.e., come to inhabit coordinates within one unit of each other), then they return to their bases and start new forays. Because both of the foragers are characterized by the same n, they begin new forays 10 units apart from one another every n time steps. Perreault and Brantingham's (2011, Fig. 7) central finding is a non-monotonic relationship between n and the mean time required for an interaction to occur between the two foragers. Mean interaction time starts relatively high with low n, decreases as n approaches half of the starting distance between the home bases (where it reaches its minimum), and then increases again as n increases. This finding is potentially of general interest because higher interaction rates between foraging groups have implications for the effective size of the population and therefore for the population-level diversity of selectively neutral culturally transmitted traits, as shown below.

5.1.3 Summary

Binford introduces a very useful, if highly stylized, way to think about hunter-gatherer mobility as an adaptive mix of residential and logistical mobility. Kelly's central-place foraging model provides an elegant way to represent the continuum of mobility strategies along the forager-collector spectrum by varying just one parameter: effective foraging radius, r_e. A series of empirical tests of the effect of mobility on toolkit diversity yield mixed results, with some finding that mobility plays an insignificant role in explaining the richness of subsistant tool classes in recent hunter-gatherer toolkits. The interesting study by Perreault and Brantingham provides a step in the right direction even though their model does not fully represent the relationship between residential and logistical mobility. By any measure, there is still much work to be done on the issue of how mobility impacts cultural diversity in hunter-gatherers. Given that many researchers identify increased total cultural diversity and increased cultural differentiation between regions as indicators of modern behavior, a better understanding of how mobility affects these variables might provide us with an additional line of evidence for explaining the appearance of archaeological indicators of modernity during the Paleolithic.

In the next section, I describe a spatially explicit agent-based model and my experimental design. The purpose of the model is to address the following question: how does length of the effective foraging radius (r_e) affect the effective size of a subdivided population composed of central-place foraging groups? The results show that increased logistical mobility inhibits interaction among groups of central-place foragers, increasing the effective size of the metapopulation. When considered within the context of other findings regarding the effects of isolation by distance, the results also have important implications for differentiation between groups and, by extension, perhaps between regional archaeological assemblages. To the extent that we can identify shifts in hunter-gatherer mobility strategies in the Paleolithic archaeological record, these findings may help us better understand and predict (or perhaps "retrodict") the appearance of modern behavior as an emergent property of a population.

5.2 The Model

The model described in this section was programmed and run in NetLogo 5.0.2 (Wilensky 1999). The source code and complete model description, following the ODD protocol for agent-based models (Grimm et al. 2006), are available upon request.

Consider a population of N agents dispersed randomly over a 250 cells × 250 cells lattice wrapped around a torus to avoid edge effects. Each agent represents a small self-sufficient group of central-place foragers. The population of N groups represents a spatially explicit subdivided population, or metapopulation. It is assumed that all groups are of the same size and consume resources at the same rate. The metapopulation of N groups is constant; foraging groups cannot suffer "local extinction," reproduce (or fission), or join together during the course of a simulation run. Each cell of the grid may contain a resource that provides enough food to support one group for one time step. Resource density is given by the parameter d. If $d = 1$, then every cell contains food. If $d = 0.5$, then only half of the cells (chosen randomly) contain food resources. Foragers deplete resources. When an agent consumes food, the food does not reappear in its cell until 800 time steps have passed. Thus, the resource in this model represents a slowly regenerating source of calories and nutrients. The rate of resource regeneration and N are held constant over all simulations.

Foraging decisions, such as how long to remain at a residential base and how far to move to a new residential base, depend upon the logic of Kelly's (1995) central-place foraging model and, more specifically, on the parameter r_e. As their name implies, central-place foragers in this model conduct logistical forays from a "central-place," or residential base, to procure food from the landscape. Each group is allowed to consume only those resources located within the effective foraging radius, r_e, of its current residential base. Groups have no information about the state of resources outside of their current foraging area except for the fact that they left the resources depleted in their previous foraging area. Just as in Kelly's mathematical model, lower values of r_e correspond to strategies that emphasize residential mobility and higher values of r_e correspond to strategies that emphasize logistical mobility. All forager groups have the same effective foraging radius during each simulation run. Effective foraging radius serves as the primary experimental parameter in this study. Data are collected to investigate the effect of r_e on the effective size of the metapopulation of $N = 25$ groups under a range of conditions.

As in Binford (1980), logistical mobility refers to the movement of people between a residential base and a logistical camp. Each time step, each group randomly chooses a cell to serve as its logistical camp from among the set of cells that satisfy two criteria: (1) the cell is located within a distance r_e of the agent's residential base and (2) the cell currently contains food. To represent the logistical foray of some of the group's members, the agent moves to the logistical camp and consumes the resource before moving back to the residential base. In the event that none of the cells within the foraging area contains food, then the group relocates its residential base. To avoid overlap between the new foraging area and recently depleted areas, residential camps are moved a distance of $2r_e + 1$ in a heading defined by adding a value chosen randomly from a uniform distribution bound by −45 and 45 to the group's previous heading. In other words, residential mobility follows a correlated random walk with a step length of $2r_e + 1$. To better isolate the effect of r_e on effective population size, I assume that all of the other needs that might require mobility (travel for water or for raw materials such as stone or firewood, or to share information with other groups) are embedded within logistical and residential mobility. Also note that a group cannot move logistically *and* residentially during the course of a single time step. Each forager moves its residential base *or* conducts a logistical foray during each time step, but it cannot do both. In the event that a residential move places a group in a foraging area that is completely devoid of food, that group will make another residential move during the subsequent time step.

For the sake of simplicity and because within-group diversity was not a central focus of this study, it is assumed that all of the members of a group display the same variant of a selectively neutral cultural trait. Different variants of the selectively neutral trait are represented by integers. Variants can be transmitted between groups that find themselves located within each other's interaction radius, r_i. In cases where ego is the only group within its interaction radius, ego retains the variant it displayed in the previous time step (or, in other words, the members of ego learn from their "former" selves). When two or more groups (including ego) are located within r_i of ego, ego chooses one of them at random to serve as its teacher group for the current time step. Ego adopts the variant that the teacher group displayed in the previous time step. Because ego can potentially choose itself to serve as its teacher group in all cases, it is possible for ego to adopt the same variant that it displayed in the previous time step even when other groups are located within ego's interaction radius. The absence of horizontal cultural transmission between groups of the same "generation" is consistent with the Wright-Fisher model assumption that generations are discrete and non-overlapping. Note that in this model, cultural transmission does not include copying error; all variants are copied with perfect fidelity.

There are a number of ways in which groups may come into contact with each other while foraging for food. Perhaps the most obvious type of interaction occurs when two or more residential bases are separated by a distance less than or equal to r_i. It seems unlikely that another group's residential base camp, even a relatively small and ephemeral one, would go unnoticed for long at relatively close range. Small dwellings, cooking fires, or perhaps even the sound of toddlers at play could betray a group's location at a distance of several kilometers or more, depending on the terrain and visibility. But it is also possible for foragers of different groups to come into contact while out on logistical forays, or on what Tostevin (2007) calls "the pathways of the landscape." Although it may be more difficult to detect a logistical camp than a residential base, cultural transmission can occur as a result of interactions between foragers who meet each other at the ends of their logistical forays. Finally, logistical forays may bring a group into contact with another group's residential base, just as a residential move might situate the new base in close proximity to another group's logistical camp. The model parameter, interaction type, regulates the kinds of interactions that can result in cultural transmission between groups. The more restrictive option allows cultural transmission to occur between residential bases only. The less restrictive option allows cultural transmission to occur between residential bases, between logistical camps, and between residential bases and logistical camps.

At the start of each simulation run, $d \times 250 \times 250$ cells are seeded with food resources. So as not to start foragers on a completely pristine landscape at the start of the simulation, $N \times 800$ (the regeneration rate) of the $d \times 250 \times 250$ cells seeded with food are chosen at random, and each value in the set [1,2,3, ... ,800] is assigned to N cells chosen at random from this subset of $N \times 800$ cells. This value corresponds to the number of time steps that must pass before the food resource reappears in the cell. Next, each group is placed on a randomly chosen cell and initialized with a unique cultural variant. Thus, the richness of cultural variants (k) displayed by the metapopulation of foraging groups is equal to N (i.e., $k = N = 25$) at the start of each simulation.

During each iteration, or time step, of the simulation, the following methods occur in the following order. First, groups forage. This involves either a logistical foray to a cell within the group's foraging area or a residential move to a cell outside of its foraging area, but not both. Affected cells are updated to reflect the fact that groups consume the resources present at logistical camps. Second, each group builds a list of potential teacher groups for the current time step. Under the more restrictive setting of interaction type, ego's potential teacher list includes the subset of groups whose residential bases are located within r_i of ego's residential base (including ego). Under the less restrictive setting of interaction type, the potential teacher list includes the subset of foragers whose residential bases or logistical camps are located within r_i of ego's residential base or logistical camp (again, including ego). Third, each forager randomly chooses a group from its list of potential teacher groups and then adopts the cultural variant (i.e., the integer) that the chosen group displayed in the previous time step. This represents cultural transmission between discrete and non-overlapping "generations" of social learners. Each time step can be thought of as the time required for every group in the population to undergo social learning once (and only once) rather than as an actual human generation of 20 years. Fourth, food resources that have been absent for 800 time steps regenerate themselves. Fifth, the richness of cultural variants, k, is calculated for the metapopulation. If $k = 1$, then data are collected and the simulation ends. If $k > 1$, then the methods described in this paragraph are iterated again. Fifty unique simulations were executed for each possible combination of parameter values (see Table 5.1), resulting in a total of 1,800 runs.

The purpose of the model is to address the following research question: how does length of the effective foraging radius (r_e) affect the effective size of a subdivided population of central-place foragers? Effective foraging radius is the experimental parameter (or independent variable) in the study. Mean time to fixation ($\bar{t}$), which serves as a proxy for effective population size (see below), is the dependent variable. In the context of my model, time to fixation is the number of time steps that elapse before one of the selectively neutral cultural variants present at the start of each simulation (i.e., when $k = N$) becomes "fixed" in the population, such that all N groups display the same variant ($k = 1$). Because drift is stronger when N_e is small, populations with smaller effective sizes require less time for one of the selectively neutral variants to evolve to fixation. Populations with a larger effective size require more time for a single variant to become fixed because drift is weaker when N_e is larger. Thus, the rate of loss of diversity in neutral variants is an inverse function of effective population size under "theoretically ideal" conditions.

Kimura and Ohta (1969) show that the average number of generations required for a single selectively neutral mutant gene to reach fixation in a diploid Wright-Fisher population is approximately $4N_e$. In the case of haploid genes (and cultural variants as they are modeled here), the mean time to fixation is approximately twice the effective size of the population, or $2N_e$. The remainder of the present paper deals only with the haploid case. Kimura and Ohta (1969, p. 767, Fig. 1) point out that, while their analytical prediction based on effective population size slightly overestimates the actual mean time to fixation, the discrepancy is small. To conduct an independent check of my agent-based model, I collected data from 50 simulations in which cultural variants are passed via global-scale unbiased cultural transmission. Although the data collected from simulations in which $N_e = N = 25$ ($\bar{t} = 45.88$, standard deviation = 24.80) did not significantly differ from the value provided by Kimura and Ohta's analytical estimate (one-sample Wilcoxon test[5]: V = 443.5, $p = 0.06$), a $\bar{t}$ of 45.88 does provide a sense of the degree to which $2N_e$ "overestimates" the actual time to fixation in the haploid case.

Table 5.1 Parameter values used in this study

Parameter	Value(s)
Number of groups, N	25
Resource regeneration rate	800
Resource density, d	0.5, 0.75, 1
Interaction radius, r_i	5, 10
Effective foraging radius, r_e	5, 10, 15
Interaction type	Residential bases only, all combinations of residential bases and logistical camps

5.3 Results

To assess how hunter-gatherer mobility strategies along Binford's forager-collector continuum affect the effective size of a spatially explicit subdivided population of central-place foragers I use the model described above to investigate the effect of r_e on mean time to fixation (where $\bar{t}/2$ is my proxy for N_e) while holding N constant. The central question is whether r_e has a significant effect on mean time to fixation and, thus, effective population size, while controlling for N, d, and r_i, where $r_i \leq 2r_e$. Finding a significant effect of r_e

[5] I employ non-parametric tests throughout this paper because in most cases, though not all, a Shapiro-Wilk test shows that my data are not distributed normally. For what it is worth, in all cases the parametric test results lead to qualitatively identical conclusions.

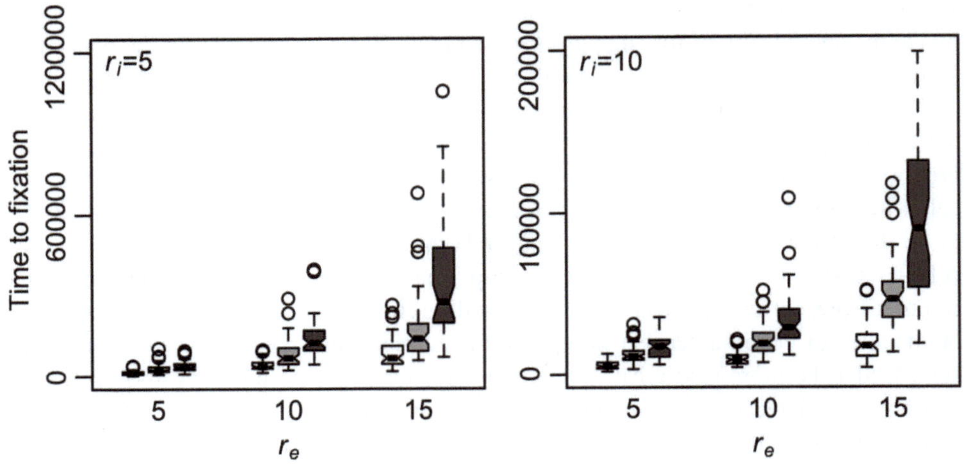

Fig. 5.1 The effect of r_e on time to fixation when cultural transmission is restricted to interactions between residential bases. Each notched *boxplot* summarizes data collected from 50 simulations. Different values of d are distinguished by color (*white*: $d = 0.5$; *light gray*: $d = 0.75$; *dark gray*: $d = 1$). r_i is interaction radius (*left*: $r_i = 5$, *right*: $r_i = 10$). Note that the scale of the y-axis differs between panels. See Table 5.2 for Mann-Whitney test results

on $\bar{t}$ while holding all other variables constant suggests that change in mobility, alone, can impact the effective size of the metapopulation, with interesting implications for total cultural diversity and regional differentiation in selectively neutral variants.

5.3.1 Time to Fixation

Figure 5.1 presents data collected from simulations in which cultural transmission can occur between residential bases only. I vary r_i and d to investigate whether the scale of the interaction sphere and the density of food resources influence the effect of r_e on time to fixation. Violating the assumption of panmixia, by limiting intergroup cultural transmission to encounters that occur while foraging, drastically increases the mean time to fixation even though the number of groups is held constant at $N = 25$. For instance, the shortest mean time to fixation observed (when $d = 0.5$, $r_i = 10$, and $r_e = 5$) provides an N_e estimate that is 113 times larger than the actual N. The most extreme case ($d = 1$, $r_i = 5$, and $r_e = 15$) yields a mean time to fixation of 350,114, which equates to an N_e estimate that is 7,000 times larger than the actual N. This finding is consistent with previous work that shows that in the absence of selection the effective size of a subdivided population can far exceed its census size as transmission between groups is reduced (Hartl and Clark 2007, p.127; Wakeley 1999, 2000).

More importantly to the present research question, the results of pair-wise Mann-Whitney tests (corrected for multiple tests while holding r_i and d constant) show that r_e has a significant and relatively large effect on time to fixation (see p and PS values in Table 5.2). The relationship between r_e and time to fixation is monotonic and positive. The magnitude of the positive effect of r_e on time to fixation is reduced under greater r_i, although here it remains statistically significant in all cases (Fig. 5.1, Table 5.2). Under these conditions, there are two ways to decrease the strength of the positive relationship between residential mobility and intergroup interaction and, thus, to dampen the effect of r_e on mean time to fixation (and, thus, N_e). The first is to allow for residential bases to interact over a greater distance, which can be accomplished by increasing the interaction radius from 5 to 10, as just shown. The second is to decrease resource density, d.

As it turns out, there is a third way to dampen the effect of r_e on mean time to fixation, and that is to allow cultural transmission to take place between residential bases, between logistical camps, and between residential bases and logistical camps. Figure 5.2 presents data collected from simulations under these less restrictive conditions. As before, I vary r_i and d to investigate whether the scale of the interaction sphere and resource density influence the effect of r_e on mean time to fixation. As before, the results reflect the fact that making cultural transmission spatially explicit increases the effective size of the population ($\bar{t}_{/2}$ >> N in all cases). However, unlike before, r_e has a significant effect on time to fixation in only one of nine cases when $r_i = 5$ and in just five of nine cases when $r_i = 10$ (Table 5.3), and this is explored further in the next section. Nevertheless, the results show that even under less restrictive conditions there is no evidence that increasing r_e (i.e., emphasizing logistical mobility over residential mobility) significantly reduces $\bar{t}$ under any of the parameter value combinations investigated here.

5.3.2 Number of Variants Remaining and Teachers Gained per Unit Time

I re-ran all of the simulations for a fixed number of time steps (in this case, 2,000) to get a better sense of the rate of loss in diversity as well as the rate that novel teachers are gained per group while controlling for time. If the effect of r_e on time to fixation is explained by the fact that increasing logistical

Table 5.2 The effect of r_e on residential moves per 100 time steps and time to fixation when cultural transmission is restricted to interactions between residential bases. The cells under "Residential moves per 100 times steps" and "Time to fixation" present the mean(sd) of 50 simulation runs. Sections labeled "p, PS" present the results of pair-wise Mann-Whitney tests. The first value in each cell is the p-value (*significant after Bonferroni correction for multiple tests; the critical value is $\alpha = 0.05/3 = 0.017$ for all tests). H_0: the distributions of the two samples differ by a location shift of 0. The second value in each cell is the probability of superiority, PS. PS is calculated by dividing the U value provided by each pair-wise Mann-Whitney test by the product of the two samples' sizes (e.g., U/(50 × 50)). PS provides a sense of the size of the effect of r_e. Here, PS should be interpreted as the probability that a value chosen at random from the sample of 50 values where r_e is smaller is superior to (higher in rank order than) a value chosen at random from the sample of 50 values where r_e is larger. A PS value of 0.5 signifies no effect, and effect size increases as the PS value approaches either 0 or 1. d is resource density and r_i is interaction radius. 5–10: comparison of $r_e = 5$ data to $r_e = 10$ data. 5–15: comparison of $r_e = 5$ data to $r_e = 15$ data. 10–15: comparison of $r_e = 10$ data to $r_e = 15$ data

	Residential moves per 100 time steps			p, PS			Time to fixation			p, PS		
$r_i = 5$												
d	$r_e = 5$	$r_e = 10$	$r_e = 15$	5–10	5–15	10–15	$r_e = 5$	$r_e = 10$	$r_e = 15$	5–10	5–15	10–15
0.5	5.77(0.04)	1.63(0.01)	0.78(0.01)	<0.001*, 1	<0.001*, 1	<0.001*, 1	13,802.08(8,042.56)	41,530.54(20,724.27)	83,493.68(54,178.09)	<0.001*, 0.06	<0.001*, 0.01	<.001*, 0.20
0.75	2.70(0.02)	0.73(0.01)	0.33(0.004)	<0.001*, 1	<0.001*, 1	<0.001*, 1	28,320.82(1,8736.16)	80,388.42(52,232.34)	168,921.10(117,644.50)	<0.001*, 0.10	<0.001*, 0.01	<0.001*, 0.18
1	1.76(0.01)	0.46(0.003)	0.20(0.002)	<0.001*, 1	<0.001*, 1	<0.001*, 1	37,576.44(20,453.88)	137,754.90(71,111.73)	350,113.90(205,600.30)	<0.001*, 0.03	<0.001*, 0	<0.001*, 0.11
$r_i = 10$												
d	$r_e = 5$	$r_e = 10$	$r_e = 15$	5–10	5–15	10–15	$r_e = 5$	$r_e = 10$	$r_e = 15$	5–10	5–15	10–15
0.5	5.80(0.07)	1.63(0.03)	0.77(0.02)	<0.001*, 1	<0.001*, 1	<0.001*, 1	5,673.52(2,851.30)	9,496.42(3,800.87)	19,046.28(10,857.27)	<0.001*, 0.19	<0.001*, 0.07	<0.001*, 0.19
0.75	2.70(0.02)	0.73(0.02)	0.32(0.01)	<0.001*, 1	<0.001*, 1	<0.001*, 1	12,447.06(5,991.18)	20,767.94(9,877.58)	48,438.00(20,608.16)	<0.001*, 0.23	<0.001*, 0.01	<0.001*, 0.08
1	1.76(0.02)	0.46(0.01)	0.20(0.003)	<0.001*, 1	<0.001*, 1	<0.001*, 1	16,834.16(6,578.90)	32,936.54(17,215.52)	96,278.02(48,768.43)	<0.001*, 0.15	<0.001*, 0.01	<0.001*, 0.08

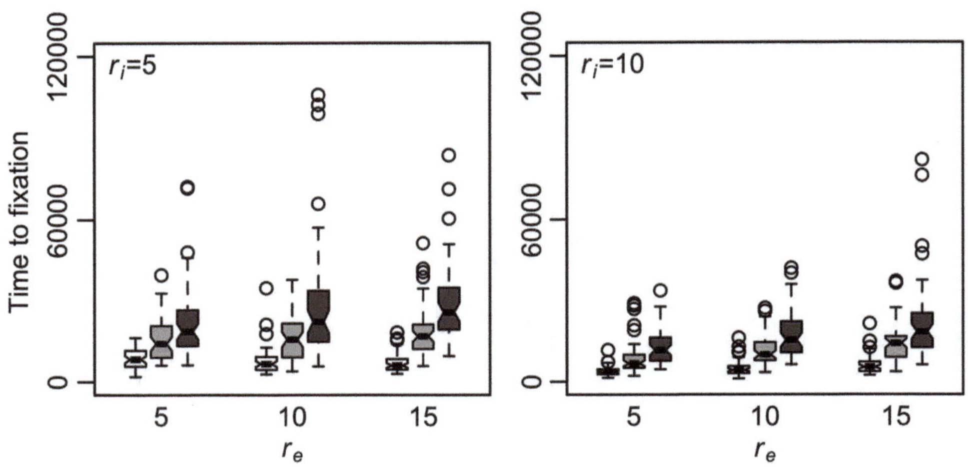

Fig. 5.2 The effect of r_e on time to fixation when cultural transmission is allowed to occur between residential bases, between logistical camps, and between residential bases and logistical camps. Each notched *boxplot* summarizes data collected from 50 simulations. Different values of d are distinguished by color (*white*: $d = 0.5$; *light gray*: $d = 0.75$; *dark gray*: $d = 1$). r_i is interaction radius (*left*: $r_i = 5$, *right*: $r_i = 10$). See Table 5.3 for Mann-Whitney test results

mobility restricts how widely groups interact with each other, then one should expect (1) a positive relationship between r_e and number of traits in the population after 2,000 time steps and (2) a negative relationship between r_e and mean number of different teachers per group after 2,000 time steps. Both relationships should be stronger for cases in which r_e has a stronger effect on time to fixation (Tables 5.2 and 5.3).

Figure 5.3 presents data collected from simulations in which cultural transmission can occur between residential bases only. I vary r_i and d to investigate whether the scale of the interaction sphere and resource density influence the effect of r_e on the variables of interest. The top row presents the number of unique variants present in the population after 2,000 times steps. Larger values are indicative of larger effective population sizes because they result from slower rates of loss (i.e., weaker drift). The bottom row presents the mean number of unique teachers (not including ego) per group after 2,000 time steps. In this case, larger values are indicative of a more freely mixing metapopulation and, thus, a smaller effective population size. Figure 5.3 shows a positive relationship between r_e and the number of variants remaining after 2,000 times steps and a negative relationship between r_e and mean number of teachers per group. The results of the Mann-Whitney tests show that these effects are significant and relatively large in all cases (Table 5.4). These results are consistent with the explanation that, when intergroup interaction is allowed to occur between residential bases only, increased logistical mobility increases the effective size of a metapopulation by decreasing the number of groups with which each group interacts.

Figure 5.4 presents data collected from simulations in which cultural transmission is allowed to take place between residential bases, between logistical camps, and between residential bases and logistical camps. It is under this condition that r_e does not have a significant effect on time to fixation in some cases (Table 5.3). I vary r_i and d as before, and again collect data after just 2,000 time steps in each simulation. The relationship between r_e and mean number of teachers per group is negative and significant in all cases, although the effect size is not as large as in the case where cultural transmission occurs between residential bases only (compare Figs. 5.3 and 5.4 or Tables 5.4 and 5.5). The weaker effect of r_e on mean number of teachers per group explains both why we do not always see a significant effect of r_e on number of traits remaining after 2,000 time steps (Table 5.5) and why r_e does not always have a significant effect on time to fixation (Table 5.3) when cultural transmission is allowed to occur between any combination of residential bases and logistical camps.

5.4 Discussion

The results show that when slowly regenerating food resources are distributed randomly and residential moves occur in a correlated random walk, increased logistical mobility (or reduced residential mobility) can increase the mean time required for a selectively neutral trait to become fixed in a spatially explicit metapopulation composed of central-place foraging groups. But why does a longer effective foraging radius often increase the effective size of the spatially explicit subdivided population, and what does this finding mean for explaining modern behavior as a population-level property?

5.4.1 How Does Increased Logistical Mobility Increase the Effective Size of a Subdivided Population?

First, recall that the Wright-Fisher model assumes panmixia, or a freely mixing population. In the context of our cultural evolutionary model, a freely mixing population is one in which every group is equally likely to learn from any other

Table 5.3 The effect of r_e on residential moves per 100 time steps and time to fixation when cultural transmission is allowed to occur between residential bases, between logistical camps, and between residential bases and logistical camps. See caption to Table 5.2 for details on cell values. d is resource density and r_i is interaction radius. 5–10: comparison of $r_e = 5$ data to $r_e = 10$ data. 5–15: comparison of $r_e = 5$ data to $r_e = 15$ data. 10–15: comparison of $r_e = 10$ data to $r_e = 15$ data

	Residential moves per 100 time steps			p,PS			Time to fixation			p,PS		
$r_i = 5$												
d	$r_e = 5$	$r_e = 10$	$r_e = 15$	5–10	5–15	10–15	$r_e = 5$	$r_e = 10$	$r_e = 15$	5–10	5–15	10–15
0.5	5.78(0.04)	1.62(0.03)	0.76(0.04)	<0.001*, 1	<0.001*, 1	<0.001*, 1	8,809.02(3,995.74)	7,858.00(5,349.48)	7,218.00(3,508.35)	0.058, 0.63	0.032, 0.61	0.807, 0.48
0.75	2.70(0.02)	0.73(0.02)	0.32(0.01)	<0.001*, 1	<0.001*, 1	<0.001*, 1	15,231.94(7,082.80)	16,422.98(7,970.74)	18,610.14(9,909.49)	0.497, 0.46	0.117, 0.41	0.414, 0.45
1	1.76(0.02)	0.46(0.01)	0.20(0.005)	<0.001*, 1	<0.001*, 1	<0.001*, 1	22,367.52(14,461.24)	29,426.76(22,585.20)	28,647.56(15,220.35)	0.078, 0.40	0.006*, 0.34	0.493, 0.46
$r_i = 10$												
d	$r_e = 5$	$r_e = 10$	$r_e = 15$	5–10	5–15	10–15	$r_e = 5$	$r_e = 10$	$r_e = 15$	5–10	5–15	10–15
0.5	5.81(.08)	1.63(0.05)	0.76(0.03)	<0.001*, 1	<0.001*, 1	<0.001*, 1	4,023.02(1,769.12)	4,821.88(2,951.67)	6,069.92(3,642.29)	0.218, 0.43	<0.001*, 0.30	0.035, 0.38
0.75	2.69(.03)	0.73(.02)	0.32(0.01)	<0.001*, 1	<0.001*, 1	<0.001*, 1	8,670.86(6,258.78)	11,873.76(5,885.57)	14,321.08(7,092.25)	<0.001*, 0.29	<0.001*, 0.23	0.050, 0.39
1	1.75(.02)	0.46(0.01)	0.19(0.01)	<0.001*, 1	<0.001*, 1	<0.001*, 1	12,535.94(6,058.13)	17,654.64(9,179.51)	22,250.36(15,557.84)	0.003*, 0.33	<0.001*, 0.25	0.151, 0.42

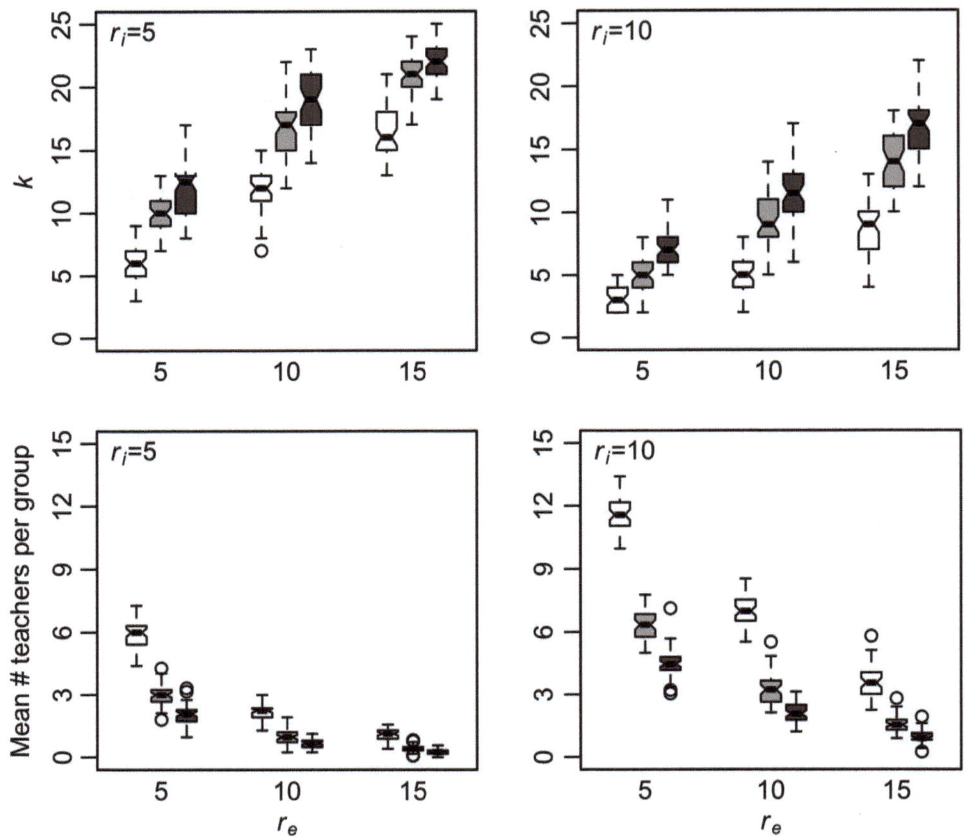

Fig. 5.3 The effect of r_e on number of variants (k) (*top row*) and mean number of teachers per group (*bottom row*) after 2,000 time steps when cultural transmission is restricted to interactions between residential bases. Each notched *boxplot* summarizes data collected from 50 simulations. Different values of d are distinguished by color (*white*: $d = 0.5$; *light gray*: $d = 0.75$; *dark gray*: $d = 1$). r_i is interaction radius (*left column*: $r_i = 5$, *right column*: $r_i = 10$). See Table 5.4 for Mann-Whitney test results

group including itself. Under such "theoretically ideal" conditions, $N = N_e$ and mean time to fixation for haploid traits is closely approximated by $2N_e$, as shown above. Second, consider that the opposite of a freely mixing population is a "non-mixing" (or asexual) population. In the context of cultural evolution, a non-mixing population of social learners can be represented by the vertical cultural transmission of haploid variants. The mean time to fixation of a population initialized with $k = N$ and in which each experienced individual passes its variant vertically to a different naïve individual (its cultural "offspring," so to speak) is infinity for $N > 1$. Obviously, the mechanism of cultural transmission can drastically affect a population's effective size. See Aoki et al. (2011) for a slightly different demonstration—using the Moran model rather than a Wright-Fisher model—of this same general point.

Now, if we return to the "theoretically ideal" case, mobility becomes a more important factor in determining the rate at which foraging groups in a spatially explicit population interact with each other as we gradually relax the assumption of panmixia by decreasing the length of the interaction radius. In other words, when r_i is relatively low, the mean number of different groups with which each group interacts per unit time is determined in large part by residential mobility. Indeed, in the simple central-place foraging model presented here, foraging groups that emphasize residential mobility (low r_e) move their residential bases significantly more often (Tables 5.2 and 5.3) and thus explore a larger proportion of the total landscape per unit time than those that emphasize logistical mobility and, as a result, repeatedly trod over the same immediate surroundings as they make longer logistical trips out from their long-term residential base. This is true despite the fact that each group consumes the same amount of resources over the course of a simulation, regardless of r_e. Naturally, seeing a greater proportion of the total landscape per unit time increases the likelihood of coming into contact with a greater number of groups. It is easiest to envision why this holds true when cultural transmission occurs between residential bases only. But the results presented in Tables 5.3 and 5.5 show that it also holds true when cultural transmission is less restricted (although the effect is not always large enough in this case to significantly affect time to fixation). In general, in the presence of local spheres of intergroup interaction, a foraging strategy that emphasizes residential mobility yields a metapopulation that "mixes" (or interacts) more freely while a foraging strategy that emphasizes logistical mobility results in a more viscous metapopulation (see also Premo 2012b). I turn now to the implications of this finding for selectively neutral cultural diversity at the level of the metapopulation and for cultural differentiation between groups in a subdivided population.

Table 5.4 The effect of r_e on number of variants (k) and mean number of teachers per group after 2,000 time steps when cultural transmission is restricted to interactions between residential bases. The cells under "Number of variants (k)" and "Mean # of teachers per group" present the mean(sd) of 50 simulation runs. Sections labeled "p, PS" present the results of pair-wise Mann-Whitney tests. The first value in each cell is the p-value (*significant after Bonferroni correction for multiple tests; the critical value is $\alpha = 0.05/3 = 0.017$ for all tests). H_0: the distributions of the two samples differ by a location shift of 0. The second value in each cell is the probability of superiority, PS. PS is calculated by dividing the U value provided by each pair-wise Mann-Whitney test by the product of the two samples' sizes (e.g., U/(50 × 50)). PS provides a sense of the size of the effect of r_e. Here, PS should be interpreted as the probability that a value chosen at random from the sample of 50 values where r_e is smaller is superior to (higher in rank order than) a value chosen at random from the sample of 50 values where r_e is larger. A PS value of 0.5 signifies no effect, and effect size increases as the PS value approaches either 0 or 1. d is resource density and r_i is interaction radius. 5–10: comparison of $r_e = 5$ data to $r_e = 10$ data. 5–15: comparison of $r_e = 5$ data to $r_e = 15$ data. 10–15: comparison of $r_e = 10$ data to $r_e = 15$ data

	Number of variants (k)			p, PS			Mean # of teachers per group			p, PS		
$r_i = 5$												
d	$r_e = 5$	$r_e = 10$	$r_e = 15$	5–10	5–15	10–15	$r_e = 5$	$r_e = 10$	$r_e = 15$	5–10	5–15	10–15
0.5	5.92(1.35)	11.80(1.71)	16.40(1.71)	<0.001*, 0.01	<.001*, 0	<0.001*, 0.02	5.92(0.61)	2.16(0.37)	1.07(0.26)	<0.001*, 1	<0.001*, 1	<0.001*, 0.99
0.75	9.68(1.45)	16.80(2.66)	21.08(1.55)	<0.001*, 0.01	<.001*, 0	<0.001*, 0.08	3.02(0.48)	0.96(0.38)	0.39(0.18)	<0.001*, 1	<0.001*, 1	<0.001*, 0.93
1	12.22(2.05)	18.94(2.09)	22.24(1.49)	<0.001*, 0.01	<.001*, 0	<0.001*, 0.11	2.07(0.47)	0.64(0.24)	0.25(0.14)	<0.001*, 1	<0.001*, 1	<0.001*, 0.91
$r_i = 10$												
d	$r_e = 5$	$r_e = 10$	$r_e = 15$	5–10	5–15	10–15	$r_e = 5$	$r_e = 10$	$r_e = 15$	5–10	5–15	10–15
0.5	2.98(0.82)	5.06(1.36)	8.62(1.86)	<0.001*, 0.10	<0.001*, 0	<0.001*, 0.06	11.58(0.80)	6.98(0.74)	3.57(0.73)	<0.001*, 1	<0.001*, 1	<0.001*, 1
0.75	5.22(1.31)	9.26(2.04)	13.98(2.21)	<0.001*, 0.05	<0.001*, 0	<0.001*, 0.05	6.33(0.68)	3.27(0.76)	1.55(0.40)	<0.001*, 1	<0.001*, 1	<0.001*, 0.99
1	7.08(1.37)	11.78(2.39)	16.68(1.91)	<0.001*, 0.04	<0.001*, 0	<0.001*, 0.06	4.51(0.69)	2.11(0.52)	0.96(0.31)	<0.001*, 1	<0.001*, 1	<0.001*, 0.98

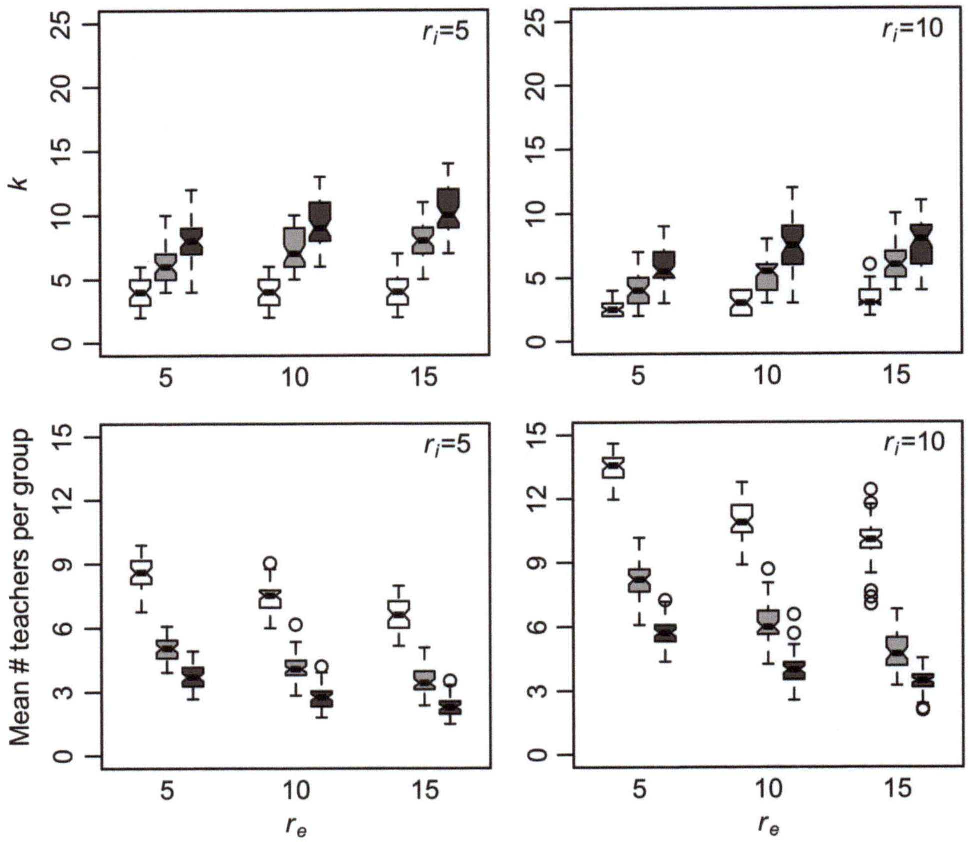

Fig. 5.4 The effect of r_e on number of variants (k) (*top row*) and mean number of teachers per group (*bottom row*) after 2,000 time steps when cultural transmission is allowed to occur between residential bases, between logistical camps, and between residential bases and logistical camps. Each notched *boxplot* summarizes data collected from 50 simulations. Different values of d are distinguished by color (*white*: $d = 0.5$; *light gray*: $d = 0.75$; *dark gray*: $d = 1$). r_i is interaction radius (*left column*: $r_i = 5$, *right column*: $r_i = 10$). See Table 5.5 for Mann-Whitney test results

5.4.2 What Are the Implications for Our Understanding of the Emergence of Modern Behavior?

My results show that a spatially explicit metapopulation of central-place foraging groups that emphasize logistical mobility will often display a larger effective population size than an equivalently sized metapopulation composed of central-place foragers who emphasize residential mobility. But Paleolithic archaeologists do not commonly study culture material at the level of the metapopulation of all Pleistocene hominin groups. They more commonly study material culture excavated from a single locality, such as a cave site, or collected from within the regional boundaries of a watershed or a current nation state. In short, the data that Paleolithic archaeologists consider are usually much closer in spatial scale to the scale of one to a few groups rather than to the entire metapopulation. The issue of scale is especially important here, because decreased intergroup interaction *decreases* the effective population size of each group even though it *increases* the effective size of the metapopulation. Thus, it is imperative to discuss how mobility affects diversity not only at the level of the metapopulation, but also at the level of the groups or regions that archaeologists more commonly study.

A discussion of the effects of mobility on differentiation between groups requires a return to theoretical population genetics. Sewall Wright (1943) shows that decreased intergroup interaction (in his case, gene flow) not only increases the effective size of a subdivided population (as shown above for the case of cultural transmission) but also increases the level of differentiation between groups while at the same time increasing the strength of drift within them (not shown above). In fact, the work of Wright and others shows that isolation by distance increases diversity at the level of the metapopulation precisely because it increases the diversity explained by differences *between* groups more than it reduces the diversity found *within* groups.

Wright's findings on effects of isolation by distance in a stepping stone model of population structure have interesting implications for understanding how shifts in Pleistocene hominin mobility might have affected the appearance of modern behavior at a scale that is more amenable to archaeological inquiry. As shown above, a shift to a mobility strategy that emphasizes logistical mobility can increase cultural diversity at the level of the metapopulation because it reduces intergroup cultural transmission. Wright's work shows that one should also expect increased differentiation in cultural variants *between* groups when intergroup interaction is reduced. Stated more generally, any process that decreases intergroup

Table 5.5 The effect of r_e on number of variants (k) and mean number of teachers per group after 2,000 time steps when cultural transmission is allowed to occur between residential bases, between logistical camps, and between residential bases and logistical camps. See caption to Table 5.4 for details on cell values. d is resource density and r_i is interaction radius. 5–10: comparison of $r_e = 5$ data to $r_e = 10$ data. 5–15: comparison of $r_e = 5$ data to $r_e = 15$ data. 10–15: comparison of $r_e = 10$ data to $r_e = 15$ data

	Number of variants (k)			p, PS			Mean # of teachers per group			p, PS		
$r_i = 5$												
d	$r_e = 5$	$r_e = 10$	$r_e = 15$	5–10	5–15	10–15	$r_e = 5$	$r_e = 10$	$r_e = 15$	5–10	5–15	10–15
0.5	3.80(1.12)	3.94(1.02)	4.20(1.07)	0.487, 0.46	0.106, 0.41	0.393, 0.45	8.59(0.68)	7.43(0.67)	6.60(0.69)	<0.001*, 0.89	<0.001*, 0.98	<.001*, 0.80
0.75	6.18(1.17)	7.40(1.37)	7.88(1.51)	<0.001*, 0.25	<0.001*, 0.19	0.108, 0.41	5.00(0.52)	4.14(0.59)	3.52(0.61)	<0.001*, 0.87	<0.001*, 0.97	<.001*, 0.77
1	8.08(1.54)	9.34(1.76)	10.32(1.83)	<0.001*, 0.29	<0.001*, 0.18	0.010*, 0.35	3.69(0.56)	2.74(0.55)	2.29(0.46)	<0.001*, 0.88	<0.001*, 0.97	<.001*, 0.73
$r_i = 10$												
d	$r_e = 5$	$r_e = 10$	$r_e = 15$	5–10	5–15	10–15	$r_e = 5$	$r_e = 10$	$r_e = 15$	5–10	5–15	10–15
0.5	2.54(0.58)	2.84(0.82)	3.20(0.97)	0.080, 0.41	<0.001*, 0.30	0.082, 0.40	13.46(0.67)	10.93(0.93)	9.98(1.07)	<0.001*, 0.99	<0.001*, 1	<0.001*, 0.77
0.75	4.06(1.15)	5.36(1.17)	6.08(1.40)	<0.001*, 0.22	<0.001*, 0.14	0.010*, 0.35	8.20(0.84)	6.15(0.88)	4.85(0.88)	<0.001*, 0.95	<0.001*, 1	<0.001*, 0.85
1	5.66(1.57)	7.46(1.88)	7.88(1.80)	<0.001*, 0.24	<0.001*, 0.19	0.292, 0.44	5.74(0.65)	4.02(0.72)	3.41(0.55)	<0.001*, 0.96	<0.001*, 1	<0.001*, 0.76

cultural transmission in a subdivided population will have the two-pronged effect of increasing neutral cultural diversity at the level of the metapopulation and increasing neutral cultural differentiation between groups. Thus, to the extent that greater regional diversification in culturally transmitted variants serves as an indicator of modern behavior, perhaps it, too, might be explained in part by increased isolation by distance due to a greater emphasis on logistical mobility.

5.5 Conclusion

The results reported here show that, holding all else constant, a shift to a more collector-like foraging strategy yields metapopulations that are characterized by increased isolation between central-place foraging groups and larger effective population sizes. Larger effective population sizes support greater neutral cultural diversity at the level of the metapopulation. Considered within the context of Wright's work on isolation by distance, the results also suggest that an emphasis on logistical mobility would result in greater cultural differentiation between groups within a spatially explicit subdivided population. Archaeologists have generally recognized greater overall diversity and increased regional differentiation in culture material as signals of modern behavior. However, as is often the case, this finding raises more questions than it answers. To what extent are signs of increased cultural differentiation between regions during the Paleolithic due to a shift in mobility rather than to some form(s) of biased cultural transmission or to different natural selective pressures in different environments? How are we to measure shifts in Pleistocene hominin mobility strategies independently of the archaeological data we would like to investigate for the effects of such shifts? In the event that evidence for shifts to a more collector-like strategy is found to regularly precede "flashes" of modern behavior in the archaeological record, what was responsible for such shifts and were those shifts reversed in cases where the archaeological record suggests that those behaviors ultimately did not persist in the region?

Although a full discussion concerning to what extent the mobility strategies of behaviorally modern humans differed from other Pleistocene hominin populations, such as Neandertals, is beyond the scope of the present study, the topic is worth visiting briefly in light of the theoretical findings presented above. Despite decades of interest in Middle Paleolithic settlement systems, the evidence for differences in land-use strategies among Pleistocene hominin populations remains spotty (many authors bemoan the small sample sizes) and contentious. As recently recounted by Wallace and Shea (2006) and Walker and Churchill (2014), archaeologists have not been of one mind in interpreting Middle Paleolithic data in the context of hominin mobility. The available data suggest to some that Neandertals emphasized logistical mobility, while to others the data suggest that Neandertals emphasized residential mobility. Lieberman and Shea (1994) and Wallace and Shea (2006) discuss evidence that might support the hypothesis that Levantine Neandertals emphasized logistical mobility (or "radiating mobility") while anatomically modern humans in the Levant some 30,000 years earlier perhaps emphasized residential mobility (or "circulating mobility"). On the other hand, Kuhn (1992) is not alone in floating the hypothesis that at least some Upper Paleolithic modern humans may have emphasized logistical mobility more than Middle Paleolithic Neandertals did (see also Riel-Salvatore and Barton 2004). These issues remain unresolved. But if there is some truth to the notion that there was a difference in the settlement systems of Neandertals and modern humans during the Late Pleistocene, with modern humans emphasizing logistical mobility more than Neandertals, then *holding all else constant* the findings presented above raise the possibility that differences in mobility rather than differences in cognition may have played an important role in explaining some of the differences between the Middle Paleolithic and Upper Paleolithic archaeological records.

Yet, given the nature of the archaeological evidence available, it is premature to rule out the possibility that all Pleistocene hominin foragers were behaviorally flexible enough to modify their mix of logistical and residential mobility in response to changing conditions in ways that are more-or-less captured by Kelly's central-place foraging model. In other words, it seems unreasonable at this time to pin Neandertals and modern humans to different places along Binford's forager-collector continuum as if their land-use strategies were not only different but also static (nor is this is a position that many researchers, if any, hold today). If one avoids rigid, monolithic classifications of Neandertals as more forager-like and modern humans as more collector-like (or vise versa) and instead allows for more fluidity in Pleistocene hominin adaptive responses to changing conditions (e.g., Delanges and Rendu 2011), then one leaves the door open to the possibility that even relatively subtle shifts in mobility strategies could modify metapopulation dynamics such that some of the archaeological signals currently recognized as hallmarks of modern behavior—increased total cultural diversity and increased cultural differentiation between regions—could appear and persist for as long as conditions allow, only to disappear when the metapopulation dynamics are modified again. Note that this statement is not a conclusion about the past but rather a hypothesis that remains to be tested against the culture material record.

In thinking about how this hypothesis might be tested empirically, I wish to reiterate that the effects of mobility on cultural *diversity* discussed above are much more likely to be seen in selectively neutral cultural variants passed via unbiased transmission than in cultural variants that are adaptive and/or passed via a form of biased cultural transmission. Because tool classes used to procure food (i.e., Oswalt's (1976) subsistants) were likely affected by cultural selection (and perhaps even natural selection working on those who employed them), their diversity is unlikely to vary with population size in accordance with the predictions of neutral theory and therefore they are unlikely to serve as a good test of the so-called "demographic hypothesis." The same holds true for cultural variants that are even further removed from the tool user's reproductive success than subsistants. While "stylistic" variants used to convey something about one's social identity or group membership may not be adaptive in the biological sense, they are likely to be affected by cultural selection—that is, they are likely to be passed via biased forms of cultural transmission. Thus, it is not enough for empirical tests of the effect of mobility on cultural diversity in the Paleolithic to target so-called non-functional aspects of technology. Empirical tests of the demographic hypothesis that focus on cultural diversity must target cultural variants that not only did not affect the reproductive fitness of those who displayed them but also—and just as importantly—were passed via unbiased cultural transmission. Identifying Paleolithic cultural variants that satisfy those requirements requires more thought and discussion.

Acknowledgements I thank Alex Mesoudi and Kenichi Aoki for the invitation to contribute a paper to this interesting volume. I thank Amelie Schmolke and my reviewers, Kenichi Aoki and Charles Perreault, for their insightful and very constructive comments. Any remaining errors in logic or presentation are mine alone.

References

Ambrose SH (1998) Chronology of the Later Stone Age and food production in East Africa. J Archaeol Sci 25:377–392

Ambrose SH (2001) Paleolithic technology and human evolution. Science 291:1748–1753

Ambrose SH (2010) Coevolution of composite-tool technology, constructive memory, and language implications for the evolution of modern human behavior. Curr Anthropol 51:S135–S147

Ambrose SH, Lorenz KG (1990) Social and ecological models for the Middle Stone Age in southern Africa. In: Mellars P (ed) The emergence of modern humans: an archaeological perspective. Edinburgh University Press, Edinburgh, pp 3–33

Aoki K, Lehmann L, Feldman MW (2011) Rates of cultural change and patterns of cultural accumulation in stochastic models of social transmission. Theor Popul Biol 79:192–202

Binford LR (1980) Willow smoke and dogs' tails: hunter-gatherer settlement systems and archaeological site formation. Am Antiq 45:4–20

Binford LR (2001) Constructing frames of reference: an snalytical method for theory building using ethnographic and environmental data sets. University of California Press, Berkeley

Boyd R, Richerson PJ (1985) Culture and the evolutionary process. University of Chicago Press, Chicago

Brown KS, Marean CW, Herries AIR, Jacobs Z, Tribolo C, Braun D, Roberts DL, Meyer MC, Bernatchez J (2009) Fire as an engineering tool of early modern humans. Science 325:859–862

Burbano HA, Hodges E, Green RE, Briggs AW, Krause J et al (2010) Targeted investigation of the Neandertal genome by array-based sequence capture. Science 328:723–725

Collard M, Kemery MD, Banks S (2005) Causes of toolkit variation among hunter–gatherers: a test of four competing hypotheses. Can J Archaeol 29:1–19

Collard M, Buchanan B, Morin J, Costopoulos A (2011) What drives the evolution of hunter-gatherer subsistence technology? A reanalysis of the risk hypothesis with data from the Pacific Northwest. Philos Trans R Soc B 366:1129–1138

Collard M, Buchanan B, O'Brien MJ (2013a) Population size as an explanation for patterns in the Paleolithic record: more caution is needed. Curr Anthropol 54:S388–S396

Collard M, Buchanan B, O'Brien MJ, Scholnick J (2013b) Risk, mobility, or population size? Drivers of technological diversity among recent western North American hunter-gatherers. Philos Trans R Soc B 368:20120412

Crow JF, Kimura M (1970) An introduction to population genetics theory. Harper and Row, New York

Delanges A, Rendu W (2011) Shifts in Neandertal mobility, technology and subsistence strategies in western France. J Archaeol Sci 38:1771–1783

d'Errico F (2003) The invisible frontier. A multiple species model for the origin of behavioral modernity. Evol Anthropol 12:188–202

d'Errico F, Henshilwood CS (2007) Additional evidence for bone technology in the southern African Middle Stone Age. J Hum Evol 52:142–163

d'Errico F, Stringer CB (2011) Evolution, revolution or saltation scenario for the emergence of modern cultures? Philos Trans R Soc B 366:1060–1069

d'Errico F, Henshilwood CS, Vanhaeren M, van Niekerk K (2005) *Nassarius kraussianus* shell beads from Blombos Cave: evidence for symbolic behaviour in the Middle Stone Age. J Hum Evol 48:3–24

Fisher RA (1930) The genetical theory of natural selection. Clarendon Press, Oxford

Foley R, Lahr MM (1997) Mode 3 technologies and the evolution of modern humans. Camb Archaeol J 7:3–36

Green RE, Krause J, Briggs AW, Maricic T, Stenzel U et al (2010) A draft sequence of the Neandertal genome. Science 328:710–722

Grimm V, Berger U, Bastiansen F, Eliassen S, Ginot V et al (2006) A standard protocol for describing individual-based and agent-based models. Ecol Model 198:115–126

Hartl DL, Clark AG (2007) Principles of population genetics, 4th edn. Sinauer Associates, Sunderland

Henrich J (2004) Demography and cultural evolution: how adaptive cultural processes can produce maladaptive losses: the Tasmanian case. Am Antiq 69:197–214

Henshilwood CS, d'Errico F, Marean CW, Milo RG, Yates R (2001) An early bone tool industry from the Middle Stone Age at Blombos Cave, South Africa: implications for the origins of modern human behaviour, symbolism and language. J Hum Evol 41:631–678

Henshilwood CS, d'Errico F, Yates R, Jacobs Z, Tribolo C, Duller GAT, Mercier N, Sealy JC, Valladas H, Watts I, Wintle AG (2002) Emergence of modern human behaviour: Middle Stone Age engravings from South Africa. Science 295:1278–1280

Henshilwood CS, d'Errico F, Vanhaeren M, van Niekerk K, Jacobs Z (2004) Middle Stone Age shell beads from South Africa. Science 304:404

Henshilwood CS, d'Errico F, Watts I (2009) Engraved ochres from the Middle Stone Age levels at Blombos Cave, South Africa. J Hum Evol 57:27–47

Hitchcock RK (1982) The ethnoarchaeology of sedentism: mobility strategies and site structure among foraging and food producing groups in the Eastern Kalahari Desert, Botswana. Unpublished Ph.D. dissertation, University of New Mexico, Albuquerque

Kelly RL (1983) Hunter-gatherer mobility strategies. J Anthropol Res 39:277–306

Kelly RL (1990) Marshes and mobility in the Western Great Basin. In: Janetski J, Madsen DB (eds) Wetlands adaptations in the Great Basin. Brigham Young University Press, Provo, pp 259–276

Kelly RL (1991) Sedentism, sociopolitical inequality, and resource fluctuations. In: Gregg S (ed) Between bands and states. Southern Illinois University Press, Carbondale, pp 135–158

Kelly RL (1995) The foraging spectrum: diversity in hunter-gatherer lifeways. Smithsonian Institution Press, Washington

Kimura M (1968) Evolutionary rate at the molecular level. Nature 217:624–626

Kimura M (1983) The neutral theory of molecular evolution. Cambridge University Press, Cambridge

Kimura M, Crow JF (1964) The number of alleles that can be maintained in a finite population. Genetics 49:725–738

Kimura M, Ohta T (1969) The average number of generations until fixation of a mutant gene in a finite population. Genetics 61:763–771

Kintigh KW, Altschul JH, Beaudry MC, Drennan RD, Kinzig AP et al (2014) Grand challenges for archaeology. Am Antiq 79:5–24

Klein RG (1992) The archaeology of modern humans. Evol Anthropol 1:5–14

Klein RG (1995) Anatomy, behavior, and modern human origins. J World Prehist 9:167–198

Klein RG (2000) Archaeology and the evolution of human behavior. Evol Anthropol 9:17–36

Klein RG (2008) Out of Africa and the evolution of human behavior. Evol Anthropol 17:267–281

Kuhn SL (1992) On planning and curated technologies in the Middle Paleolithic. J Anthropol Res 48:185–214

Kuhn SL, Stiner MC, Reese DS, Gulec E (2001) Ornaments of the earliest Upper Paleolithic: new insights from the Levant. Proc Natl Acad Sci U S A 98:7641–7646

Lieberman DE, Shea JJ (1994) Behavioral differences between archaic and modern humans in the Levantine Mousterian. Am Anthropol 96:300–332

MacDougall I, Brown FH, Fleagle JG (2005) Stratigraphic placement and age of modern humans from Kibish, Ethiopia. Nature 433:733–736

Marean CW, Bar-Matthews M, Bernatchez J, Fisher E, Goldberg P et al (2007) Early human use of marine resources and pigment in South Africa during the Middle Pleistocene. Nature 449:905–908

Maricic T, Gunther V, Georgiev O, Gehre S, Curlin M et al (2013) A recent evolutionary change affects a regulatory element in the human FOXP2 gene. Mol Biol Evol 30:844–852

McBrearty S, Brooks AS (2000) The revolution that wasn't: a new interpretation of the origin of modern human behavior. J Hum Evol 39:453–563

Mellars P (1996) The emergence of biologically modern populations in Europe: a social and cognitive 'revolution'? In: Runciman WG, Maynard Smith J, Dunbar RIM (eds) Evolution of social behaviour patterns in primates and man. Clarendon Press, Oxford, pp 179–202

Neiman FD (1995) Stylistic variation in evolutionary perspective: inferences from decorative diversity and interassemblage distance in Illinois woodland ceramic assemblages. Am Antiq 60:7–36

Nowell A (2010) Defining behavioral modernity in the context of Neandertal and anatomically modern human populations. Annu Rev Anthropol 39:437–452

Oswalt WH (1976) An anthropological analysis of food-getting technology. Wiley, New York

Perreault C, Brantingham PJ (2011) Mobility-driven cultural transmission along the forager-collector continuum. J Anthropol Archaeol 30:62–68

Powell A, Shennan S, Thomas MG (2009) Late Pleistocene demography and the appearance of modern human behavior. Science 324:1298–1301

Premo LS (2012a) Local extinctions, connectedness, and cultural evolution in structured populations. Adv Complex Syst 15:1150002

Premo LS (2012b) The shift to a predominantly logistical mobility strategy can inhibit rather than enhance forager interaction. Hum Ecol 40:647–649

Premo LS, Kuhn SL (2010) Modeling effects of local extinctions on culture change and diversity in the Paleolithic. PLoS One 5:e15582

Premo LS, Scholnick JB (2011) The spatial scale of social learning affects cultural diversity. Am Antiq 76:163–176

Prufer K, Racimo F, Patterson N, Jay F, Sankararaman S et al (2014) A complete genome sequence of a Neanderthal from the Altai Mountains. Nature 505:43–49

Read D (2008) An interaction model for resource implement complexity based on risk and number of annual moves. Am Antiq 73:599–625

Reich D, Green RE, Kircher M, Krause J, Patterson N et al (2010) Genetic history of an archaic hominin group from Denisova Cave in Siberia. Nature 468:1053–1060

Riel-Salvatore J, Barton CM (2004) Late Pleistocene technology, economic behavior, and land-use dynamics in southern Italy. Am Antiq 69:257–274

Shennan SJ (2001) Demography and cultural innovation: a model and its implications for the emergence of modern human culture. Camb Archaeol J 11:5–16

Shennan SJ, Wilkinson JR (2001) Ceramic style change and neutral evolution: a case study from Neolithic Europe. Am Antiq 66:577–593

Shott M (1986) Technological organization and settlement mobility: an ethnographic examination. J Anthropol Res 42:15–51

Soressi M, McPherron SP, Lenoir M, Dogandzic T, Goldberg P et al (2013) Neandertals made the first specialized bone tools in Europe. Proc Natl Acad Sci U S A 110:14186–14190

Texier P-J, Porraz G, Parkington J, Rigaud J-P, Poggenpoel C et al (2010) A Howiesons Poort tradition of engraving ostrich eggshell containers dated to 60,000 years ago at the Diepkloof Rock Shelter, South Africa. Proc Natl Acad Sci U S A 107:6180–6185

Torrence R (1983) Time budgeting and hunter-gatherer technology. In: Bailey G (ed) Hunter-gatherer economy in prehistory. Cambridge University Press, Cambridge, pp 11–22

Torrence R (1989) Re-tooling: towards a behavioral theory of stone tools. In: Torrence R (ed) Time, energy and stone tools. Cambridge University Press, Cambridge, pp 57–66

Torrence R (2000) Hunter-gatherer technology: macro- and microscale approaches. In: Panter-Brick C, Layton RH, Rowley-Conwy P (eds) Hunter-gatherers: an interdisciplinary perspective. Cambridge University Press, Cambridge, pp 99–143

Tostevin GB (2007) Social intimacy, artefact visibility, and acculturation models of neanderthal-modern human interaction. In: Mellars P, Boyle K, Bar-Yosef O, Stringer C (eds) Rethinking the human revolution: new behavioural and biological perspectives on the origins and dispersal of modern humans. Cambridge University Press, Cambridge, pp 341–357

Tryon CA, Faith JT (2013) Variability in the Middle Stone Age of Eastern Africa. Curr Anthropol 54:S234–S254

Vaesen K (2012) Cumulative cultural evolution and demography. PLoS One 7:e40989

Vanhaeren M, d'Errico F, Stringer C, James SL, Todd JA, Mienis HK (2006) Middle Paleolithic shell beads in Isreal and Algeria. Science 312:1785–1788

Wadley L (2013) Recognizing complex cognition through innovative technology in Stone Age and Palaeolithic sites. Camb Archaeol J 23:163–183

Wakeley J (1999) Nonequilibrium migration in human history. Genetics 153:1863–1871

Wakeley J (2000) The effects of subdivision on the genetic divergence of populations and species. Evolution 54:1092–1101

Walker CS, Churchill SE (2014) Territory size in *Canis lupus*: implications for neandertal mobility. In: Carlson K, Marchi D (eds) Reconstructing mobility: environmental, behavioral, and morphological determinants. Springer Press, New York, pp 209–226

Wallace IJ, Shea JJ (2006) Mobility patterns and core technologies in the Middle Paleolithic of the Levant. J Archaeol Sci 33: 1293–1309

White AA (2013) An abstract model showing that the spatial structure of social networks affects the outcomes of cultural transmission processes. J Artif Soc Soc Simul 16(3):9

Wilensky U (1999) NetLogo. http://ccl.northwestern.edu/netlogo/. Center for Connected Learning and Computer-Based Modeling, Northwestern University, Evanston

Wright S (1931) Evolution in Mendelian populations. Genetics 16:97–159

Wright S (1943) Isolation by distance. Genetics 28:114–138

Wurz S (2013) Technological trends in the Middle Stone Age of South Africa between MIS 7 and MIS 3. Curr Anthropol 54:S305–S319

Yellen JE, Brooks AS, Cornelissen E, Mehlman MJ, Stewart K (1995) A Middle Stone Age worked bone industry from Katanda, Upper Semliki Valley, Zaire. Science 268:553–556

Behavioral Modernity and the Cultural Transmission of Structured Information: The Semantic Axelrod Model

6

Mark E. Madsen and Carl P. Lipo

Abstract

Cultural transmission models are coming to the fore in explaining increases in the Paleolithic toolkit richness and diversity. During the later Paleolithic, technologies increase not only in terms of diversity but also in their complexity and interdependence. As Mesoudi and O'Brien (Biolog Theory 3:63–72, 2008) have shown, selection broadly favors social learning of information that is hierarchical and structured. We believe that teaching provides the necessary scaffolding for transmission of more complex cultural traits. Here, we introduce an extension of the Axelrod (J Confl Resolut 41:203–226, 1997) model of cultural differentiation in which traits have prerequisite relationships, and where social learning is dependent upon the ordering of those prerequisites. We examine the resulting structure of cultural repertoires as learning environments range from largely unstructured imitation, to structured teaching of necessary prerequisites, and we find that in combination with individual learning and innovation, high probabilities of teaching prerequisites leads to richer cultural repertoires. Our results point to ways in which we can build more comprehensive explanations of the archaeological record of the Paleolithic as well as other cases of technological change.

Keywords

Structured trait model • Axelrod model • Unbiased transmission • Knowledge prerequisites • Cumulative cultural transmission

6.1 Introduction

Although humans and our hominin ancestors have been cultural animals throughout our evolutionary history, an important change occurred in our lineage during the Middle and Upper Paleolithic. For millennia our ancestors manufactured relatively small toolkits and their material culture was remarkably similar across continental distances and over many generations. Beginning in the Middle Paleolithic and continuing through the Upper Paleolithic, the archaeological record reflects an explosion in our cultural repertoire. Over tens of thousands of years, artifactual toolkits shift from sets of relatively few objects with multiple uses to large collections of functionally-specialized tools that employed increasingly complex technologies and that were manufactured from an enriched range of materials. The changes in artifacts suggest that human solutions to the problems of everyday life became regionalized and differentiated. Further, the economic basis of our lives began to broaden and also, in many areas, to become specialized (Bar-Yosef 2002; d'Errico and Stringer 2011; Straus 2005).

M.E. Madsen (✉)
Department of Anthropology, University of Washington, Box 353100, Seattle, WA 98195, USA
e-mail: mark@madsenlab.org

C.P. Lipo
Department of Anthropology and IIRMES, California State University at Long Beach, 1250 Bellflower Blvd, Long Beach, CA 90840, USA
e-mail: Carl.Lipo@csulb.edu

A. Mesoudi and K. Aoki (eds.) *Learning Strategies and Cultural Evolution during the Palaeolithic*, Replacement of Neanderthals by Modern Humans Series, DOI 10.1007/978-4-431-55363-2_6,

While early researchers believed that the Upper Paleolithic resulted from a singular "revolution" in human evolution leading to behaviorally modern homo sapiens, this view is held by a minority of paleoanthropologists and archaeologists today (e.g., Klein 2009). Careful examination of the Middle Paleolithic archaeological record especially in Africa and the Near East suggests that this change in behavior did not occur as a single distinct event, instead occurring over a long period of time since much of the enriched material culture we later characterize as the "Upper Paleolithic" had precursors. In addition, this change now appears to be patchy and fitful, with modern features appearing and frequently being lost again (Bouzouggar et al. 2007; d'Errico and Henshilwood 2007; d'Errico and Stringer 2011; Straus 2005; McBrearty and Brooks 2000; McBrearty 2007). Nor does behavioral modernity map neatly to biological taxa and their movements, given that evidence for the precursors of fully modern behavior is abundant in deposits associated with Neaderthals in addition to modern *Homo sapiens* (Villa and Roebroeks 2014).

The "learning hypothesis" studied in this series of volumes makes the plausible claim that behavioral modernity is the product of cumulative changes in the way cultural information was acquired and retained across generations (Nishiaki et al. 2013), thus providing a potential explanation for the slow evolution of "modern" features, its patchiness in space and time, and the lack of a neat mapping between hominin taxonomy and material culture. In short, according to the learning hypothesis, behavioral modernity arose through a change or changes in the way social learning operated within hominin groups, with those groups adopting richer modes of cultural learning surviving and spreading compared to those who retained simpler forms of social learning.

Within the umbrella of the learning hypothesis, there are many ways in which social learning and thus intergenerational cultural transmission could have changed, and an increasing amount of research is focused upon formulating and testing different models. One class of studies is focused upon factors exogeneous to the learning or imitation process itself. Shennan (2000, 2001) proposed that population size has a powerful effect on diversity within cultural transmission processes, which Henrich showed in the case of toolkit element loss during a Tasmanian population bottleneck (Henrich 2004). In a similar line of reasoning, Kuhn (2013) argues that low population size and density put Neanderthals in a situation where innovations spread slowly and ultimately led to their demise relative to modern humans. Furthermore, a growing set of experimental studies clearly show a relationship between accumulation of complex cultural traits and the number of cultural "models" from whom individuals can learn (Muthukrishna et al. 2014; Derex et al. 2013; Kempe and Mesoudi 2014). Not all studies have shown a strong association between population size and cultural diversity, however. Collard and colleagues, find little association in a linked series of comparative studies (Collard et al. 2011, 2013a,b,c). Finally, in his analysis of the overall evolutionary rate, Aoki (2013) found that innovation rates were more important than population size to determining the rate of evolution in a population.

To us, this body of work indicates that while population size is an important parameter in mathematical models, it may be better understood as a second-order effect in the real world, interacting with a myriad of other factors and thus often dominated by those factors. Another important factor is the structure of bands or demes into larger regional metapopulations. Network topology, for example, is known to have a substantial effect upon contagion or diffusion processes (e.g., Castellano et al. 2009; Smilkov and Kocarev 2012). Thus, it is likely that regional structure has critical effects on the outcomes we can expect from a single social "learning rule." Along these lines, Premo (2012) has examined whether metapopulation dynamics that include local extinction and recolonization might provide an improved account for the retention and expansion of diversity.

A second group of studies has focused upon endogeneous changes to social learning processes. Many authors in this volume series, for example, have looked at aspects of the way individuals learn skills and acquire information (Aoki 2013; Nishiaki et al. 2013). We know that learning and teaching styles vary across human groups, and formal modeling efforts are beginning to make clear that such variation has evolutionary consequences that might lead to a rapid expansion of the human cultural repertoire (Nakahashi 2013). Those populations which increased the amount or effectiveness of teaching would have a fitness advantage over those who relied upon imitation and "natural pedagogy" in passing along technological and foraging knowledge (Csibra and Gergely 2011; Fogarty et al. 2011; Terashima 2013). Demography and population structure would then play an important role in reinforcing the fitness differences which different learning strategies would create, as pointed out by Kuhn (2013).

Ultimately, a full "learning explanation" for behavioral modernity will be multifacted, including demographic and spatial changes as well as changes to the mechanisms of social learning and technological innovation themselves. Sterelny (2012, p. 61) sums up this kind of multifactorial approach to behavioral modernity well:

> . . . the cultural learning characteristic of the Upper Paleolithic transition and later periods of human culture—social transmission with both a large bandwidth and sufficient accuracy for incremental improvement—requires individual cognitive adaptations for cultural learning, highly structured learning environments, and population structures that both buffer existing resources effectively and support enough specialization to generate a supply of innovation.

In research designed to explore how the structure of a learning environment affects the results of social learning, Creanza and colleagues (2013), Aoki (2013), Nakahashi (2013), and Castro and colleagues (2014) developed models that examine how explicit teaching (as opposed to simple imitation) affects the overall evolutionary rate or cultural diversity in a population. Castro et al. for example, find that cumulative cultural transmission requires active teaching in order to achieve fidelity across generations. Our work in this chapter follows these authors, focusing on the nature of transmitted information itself and the effects of teaching upon the richness of structured technological knowledge.

In particular, we suggest that when knowledge is structured such that skills and information must be learned in sequences, high fidelity learning environments are critical to evolving ever-richer cultural repertoires, of the type seen in behaviorally modern assemblages. To formalize this idea, we construct a model which:

- Represents cultural traits as hierarchically structured, in order to study increases in complexity,
- Has a learning rule sensitive to the order in which cultural traits are acquired, with multiple levels of fidelity, and
- Has a mechanism (such as homophily) that allows cultural differentiation endogeneous to the model.

As we alter the "learning environment" in our models from less to more frequent teaching of traits and their prerequisites, we expect to see greater diversity, larger structured sets of traits persisting in the population, and greater differentiation of the population into "different" cultural configurations. We also expect that individual innovation, independent of the social learning context, will play a role in the accumulation of cultural complexity by allowing a population to explore increasingly large spaces of technological design possibilities; this expectation is concordant with Aoki's (2013) result in Volume I of this series.

In this chapter, we introduce a simulation model which combines a hierarchical trait space capable of expressing dependencies or semantic relationships between skills and information (Mesoudi and O'Brien 2008), and a modified version of Robert Axelrod's (1997) homophilic social learning model which allows us to examine the conditions under which evolution in a hierarchical design space leads to cultural differentiation. After describing the model, we study its dynamics and provide an initial assessment of its suitability for studying the onset of behavioral modernity in the later Paleolithic. Models like this begin to move beyond diffusion dynamics, bringing the actual meaning and relations of traits into the modeling process. Hence, we call these "semantic Axelrod" models, and believe that such models form a platform for formalizing the type of multi-factor hypotheses necessary to examine major transitions in human evolution, such as "behavioral modernity."

6.2 The Semantic Axelrod Model for Trait Prerequisites

Much of our technical knowledge, whether of stone tool manufacture, throwing clay pots, or computer repair, is built from simple tasks, bits of background knowledge, and step-by-step procedures (Neff 1992; Schiffer and Skibo 1987). These pieces of cultural information are not simply a set of alternative options, which can be mixed and matched in any combination. Instead, there are dependencies and relationships between items which affect how skills and information are learned and passed on between individuals. Some items will be related in time, as steps in a process. Others will be related by subsumption: arrowheads are a subclass of bifacial stone tools, and require many of the same production techniques as bifaces used in other projectiles. Still others will be related as sets of alternatives: choices of surface treatment for a given ceramic paste, given the firing regime selected, for example. To date, most archaeological models of tool production have focused upon temporal relations in the construction of an artifact, as in "sequence models" or "chaîne opératoire," but it is important to remember that other representations are possible, including trees and more general graphs to capture relations of use, reworking, or discard (Bamforth and Finlay 2008; Ferguson 2008; Högberg 2008; Bleed 2001, 2002, 2008; Schiffer and Skibo 1987; Stout 2002).

Given conscious reflection, we describe and organize our knowledge and skills in many ways, but it is common (especially while learning a new skill) to think of a complex process as a "script" or "recipe" (Schank and Abelson 1977). Experts in a task or field may not represent their knowledge this way, having internalized such structures below the conscious level. Experts will often know more than one way to accomplish any given goal, and be able to repurpose and recombine methods and tools, as opposed to the simpler, more linear or tree-based recipes of the novice or student (e.g., Bleed 2002, 2008; Stout 2002). Nevertheless, it is common to teach or learn new information and skills in a stepwise manner.

In this chapter, we focus not on the execution steps of a recipe (and thus not on sequence models), but the relations between skills and information *during the learning process*. In specific, we focus upon the *prerequisite* relationships that exist between cultural traits, since the ordered dependencies between skills and information form one of the structures within social learning occurs during development (and into adulthood). Some pieces of information or skills must be in place before a person can effectively learn or practice others. Examples from our own childhoods abound: one needed to understand addition and subtraction and multiplication before learning long division; in order to make soup, we need to understand how to simmer rather than boil, how to chop and slice, what ingredients might be combined, and so on. The fact that knowledge and skills build upon one another make prerequisite relations between cultural traits ubiquitous. In this chapter, we represent prerequisite relations as trees in the graph-theoretic sense (Diestel 2010), replacing the "nominal scale" structure of "locus/allele" models or paradigmatic classifications and some typologies (Dunnell 1971), but we emphasize that the tree models we discuss here are still classifications and thus analytic tools, designed to allow us to measure variation in the archaeological record, not reconstruct emic models of Paleolithic technologies.

Our model also requires a way of representing a changing learning environment, in ways that create higher fidelity and greater possibility for building cumulative knowledge. In real learning environments, there are many possibilities, but deliberate teaching and apprentice learning are repeatedly seen across human groups as ways that naive individuals can reliably learn the complex skills and information needed for foraging, artifact production and maintenance, and navigating an increasingly rich social world. The point of structuring the learning environment with teaching and/or apprenticeship is to give the learner skilled models to imitate, shortcut trial and error when acquiring a skill, provide a reference for needed information, and to guide individuals to put their information and skills together into appropriate sequences to accomplish an overall goal. Apprenticeship and formalized teaching provide a social learning "scaffold," helping to lower the amount of individual trial and error learning needed to master a body of material (Wimsatt and Griesemer 2007; Wimsatt 2007).

Within a standard discrete-time simulation model of a social learning process, we can model this type of learning environment with the following modifications:

1. Represent the order in which skills and information need to be acquired as a series of trees, with vertices representing traits (either a skill or piece of information), and edges the prerequisite relations between them.
2. Disallowing individuals the ability to copy traits from a cultural model for which they do not have necessary background or prerequisites, given the relations in the applicable tree model.
3. Creating a probability that individuals, if disallowed a trait, can be taught one of the needed prerequisites instead by that cultural model, leading to the potential accumulation of fuller knowledge and skills over time.

By changing the probability that individuals learn a missing prerequisite trait, we can "tune" the learning environment. Low probabilities might correspond, for example, to a learning environment where individuals can observe others executing a production step, but are given little or no instruction or guidance on what they need to know in order to successfully master it. High probabilities of learning prerequisites would correspond, on the other hand, to environments where individuals receive instruction, or work together with a more skilled individual who guides them toward learning the information and skills they lack. In the next section, we discuss our model of trait relationships and the learning environment in more detail.

6.2.1 Representation of Traits and Their Prerequisites

In order to represent the "prerequisite" relations between a number of cultural traits, we organize the traits into trees,[1] where nodes higher in the tree represent knowledge, skills, or concepts which are necessary for traits further down the tree. Let us consider the different skills and information necessary for the construction of a single artifact, say a dart thrown by an atlatl. An artisan will possess information about different raw materials, an understanding of what materials are suitable for specific purposes, skills and information concerning the knapping of different types of bifaces, methods of hafting bifaces into different kinds of shafts, and so on. Stout (2011) organized such knowledge into "action hierarchies," which represent sequences of actions, sets of choices, and optional elements for the construction of a class of stone tools, drawing the representation from Moore's (2010) graphical notation.

We should emphasize that employing tree structures to represent learning dependencies is a modeling choice. Other choices may be sensible as well. General graphs could represent webs of relations between concepts or skills, and multigraphs (replacing adjacency matrices with tensors) can represent different types of relations in a single structure (Nickel et al. 2011). For purposes of the present chapter, we are interested in the order in which people usually *learn* skills and information, rather than the order in which steps are executed. The difference is potentially significant, in that two adjacent steps in a sequence might involve very different information, tools, or skills, which can be learned in parallel without dependencies. Because, in our model, traits cannot be learned unless an individual possesses the necessary prerequisites, we introduce the idea of a "learning hierarchy," which is a division of Stout's action hierarchy into components which are learned with ordered dependencies, and independent components represented in separate trees. For example, one might learn about the sources of good

[1] A tree is a graph with no cycles or loops. That is, a tree is a connected graph on n vertices that possesses at most $n - 1$ edges (Diestel 2010). Furthermore, in this chapter we are concerned with *rooted* trees, in which one vertex is distinguished as the "origin" of the tree, giving rise to a hierarchical structure.

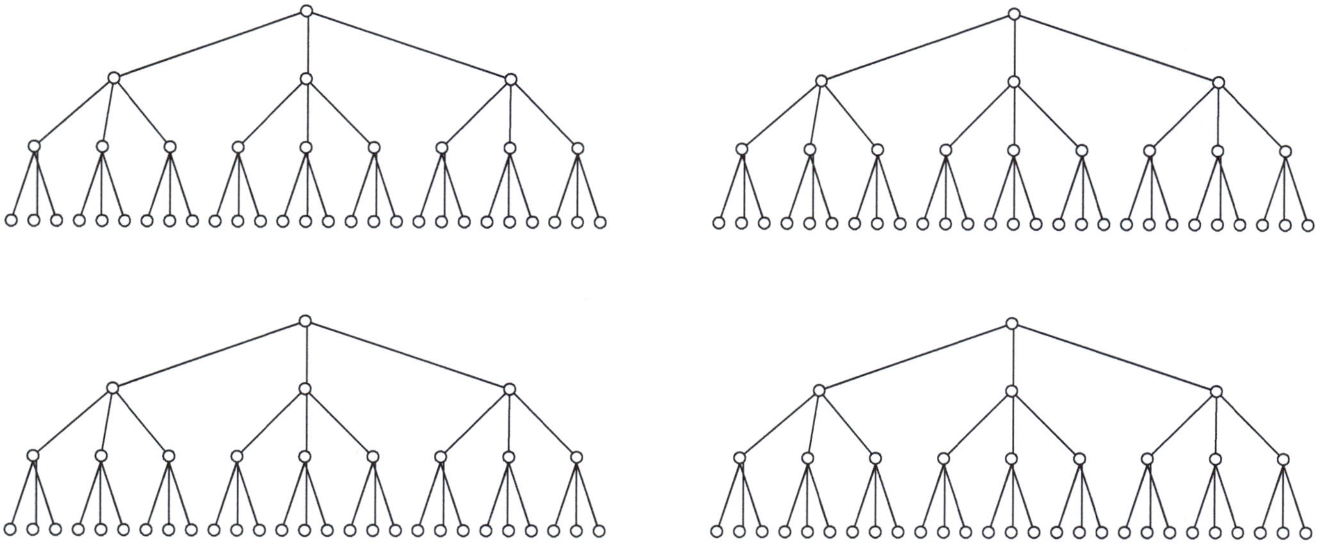

Fig. 6.1 A single trait tree, represented by a balanced tree with branching factor 3 and depth factor 3, order 40. In our model, nodes higher in the tree represent prerequisites for nodes lower down the tree. Each instance of the model will have several or many of these trees in the design space

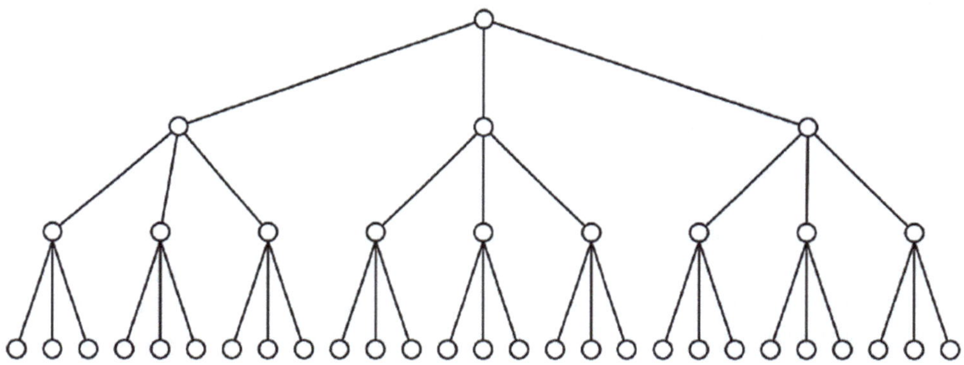

Fig. 6.2 A design space composed of 4 independent trees, each tree with branching factor 3 and depth factor 3, order 40. We also studied larger design spaces with 16 independent trees, and with larger branching and depth factors

lithic raw materials, independent of learning how to perform different percussion techniques. In our model, each of these independent areas is represented by a separate tree of traits.

In each simulation model, we begin with a trait or "design space" that incorporates several independent sets of traits (O'Brien et al. 2010). The overall design space of a simulation model is thus a forest,[2] composed of several trees (Fig. 6.1). For each tree in a learning hierarchy, we employ balanced trees which have the same number of nodes at each level, to provide a simplified model of a design space with which to begin our exploration of this class of social learning model, although real design spaces are undoubtedly more complex in their geometry. Each tree in our model is specified by a branching factor r and depth h. As a result, each trait tree in the design space has $\sum_{i=0}^{h} r^i$ traits.

The tree depicted in Fig. 6.2 thus has 40 vertices, for example. In this chapter, we examine both small (4 trees) and larger (16 trees) design spaces, to see how learning may differ in problems involving design spaces of different size and complexity. We examine trees with combinations of branching and depth factors of 3 and 5. Thus, a design space with 4 trees with branching and depth factors of 3 (as in Fig. 6.2) would have 160 traits, whereas a design space with 16 trees of branching and depth factors of 5 would have a total of 62,496 traits.[3] Even the small design spaces we consider here create a large space for cultural change and differentiation, given the number of possible trees one can construct on even 40 vertices.[4] In the experiments reported here, the overall

[2]A forest is a graph composed of multiple components, each of which is a tree.

[3]We initially chose 6 as the limit on branching and depth factors, but found that we cannot calculate certain symmetry statistics, such as the size of the automorphism group, on trees that large using existing tools. Even a tree with $r = 5, h = 6$ has over $10^{1,623}$ possible symmetries, and an attempt to calculate the symmetries for $r = 6, h = 6$ did not complete given the memory limits of the computers we had available.

[4]If we consider each trait to be unique and non-interchangeable, the number of unique trees with unique vertex labels is n^{n-2} by Cayley's theorem (Diestel 2010). For example, for each trait tree of 40 vertices, there are roughly 10^{60} possible trees. Even if we consider traits to be interchangeable (e.g., we look at the abstract topology of trees rather than the details of individual traits), there are *at least* 10^{16}

size of the design space remains constant over time, which is a simplifying assumption as we develop this class of structured information models. In future work, we will explore the role of invention in episodically creating large new regions of design space for the evolving population to explore.

Given the total "design space" represented by a forest of trait trees, each individual in our model is initialized with a small number of "initial" traits. Initial traits are chosen randomly but heavily weighted towards the roots of the trees to represent the fact that our knowledge starts out basic and sparse. In general, all of the design spaces modeled here are larger than populations will explore within the bounds of a simulation run. In the next sections we describe the social learning model, modified from Robert Axelrod's original, by which each simulated population evolves within this tree-structured design space, and will return to the specifics of how an initial culture repertoire is chosen.

6.2.2 The Axelrod Model of Social Learning and Differentiation

Robert Axelrod (1997) formulated a model aimed at studying the conditions under which simple learning rules could lead to cultural differentiation, rather than a single fixed state (which is the result of simpler neutral or diffusion models). This makes it useful as a starting point for understanding phenomena such as behavioral modernity, in our view. Axelrod's model combines social learning, in the form of random copying, a spatial structure to interaction, in the form of localized copying of neighbors on a lattice, and the tendency to interact most strongly with those to whom we are already culturally similar (homophily). The model displays a rich and interesting set of behaviors, and has been extensively studied by social scientists and physicists (Castellano et al. 2009). First we review the basic model, and in the following section our modified algorithm.

6.2.2.1 Axelrod's Original Model

The original model locates N individuals on the nodes of a regular lattice or grid, but various network structures have also been studied. Each individual is endowed with F integer variables $(\sigma_1, \ldots, \sigma_F)$, that can each assume q values. In the original model, each variable is a "cultural feature" each of which can assume q "traits." In each step, a randomly chosen individual i and a random neighbor j are selected, and "interact" with probability equal to the overlap between their cultural repertoire. Overlap, in the basic model, is simply the fraction of features for which i and j possess the same trait value:

$$p(i,j) = \frac{1}{F}\sum_{f=1}^{F} \delta_{\sigma_f(i)\sigma_f(j)} \tag{6.1}$$

where $\delta_{i,j}$ is Kronecker's delta function, taking the value 1 when its two arguments are equal and 0 otherwise. When individuals interact, the focal individual i takes the trait value of its neighbor for one of the features where the two individuals differ.

Interaction has no effect when two individuals already possess identical cultural repertoires, and there is no probability of interaction if individuals have no traits in common. This eventually causes the model to reach an absorbing state where no further changes are possible. Instances of the model are initialized with a random distribution of traits among individuals, and left to update until the steady state is reached. The evolution of the population leads to two classes of absorbing states: (a) a "monocultural" state in which all individuals share the same set of variables, and (b) a "polycultural" state in which subpopulations exist which share the same set of variables within the group, and are completely different from their neighbors.

Which of the two results is reached, and the statistical character of "polycultural" states when they exist, depends mainly upon the number of traits possible q for each cultural feature. For small values of q, individuals share many traits with their neighbors, interactions are thus frequent, and one domain comprising a single set of traits will grow to become fixed within the entire population. In contrast, when the value of q is high, individuals start out sharing very few traits, with interactions that are correspondingly less frequent. Regions of uniform cultural variation do grow, but as they do, sets of individuals who share no traits at all (and thus do not interaction) grow as well, and often prevent any single regional culture from expanding to fix within the population.

Many variants of the basic Axelrod model have been studied, including the addition of "drift" via the introduction of copying error, situating agents on different types of complex networks, the addition of an external "field" to simulate the effects of mass media, and copying that obeys a "conformist" or majoritarian rule by selecting the most common trait among the neighbor set (Castellano et al. 2000; De Sanctis and Galla 2009; Flache and Macy 2006; González-Avella et al. 2005, 2006, 2007a,b; Klemm et al. 2003a,b, 2005; Lanchier et al. 2010; Lanchier 2012). In general, modifications of the basic model can reduce the tendency of the model to produce polycultural solutions, or change the time scale or location of the critical point.

6.2.2.2 Semantic Extensions to the Axelrod Model

We begin each simulation with N (100, 225, or 400) agents, arranged on a square grid. A design space is created, with some number of trait trees (4 or 16), with uniform branching

possible unlabelled rooted trees on 40 vertices (using Otter's (1948) approximation).

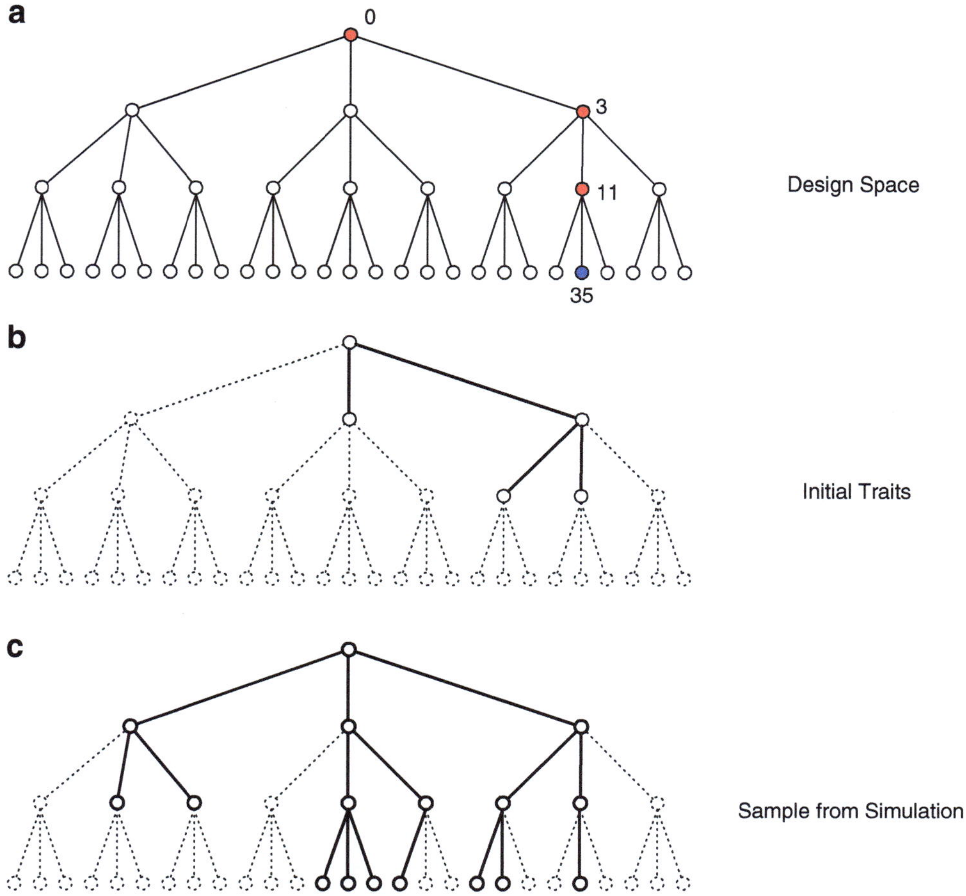

Fig. 6.3 Illustration of a design space composed of a single trait tree, along with a random initial trait chosen from the design space, and a final sample from a simulation run, showing the evolution of traits within the design space. Also shown in the (**a**) are the "prerequisites" for a cultural trait (35), as an example

factors and depth factors (3 or 5). An example of such a tree is shown in panel A of Fig. 6.3. Initial traits (and their prerequisites) are chosen randomly across the configured number of trait trees, as follows. For each individual, we select a random number t between 1 and 4, and repeat the trait selection process t times for that individual. In each selection, we choose a random tree in the design space, and then select a depth in the tree for the trait, given by $d \sim \text{Poisson}(0.5)$. This biases trait selection towards the root of the tree, as one would expect in young or inexperienced individuals. We then walk d steps into the tree, making uniform random selections for the children of each vertex. The path of vertices thus constructed is added to the individual's trait set, giving them an initial trait and its necessary prerequisites. One such initial trait is shown in Panel B of Fig. 6.3. Given that individuals begin with a small number of initial traits (between 1 and 4, selected randomly), and their prerequisites, the initial trait endowment of an individual is between 1 and $4h$, where h is the maximum depth of the design space (either 3 or 5 in the experiments reported here).[5]

Once the population is initialized, the simulation runs a discrete approximation to a continuous-time model. In other words, only one agent changes at each elemental time step, as in the original Axelrod model and the Moran model of population genetics and its cultural version (Aoki et al. 2011; Moran et al. 1962; Moran 1958). At each step, an agent (A) is chosen at random, and a random neighbor of A is then selected (agent B). Their probability of interaction is given by the overlap of trait sets, which is most simply calculated as the Jaccard overlap between the set of tree vertices each possesses, thus replacing Eq. 6.1 with:

$$J(A, B) = \frac{|V(A) \cap V(B)|}{|V(A) \cup V(B)|} \tag{6.2}$$

where $V(i)$ represents the vertex list for trait trees held by individual i in the population.

If the agents end up interacting, agent A observes the traits currently possessed by B, and selects a trait (T) that it does not already possess to learn. If agent A has the necessary prerequisite traits for the selected trait, it can learn

[5]At maximum, this yields some individuals who begin the simulation with up to 20 traits. The median number of traits in samples taken after 6–10 million time steps is considerably higher – 259 traits per cultural configuration or region. Thus, cultural repertoires in the simulation grow through copying and innovation, as expected.

trait T. If not, there is a probability $\mathbb{P}(l)$ that B can teach A a necessary prerequisite for T instead. This simulates the process of agent B structuring the learning environment of A through formal instruction or apprenticeship, for example. If such a prerequisite learning event occurs given $\mathbb{P}(l)$, agent A learns the most fundamental of T's prerequisites that it does not already possess. For example, agent A might require the trait closest to T (e.g., trait 11 in Fig. 6.3, if the original trait targeted was 35).

Additionally, at each time step, there is a probability $\mathbb{P}(m)$ that one random individual in the population will learn a new trait (and necessary prerequisites) that it does not already possess. For example, if an innovation event occurs and an agent discovers trait 35 by individual trial and error learning, we assume that the agent also discovered traits 0, 3, and 11. Thus innovation can introduce one trait to the population, or a linked set depending upon its prerequisites and what the innovating individual already "knows." This model of innovation simulates an ongoing process of individual learning unconnected to social learning or teaching within the population. Because this functions much like "infinite-alleles mutation" in the classical Wright-Fisher neutral models (Ewens 2004), or like noise terms in Axelrod, Ising, or Potts models (Castellano et al. 2009), we will refer to this as the "global innovation rate" in this chapter.

One of the editors noted that this model of innovation may not be as realistic as an alternative, where random innovations would be "discoverable" only with the correct prerequisites in place. We believe that innovation in the face of skill or knowledge prerequisites is continuous between these two models. Occasionally one will discover a new piece of knowledge or develop a skill, having learned surrounding and related knowledge. In other situations, individuals may learn sequences and sets of information or skills by trial and error and "tinkering." The "size" of innovations that can be learned purely by individual trial and error should thus vary between these extremes, biased towards the "small" end of the range. Our selection of an innovation model where individuals discover a trait and its prerequisites thus potentially overestimates the effect of individual learning, but it made certain graph operations easier, and can be relaxed in future models.

Each simulation run lasts 10^7 steps, which yields between 10^4 and 10^5 copying events per individual, depending upon population size.[6] Since we do not explicitly model the interaction between cultural transmission and biological reproduction here, we can interpret the model as representing either fine-grained learning within individuals over the course of their lifetimes, or long-term cultural evolution within a fixed size population where we are not modeling fitness. We felt this simplification was appropriate in a pilot study exploring structured information models, but a more detailed study would include dynamics on two time scales: developmental learning and evolutionary dynamics given birth and death. Samples are taken beginning at 6 million steps, and sampling at an interval of 1 million steps, and record the trait trees seen in the population. An example of such a sampled tree is shown in Panel C of Fig. 6.3. For reference, the full algorithm for each copying step is given in the Appendix as Algorithm 1.

6.3 Measuring Cultural Diversity and the Results of Structured Learning

Each sample from a simulation run is composed of the distinct sets of trait trees possessed by individuals in the population, along with summary statistics. If a simulation run converges to a monocultural solution, the sample will have one set of trait trees, which are shared across the entire population. In other cases, there will be clusters of cultural configurations which might be unique to a single individual, or shared by some number of agents. Each cluster will be composed of some number of trait trees (typically, the number configured for the simulation run: 4 or 16, but perhaps a subset), and each trait tree will be the result of many agents learning traits and their prerequisites socially, and for runs with a non-zero mutation rate, by individual learning or innovation. Each cluster will thus have some number of traits, typically higher (often much higher) than the initial endowment given to the population.

From the sampled trait trees, we calculate summary statistics as follows. The ratio of the number of traits in the sample to the full design space size (or "remaining density" of traits) is one measure of trait richness. The radius of a rooted tree is the number of edges in the path from root to the furthest edge. The average radius of trees in a sample (or its ratio to the depth of the design space) is another richness measure, aimed at measuring whether knowledge with multiple prerequisites is being learned within the simulated population. Similarly, in the original design space, the branching factor describes how many children each node in the tree started with, so measuring the average vertex degree gives us a rough measure of how broad a cultural repertoire is. Each of these measures is illustrated in Fig. 6.4 for an example tree selected from our data.

[6]One hundred thousand was chosen as a compromise for running large batches of simulations in parallel. Some simulation runs, especially in small design spaces with very high prerequisite learning rates, can converge to a monocultural solution and quasi-stable equilibrium quite quickly; in the largest design spaces and low learning rates, convergence may never occur even though the process is well-mixed. However, the processes have reached a quasi-stable equilibrium, verified by examining samples at different times for secular trends in median and mean values, which were not found.

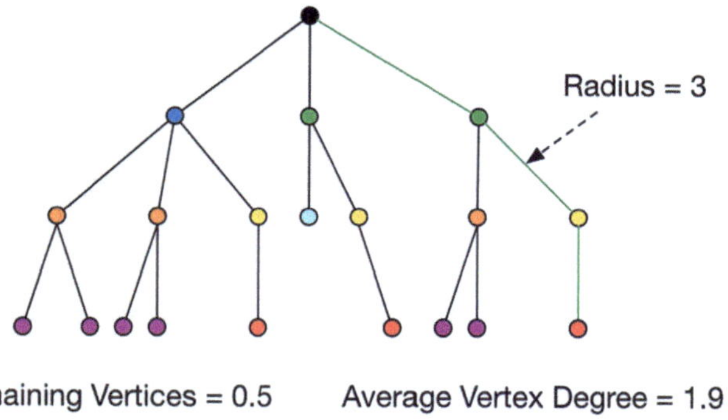

Fig. 6.4 An example set of traits at the conclusion of a simulation run, extracted from a simulation with branching factor 3 and depth factor 3, and a single trait tree as the trait space. The remaining density of vertices, mean vertex degree, and radius of the tree are noted. Vertex colors denote "structural equivalence" classes or "orbit structure," as measured by adjacency patterns, and is one measure of the symmetries present in the tree

In addition to these simple numerical measures comparing final trees to the original design space, it is useful to measure something about the overall "shape" of the trees themselves. One way of formalizing this notion is to examine the *symmetries* of the final trait trees. Examining Fig. 6.4, if we ignore the exact identities of traits for the moment, it is apparent that there are repeating patterns. For example, the left-most branches each terminate in a pair of leaves. This pattern is repeated on the second right-most branch. These types of repeating patterns are computationally expensive to search for in large sets of trees, but we can summarize them by considering trait trees as algebraic objects and examining their *automorphisms*.

An automorphism is a function which maps an object to itself, in such a way that the structure of the object is preserved (Rotman 1995). Graph automorphisms map vertices in a graph to each other, preserving properties such as the adjacency pattern of edges. The six vertices which mark the repeating pattern of leaf-pairs in Fig. 6.4 are an automorphism of the tree, and thus are symmetries we can measure. An overall measure of "how symmetrical" (or "how many interchangeable patterns") there are in a graph possesses given by the total number of automorphisms found, called the size of the automorphism group or $|Aut(G)|$ (Godsil and Royle 2001). A tree with no repeating patterns will thus have an automorphism group size of 1, indicating that the only symmetry is the entire tree itself. A balanced tree with branching and depth factors of 3, as depicted in Fig. 6.2, has approximately 1.3×10^{10} automorphisms. The more repeating patterns there are in trait trees, the more automorphisms they will possess.

Because group sizes grow quickly and the accuracy of performing calculations with truly astronomical numbers is low, another possible measure of the symmetries present is to count the *classes* of equivalences into which vertices fall. The *orbits* of the automorphism group are the sets of vertices which are interchangable by some permutation that preserves structure. For example, the graph in Fig. 6.2 has five orbits, with each vertex at a given level interchangable (in a structural sense). Similarly, the six leaf vertices that are part of pairs in Fig. 6.4 are part of the same orbit; in this illustration, each orbit is given a different color to highlight their equivalence. For each cultural region found when sampling a simulation, we calculate the size of the automorphism group and the number and multiplicity (frequency) of orbits. For this analysis, we employ the *nauty + Traces* software by Brendan McKay and Adolfo Piperno (McKay and Piperno 2014).[7]

6.4 Experiments

Given a modified Axelrod model on a tree-structured trait space, we expect to see greater cultural diversity, differentiation among groups of individuals, and larger sets of traits as the "learning environment" is tuned from a low to high probability of teaching and learning among individuals. We also expect that individual innovation, independent of the social learning context, will increase the amount of the technological design space that a population explores, which leads to enhanced opportunities for differentiation even through simple random copying. Here we measure cultural differentiation by the number of clusters of individuals who share the same trait trees when we sample the population.

Second, we looked at whether highly structured learning environments, represented here by higher probabilities of naive individuals gaining the prerequisites for the skills and information they encounter with peers, led to deeper and richer cultural repertoires. We explore a number of ways of measure the richness of a cultural repertoire in a model with structured relations between traits, through the use of graph properties and symmetry measures. The measures used are those described above: the tree radius (or depth), mean vertex degree, the fraction of remaining vertices, and the size of the automorphism group of sampled trait forests. Finally, we began to examine how the structured learning environment might interact with demography, by simulating the same parameters across two sizes of population.

For this chapter, we examined populations of size 100, 225 and 400, to begin to examine the effects of population size. For these populations, we examined design spaces that were small (4 trait trees) and large (16 trait trees). Within each size, we further examined combinations of branching factor and depth factor with values of 3 and 5, thus yielding 8 total sizes of design space (Table 6.1).

[7] Nauty+Traces can be downloaded at http://pallini.di.uniroma1.it/. We employed version 2.5r7 for this research.

Table 6.1 Size of design space for different trait tree configurations

Branching factor	Depth factor	Number of trait trees	Size of design space
3	3	4	160
5	3	4	624
3	5	4	1,456
5	5	4	15,624
3	3	16	640
5	3	16	2,496
3	5	16	5,824
5	5	16	62,496

Table 6.2 Parameter space for simulations described in this chapter

Simulation parameter	Value or values
Population rate at which new traits arise by individual learning	0.0, 5e−05, 0.0001
Maximum number of initial traits (not including their prerequisites) each individual is endowed with	4
Number of distinct trees of traits and prerequisites	4, 16
Population sizes	100, 225, 400
Replicate simulation runs at each parameter combination	25
Maximum time after which a simulation is sampled and terminated	10,000,000
Individual probability for being taught a missing prerequisite	0.05, 0.1, 0.2, 0.3, 0.4, 0.5, 0.6, 0.7, 0.8, 0.9
Number of branches at each level of a trait tree	3, 5
Depth of traits in each trait tree	3, 5

Further, we examined three levels of global mutation or innovation rate: zero, or no mutation, and 0.00005 and 0.0001. Such rates created a constant supply of new innovations, but several orders of magnitude less frequent than copying and prerequisite learning events. The full set of parameters are given in Table 6.2. In this pilot study, for each combination of all of the above parameters, we performed 25 replications. With 5 samples per simulation run, this yielded 10,963,691 samples of cultural regions.

6.5 Results

We begin by noting that compared to the original Axelrod model, or neutral and biased copying models, the dynamics of our semantic Axelrod model are highly variable. A very wide range of outcomes is possible for each parameter combination, especially when the size of the design space is large. Some variables, such as the average vertex degree of sampled trait trees, are strongly overlapping across all learning rates and do not appear diagnostic of different learning environments, at least in these initial experiments. Given the large amount of variability in the dynamics, larger numbers of replications would be useful, although this is computationally quite expensive at present.[8] That said, several features of the data are strongly suggestive that hierarchical trait models have potential in modeling cumulative technological evolution, making the computational expense worthwhile.

6.5.1 Cultural Diversity

Variation among individuals is foundational to evolutionary processes, and is the raw material from which differentiation between regions and cultural groups is constructed. Figure 6.5 depicts the number of cultural configurations (i.e., trait trees) in a population of size 100, for the smallest trait space with only 160 total traits, and relatively high levels of individual innovation. For example, in the left-most panel the large peak just above zero indicates that most simulated populations are characterized by one or a few sets of trait trees. Five learning rates are depicted, increasing from left to right across the panels. At the very lowest rate of learning fidelity, with only a 10 % chance of being taught a needed prerequisite for knowledge being copied, most of the populations simulated share a single set of traits, and even individual innovation does not drive significant exploration of the space of structured traits. With increased fidelity in teaching needed prerequisites, however, simulated populations begin exhibiting marked differentiation, with individuals possessing more unique configurations of traits from the overall design space.

[8] The simulations reported here ran on a cluster of 6 compute-optimized "extra large" Linux instances on Amazon's EC2 computing cloud, for a total of 17 days of wall clock time and 2,075 CPU hours. We plan further optimizations to the simulation code to make larger samples economically feasible.

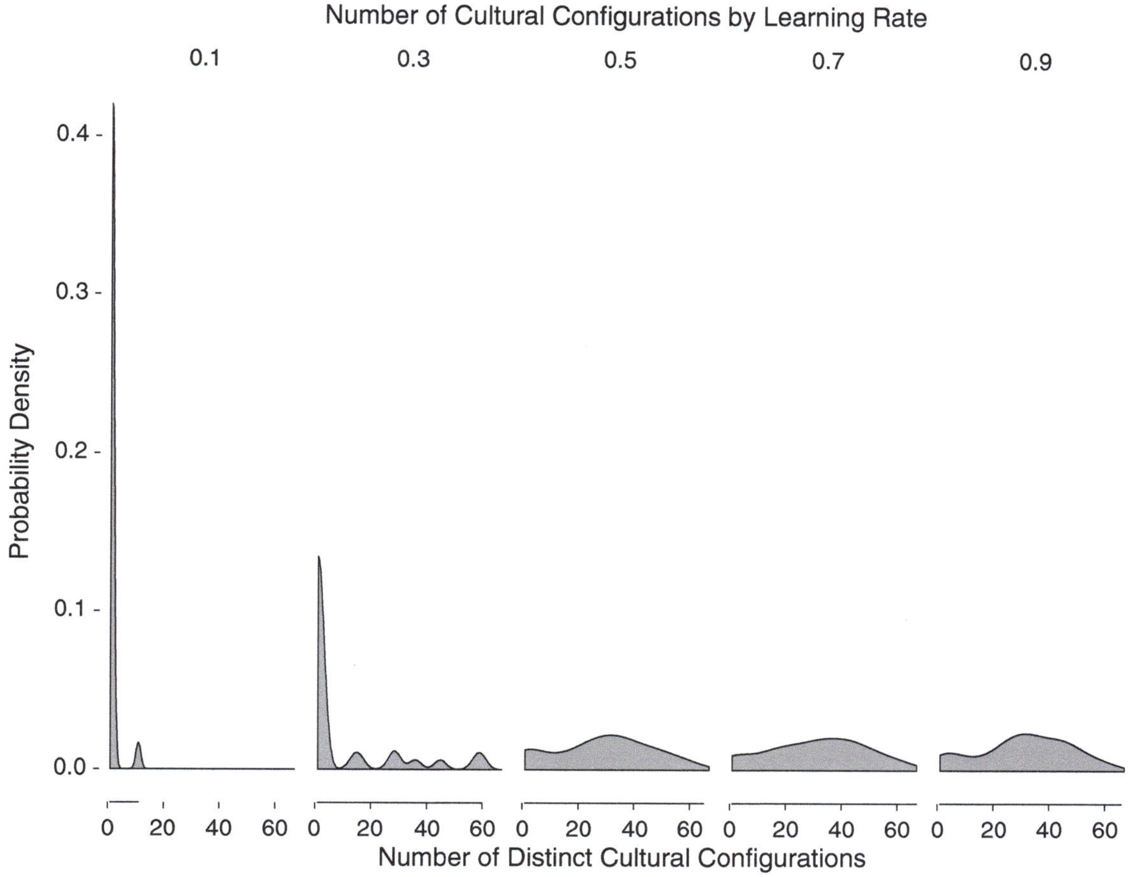

Fig. 6.5 Number of cultural configurations in simulations with the smallest trait space (160 total traits in 4 trees), and a high individual innovation rate (10^{-4})

Looking at the data from another perspective, we can hold the fidelity of learning constant (say, at a 40 % chance of being taught a needed prerequisite), with the same global innovation rate (10^{-4}) as Fig. 6.5, and examine the effect of different size design spaces (Fig. 6.6). In general, populations exhibit greater differentiation between individuals as the design space gets larger, as prerequisite learning helps individuals acquire adjacent traits, and individual innovation randomly explores more distant portions of the design space.

Given the structure of the Axelrod model, with the strong tendency towards cultural uniformity given homophily, all simulated populations converged to a single cultural configuration in the absence of a global innovation rate. This highlights the importance of various "innovation" and "invention" processes in the creation and maintenance of cultural differentiation and diversity (Eerkens and Lipo 2005; O'Brien and Shennan 2010), and suggest that highly conservative cultural repertoires, such as those posited to precede behavioral modernity in hominin populations, occur whenever individuals engage in social learning in small technological design spaces, in the absence of strong and regular individual innovation.

6.5.2 Trait Richness and Knowledge Depth

Cumulative evolution of technology is represented in our model by the population learning its way *down* the trees which compose the design space. Possession of traits deeper in the trees represents skills or information which is more specific, possessing more prerequisites. Thus, we expect that the depth (or "radius," see Fig. 6.4) of trees would increase with the prerequisite learning rate, representing a learning environment which is structured to ensure such acquisition.

Figure 6.7 gives the *normalized* mean radius of cultural regions, broken out by the prerequisite learning rate along the horizontal axis, and each group of 3 boxplots displays the differing global innovation rates studied. Radii are normalized to the depth of their design space, to facilitate comparison. The results indicate that essentially two regimes exist: shorter trees, which do not grow much beyond their initialized size, and larger trees. The mean radius has an asymptote just above 0.75, achieved with the prerequisite learning rate is approximately 0.4 or higher. Further increases do not seem to matter. Additionally, the difference between the two global innovation rates is small – what matters most in terms of

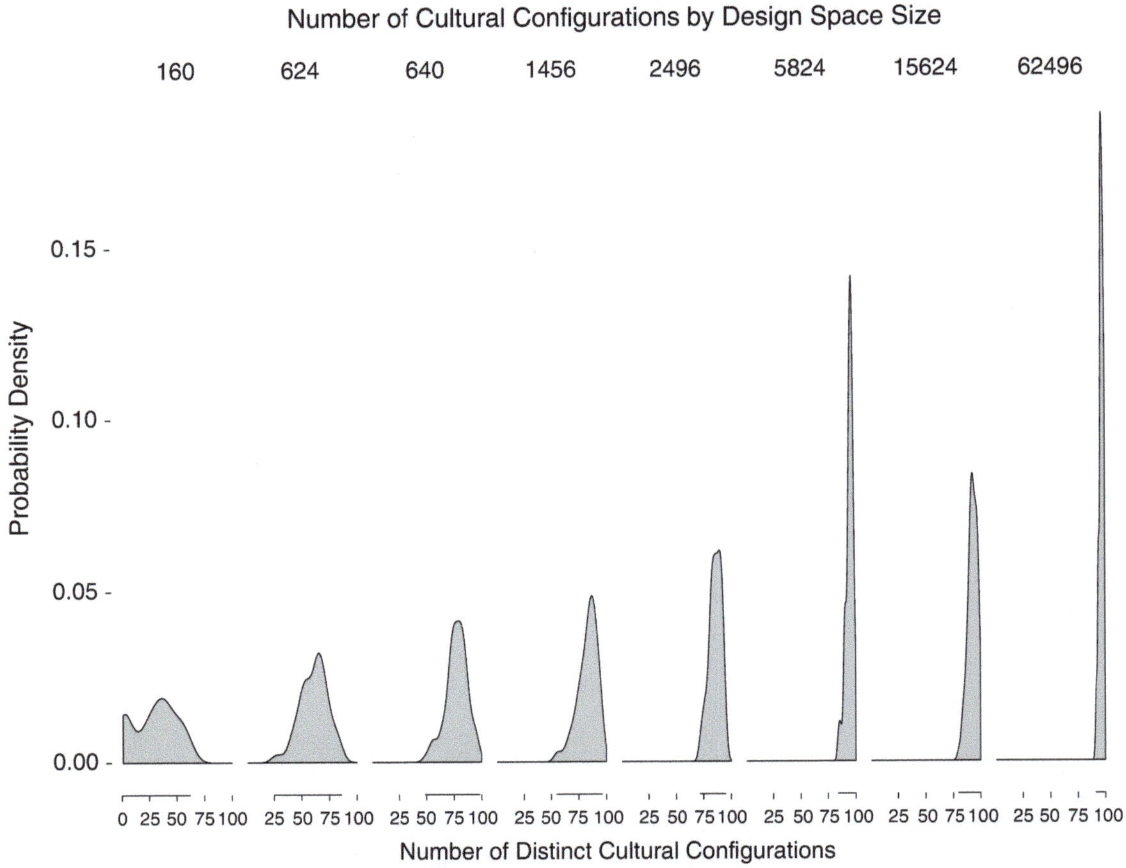

Fig. 6.6 Number of cultural configurations in simulations with an intermediate learning rate (0.4), across different sizes of trait space

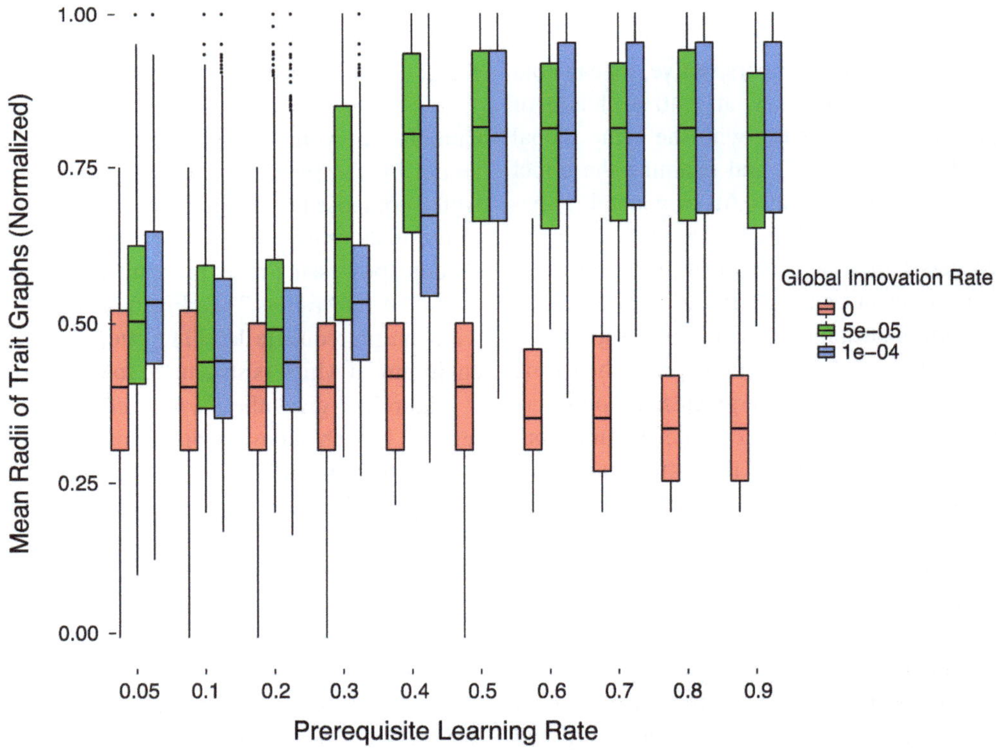

Fig. 6.7 Mean depth of trait sets, by prerequisite learning rate and global innovation rate, for population size 100

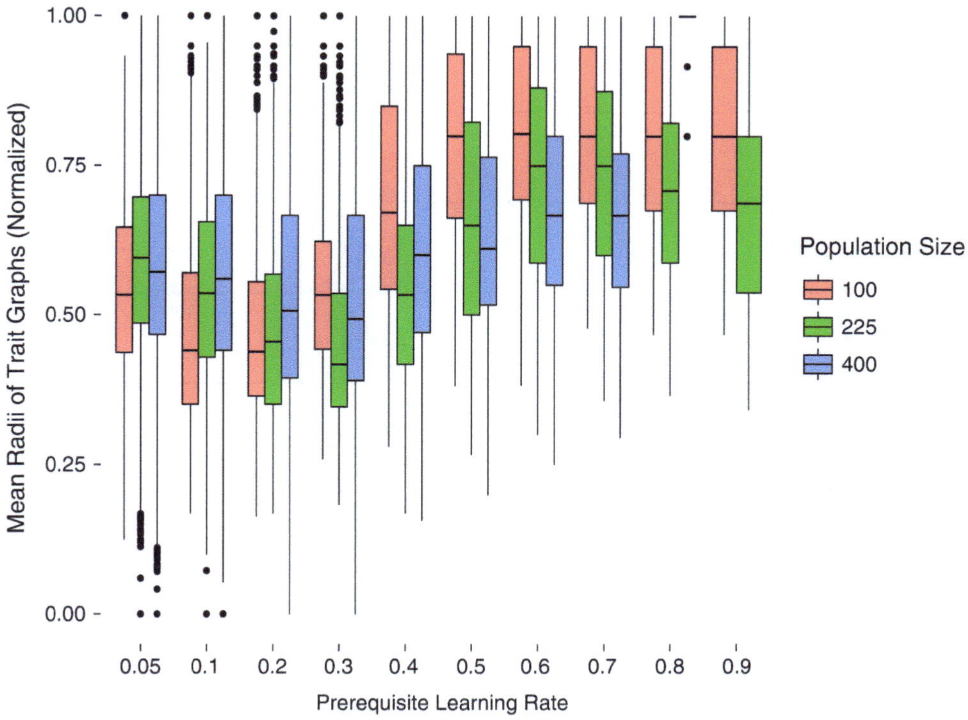

Fig. 6.8 Mean depth of trait sets, by prerequisite learning rate and population sizes of 100, 225 and 400

qualitative behavior is the presence of global innovation outside the teaching or learning of prerequisites themselves.

6.5.3 Population Size

Earlier, we mentioned that population size does not seem to be a primary factor in explaining the measured diversity in cultural transmission models, except perhaps in bottleneck situations like the one Henrich analyzes in Tasmania (2004). Instead, population size may have an interaction effect with other factors, yielding smaller second-order effects. We examined the effect of population size in the research reported here, repeating the entire set of simulation runs for populations of 100, 225, and 400.[9]

Figure 6.8 displays the relationship between mean radius (or depth) of the cultural traits in each cultural sample, as in Fig. 6.7 above, but the boxplots are instead colored by population size. At least over a range of group or deme sizes likely to be relevant to Paleolithic archaeology, population size makes no difference to the qualitative behavior of the model. There is, however, a very slight decrease in mean radius of trait sets with larger population size, which is likely a consequence of a larger population spreading out over the trait space.

[9]We should note that learning rates of 0.8 and 0.9 for population size 400 were cut short due to budget constraints, but this does not appear to affect the pattern in our dataset.

6.5.4 Trait Tree Symmetries

Finally, we examined the algebraic properties of the trait trees composing cultural regions, examining both the number of vertex equivalence classes (orbits) and the size of the automorphism group of the trait forests. We examined the raw metrics, and versions normalized by the size of the maximally symmetric forest with the same number of traits, branching factor, and depth factor. The latter proved difficult and led to serious overflow problems even with 64 bit arithmetic, so we focus here on the raw automorphism group size.

The logarithm of the automorphism group size does hint at interesting structure (Fig. 6.9). In the presence of mutation, the learning of prerequisites narrows the range of variability for the automorphism group size, and at higher learning rates renders the distribution multimodal. The modality arises because of the different combinations of branching factor and depth factor we employed for design spaces – i.e., some design spaces are "wide" and some are "narrow," while also being "shallow" or "deep." This gives rise to different modes in the measured symmetries, but overall the reduction in variability in symmetry is the most important qualitative effect seen in our data.

We do not fully understand the "shapes" of cultural regions to which the model appears to converge, but it appears that there is a tendency for trait graphs to converge towards shapes which have moderate numbers of symmetries. This graph is on a logarithmic scale, so a peak at 50

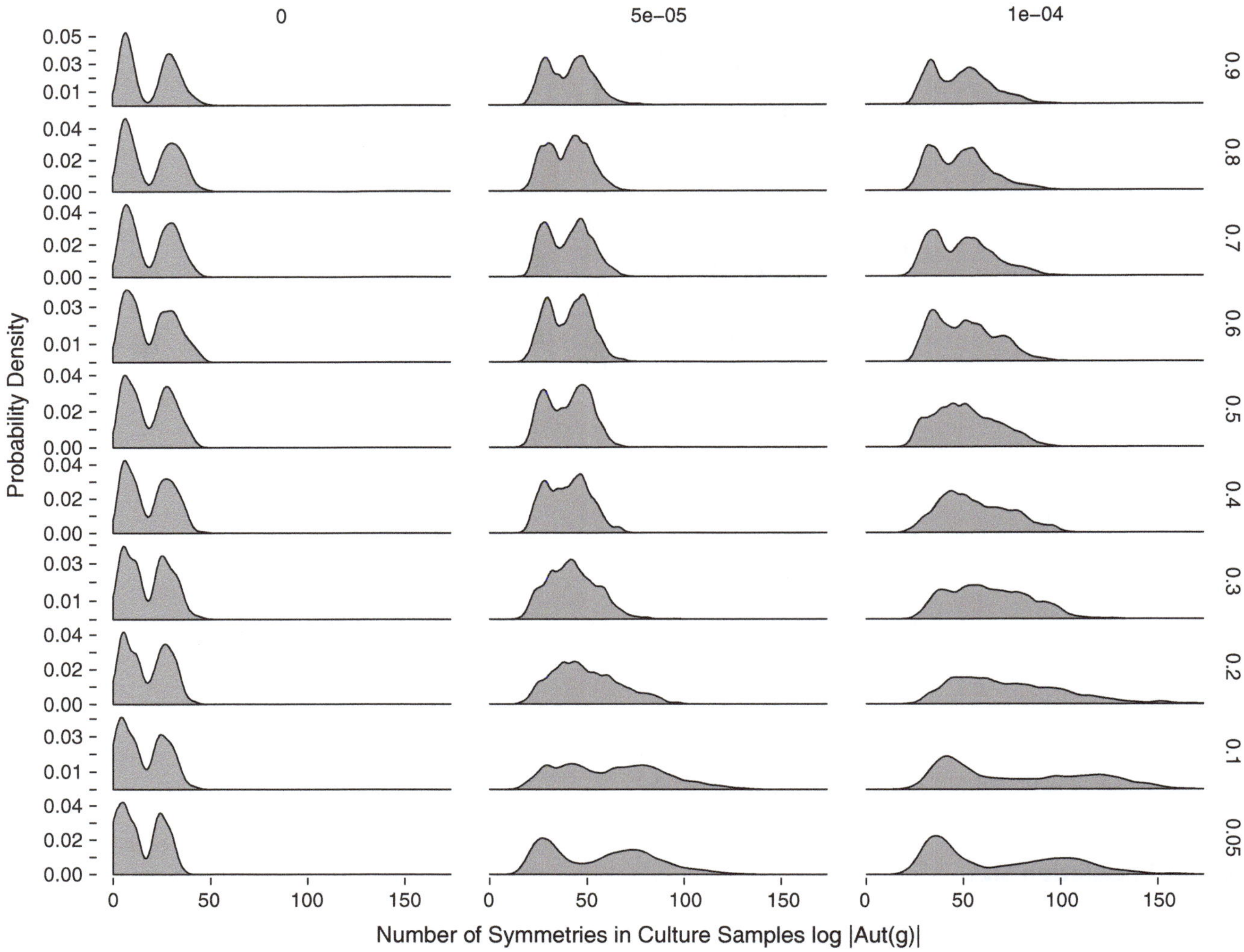

Fig. 6.9 Number of symmetries in trait tree samples, measured as the log of the order of the automorphism group of the trait graphs, broken down by prerequisite learning rate (*rows*) and global innovation rate (*columns*)

along the horizontal axis correponds to a trait graph with approximately 5×10^{21} symmetries. This is a fairly small number, compared to the original design spaces, which have symmetries ranging from approximately 10^{41}–$10^{6,496}$. Thus, the geometry of cultural traits in our hierarchical design spaces are fairly asymmetric and represent small and very specific segments of the total design space.

Further analysis of trait graph "shapes" is needed to tell whether there are repeating patterns or graph "motifs" which characterize a social learning model in a graph-structured trait space. The results here are suggestive of such a phenomenon, but inconclusive given just the bulk algebraic properties of cultural regions, since the size of the automorphism group (or the number of orbits) tells only *how many* symmetries there are, not what types of symmetries exist. The next step in our analysis of shape is to pursue a geometric decomposition of the graph following Ben MacArthur and Rubén Sánchez-García's (2008) work on the symmetries of complex networks.

6.6 Discussion

The "semantic Axelrod" model described here specifically addresses social learning of knowledge with "prerequisite" structure, and a learning environment which is tunable from low to high fidelity, simulating the intensity with which "teaching" occurs in addition to imitative copying. The model displays a characteristic increase in the cultural repertoires of individuals, as they learn in environments of higher fidelity. At the individual level, an increase in higher fidelity learning within structured information environments both creates path-dependency in what is learned, and increases the chances for specialization among individuals. Hominin populations in which complex knowledge is taught systematically along with prerequisites will accumulate and retain skills and technology faster and to a greater extent than those groups which rely upon natural pedagogy and imitation for social learning.

Previous research had established the importance of teaching and learning environments for cumulative cultural evolution and cultural diversity (Aoki 2013; Castro and Toro 2014; Creanza et al. 2013; Nakahashi 2013). Our contribution in this paper is a model capable of connecting the fact of teaching with the actual structure and content of cultural knowledge. Such models, we believe, are important in explaining the explosion of cumulative material culture that accompanies behavioral modernity. The model described here only makes a start on modeling the additive and recombinative complexity of real technologies, but it does display accumulated depth of "knowledge" or "skills," as represented by the radius or depth of trait trees. In combination with realistic models of technology – such as the production sequences studied by experts on stone tools – we believe that empirically sufficient models of the evolution of specific technologies are possible and within reach.

Several areas suggest themselves for future research in structured information or "semantic" cultural transmission models. Some we are pursuing, others remain open questions and we invite collaboration towards their solution.

- Regional scale cultural differentiation given a metapopulation embedding of the basic model.
- Additional trait relations (e.g., class subsumption, functional equivalencies).
- Realistic technology models for key artifact classes (e.g., bifaces, scrapers, pottery).
- Incorporation of trait fitness in order to study directional change.

Models of the class introduced here are "thicker" descriptions of how humans acquire skills and information in real learning environments, and thus complement existing models which describe the conditions under which teaching and structured learning might evolve and spread. We believe models of this type make a needed "downpayment" on cultural transmission models which can substantively incorporate specialties such as archaeometry, the technnological analysis of lithics and pottery (Tostevin 2012), and studies of how innovation occurs in various tool classes (e.g., O'Brien and Shennan 2010). Bringing cultural transmission modeling together with the details of technologies will be a crucial component in multifactor evolutionary explanations for the complex of changes seen in modern *Homo sapiens* and some Neanderthal populations in the later Paleolithic.

Acknowledgements The authors wish to thank Briggs Buchanan and Mark Collard for the invitation to participate in the symposium "Current Research in Evolutionary Archaeology," at the 79th Annual Meeting of the Society for American Archaeology in Austin, TX. A summary of this research was presented in that session, and Alex Mesoudi provided valuable comments on an early post-conference draft. Kenichi Aoki and an anonymous reviewer provided feedback prior to publication, and although we did not take all of their suggestions, the comments led to a number of improvements. Madsen wishes to thank Frédéric Chapoton of the Institut Camille Jordan for answering a question about the maximal automorphism group of trees.

Appendices

Algorithm Description

Algorithm 1 describes the "semantic" Axelrod model variant studied in this chapter. Within the algorithm, there are several functions which find traits with particular properties. Some, like **GetTraitUniquetoFocal**(), are fairly simple set operations but were abbreviated to clarify the notation.

Algorithm 1

```
Require: innovrate is the population rate at which individuals randomly learn a trait
Require: learningrate is the probability of learning a missing prerequisite during a learning interaction
focal ← GetRandomAgent()
neighbor ← GetRandomNeighbor(focal)
if focal = neighbor ∨ focal ∩ neighbor = ∅ ∨ neighbor ⊊ focal then
    exit { No interaction is possible, move on to next agent }
end if
prob ← (focal ∪ neighbor − focal ∩ neighbor)/focal ∪ neighbor
if RandomUniform() < prob then
    differing ← neighbor \ focal
    newtrait ← GetRandomChoice(differing)
    if hasPrerequisiteForTrait(focal, newtrait) = True then
        replace ← GetTraitUniquetoFocal(focal,neighbor)
        focal ← focal \ replace
        focal ← focal ∪ newtrait
    else
        if RandomUniform() < learningrate then
            prereq ← GetDeepestMissingPrerequisite(newtrait, focal)
            focal ← focal ∪ prereq
        end if
    end if
end if
if RandomUniform() < innovrate then
    focal3 ← GetRandomAgent()
    innovation ← GetRandomTraitNotInFocal(focal3)
    focal3 ← focal3 ∪ innovation
end if
```

GetDeepestMissingPrerequisite() is a procedure which takes the trait set of an individual, and a trait for which the individual is known to be missing necessary prerequisites, and returns the "most basic" missing prerequisite for that trait (i.e., closest to the root). This is done by finding the path which connects the root and desired trait, and walking its vertices from the root downward, checking to see if each vertex is part of the individual's trait set. The first trait not found in the individual's repertoire is returned.

Availability of Software and Analysis Code

The simulation software used in this chapter is available under an open-source license at Mark Madsen's GitHub repository https://github.com/mmadsen/axelrod-ct. Required libraries and software are listed in the source archive itself, and include Python 2.7 and the open-source MongoDB database engine to store simulation output.

The codebase consists of a set of library modules which implement the shared and unique aspects of each model, unit tests to verify the basic functionality of the code, and scripts which execute each model. The **axelrod-ct** repository contains three models:

- An implementation of the original Axelrod model using the **axelrod-ct** libraries.
- A basic model with an "extensible" trait space but no relations between traits.
- A "semantic" Axelrod model with tree-structured trait space representing prerequisite relationships between traits.

Stepwise extension from the original Axelrod to the semantic models on the same code library allowed a degree of verification, which is difficult in a situation where there is no existing mathematical theory against which to compare the code implementation (Committee on Mathematical Foundations of Verification Validation and Uncertainty Quantification, National Research Council 2012).

The analysis and final dataset reported here are available, along with the source of this paper and associated presentations, in an associated GitHub repository: https://github.com/mmadsen/madsenlipo2014. Statistical analyses of the final dataset were performed in R, rendering our results reproducible given simulated data from the "axelrod-ct" software linked above.

References

Aoki K (2013) Determinants of cultural evolutionary rates. In: Akazawa T, Nishiaki Y, Aoki K (eds) Dynamics of learning in Neanderthals and modern humans. Replacement of Neanderthals by modern humans series, vol 1. Springer, Tokyo, pp 199–210

Aoki K, Lehmann L, Feldman MW (2011) Rates of cultural change and patterns of cultural accumulation in stochastic models of social transmission. Theor Popul Biol 79:192–202

Axelrod R (1997) The dissemination of culture: a model with local convergence and global polarization. J Confl Resolut 41:203–226

Bamforth DB, Finlay N (2008) Introduction: archaeological approaches to lithic production skill and craft learning. J Archaeol Method Theory 15:1–27

Bar-Yosef O (2002) The upper paleolithic revolution. Ann Rev Anthropol 31:363–393

Bleed P (2001) Trees or chains, links or branches: conceptual alternatives for consideration of stone tool production and other sequential activities. J Archaeol Method Theory 8:101–127

Bleed P (2002) Obviously sequential, but continuous or staged? Refits and cognition in three late paleolithic assemblages from Japan. J Anthropol Archaeol 21:329–343

Bleed P (2008) Skill matters. J Archaeol Method Theory 15:154–166

Bouzouggar A, Barton N, Vanhaeren M, d'Errico F, Collcutt S, Higham T, Hodge E, Parfitt S, Rhodes E, Schwenninger JL et al (2007) 82,000-year-old shell beads from North Africa and implications for the origins of modern human behavior. Proc Natl Acad Sci 104:9964–9969

Castellano C, Fortunato S, Loreto V (2009) Statistical physics of social dynamics. Rev Modern Phys 81:591

Castellano C, Marsili M, Vespignani A (2000) Nonequilibrium phase transition in a model for social influence. Phys Rev Lett 85:3536

Castro L, Toro MA (2014) Cumulative cultural evolution: the role of teaching. J Theor Biol 347:74–83

Collard M, Buchanan B, Morin J, Costopoulos A (2011) What drives the evolution of hunter-gatherer subsistence technology? A reanalysis of the risk hypothesis with data from the Pacific Northwest. Philos Trans R Soc B Biol Sci 366:1129–1138

Collard M, Buchanan B, OBrien MJ (2013a) Population size as an explanation for patterns in the paleolithic archaeological record. Curr Anthropol 54:S388–S396

Collard M, Buchanan B, O'Brien MJ, Scholnick J (2013b) Risk, mobility or population size? Drivers of technological richness among contact-period western North American hunter-gatherers. Philos Trans R Soc B Biol Sci 368:20120412

Collard M, Ruttle A, Buchanan B, OBrien MJ (2013c) Population size and cultural evolution in nonindustrial food-producing societies. PLoS One 8:e72628

Committee on Mathematical Foundations of Verification Validation and Uncertainty Quantification, National Research Council (2012) Assessing the reliability of complex models: mathematical and statistical foundations of verification, validation, and uncertainty quantification. The National Academies Press, Washington, DC

Creanza N, Fogarty L, Feldman M (2013) Exploring cultural niche construction from the paleolithic to modern hunter-gatherers. In: Akazawa T, Nishiaki Y, Aoki K (eds) Dynamics of learning in Neanderthals and modern humans. Replacement of Neanderthals by modern humans series, vol 1. Springer, Tokyo, pp 211–228

Csibra G, Gergely G (2011) Natural pedagogy as evolutionary adaptation. Philos Trans R Soc B Biol Sci 366:1149–1157

De Sanctis L, Galla T (2009) Effects of noise and confidence thresholds in nominal and metric Axelrod dynamics of social influence. Phys Rev E 79:046108

Derex M, Beugin MP, Godelle B, Raymond M (2013) Experimental evidence for the influence of group size on cultural complexity. Nature 503:389–391

d'Errico F, Henshilwood CS (2007) Additional evidence for bone technology in the Southern African middle stone age. J Hum Evol 52:142–163

d'Errico F, Stringer CB (2011) Evolution, revolution or saltation scenario for the emergence of modern cultures? Philos Trans R Soc B Biol Sci 366:1060–1069

Diestel R (2010) Graph theory. Graduate texts in mathematics, vol 173, 4th edn. Springer, Heidelberg

Dunnell RC (1971) Systematics in prehistory. Free Press, New York

Eerkens J, Lipo C (2005) Cultural transmission, copying errors, and the generation of variation in material culture and the archaeological record. J Anthropol Archaeol 24:316–334

Ewens WJ (2004) Mathematical population genetics. Theoretical introduction, vol 1, 2nd edn. Springer, New York

Ferguson JR (2008) The when, where, and how of novices in craft production. J Archaeol Method Theory 15:51–67

Flache A, Macy MW (2006) What sustains cultural diversity and what undermines it? Axelrod and beyond. arXiv preprint physics/0604201

Fogarty L, Strimling P, Laland KN (2011) The evolution of teaching. Evolution 65:2760–2770

Godsil CD, Royle G (2001) Algebraic graph theory, vol 8. Springer, New York

González-Avella J, Cosenza M, Klemm K (2007a) Information feedback and mass media effects in cultural dynamics. J Artif Soc Soc Simul 10(3):9

González-Avella J, Eguíluz V, San Miguel M (2007b) Homophily, cultural drift, and the co-evolution of cultural groups. Phys Rev E 72(6):065102–065106

González-Avella JC, Cosenza MG, Tucci K (2005) Nonequilibrium transition induced by mass media in a model for social influence. Phys Rev E 72:065102

González-Avella JC, Eguíluz VM, Cosenza MG, Klemm K, Herrera J, San Miguel M (2006) Local versus global interactions in nonequilibrium transitions: a model of social dynamics. Phys Rev E 73: 046119

Henrich J (2004) Demography and cultural evolution: how adaptive cultural processes can produce maladaptive losses: the tasmanian case. Am Antiq 69:197–214

Högberg A (2008) Playing with flint: tracing a child's imitation of adult work in a lithic assemblage. J Archaeol Method Theory 15: 112–131

Kempe M, Mesoudi A (2014) An experimental demonstration of the effect of group size on cultural accumulation. Evol Hum Behav 35:285–290

Klein RG (2009) The human career: human biological and cultural origins. University of Chicago Press, Chicago

Klemm K, Eguíluz V, Toral R, San Miguel M (2003a) Global culture: a noise-induced transition in finite systems. Phys Rev E 67:045101

Klemm K, Eguíluz V, Toral R, San Miguel M (2003b) Nonequilibrium transitions in complex networks: a model of social interaction. Phys Rev E 67:026120

Klemm K, Eguíluz V, Toral R, San Miguel M (2005) Globalization, polarization and cultural drift. J Econ Dyn Control 29:321–334

Kuhn S (2013) Cultural transmission, institutional continuity and the persistence of the Mousterian. In: Akazawa T, Nishiaki Y, Aoki K (eds) Dynamics of learning in Neanderthals and modern humans. Replacement of Neanderthals by modern humans series, vol 1. Springer, Tokyo, pp 105–113

Lanchier N (2012) The Axelrod model for the dissemination of culture revisited. Ann Appl Probab 22:860–880

Lanchier N, Deijfen M, Häggström O, Connor S (2010) Opinion dynamics with confidence threshold: an alternative to the Axelrod model. Alea. Arxiv.org 1003.0115v2. http://arxiv.org/abs/1003.0115

MacArthur BD, Sánchez-García RJ, Anderson JW (2008) Symmetry in complex networks. Discret Appl Math 156:3525–3531

McBrearty S (2007) Down with the revolution. In: Mellars P, Boyle K, Bar-Yosef O, Stringer C (eds) Rethinking the human revolution. MacDonald Institute for Archaeological Research Monographs, Cambridge, pp 133–152

McBrearty S, Brooks AS (2000) The revolution that wasn't: a new interpretation of the origin of modern human behavior. J Hum Evol 39:453–563

McKay BD, Piperno A (2014) Practical graph isomorphism, {II}. J Symb Comput 60:94–112

Mesoudi A, O'Brien MJ (2008) The learning and transmission of hierarchical cultural recipes. Biolog Theory 3:63–72

Moore MW (2010) "Grammars of action" and stone flaking design space. In: Nowell A, Davidson I (eds) Stone tools and the evolution of human cognition. University of Colorado Press, Boulder, pp 13–43

Moran P (1958) Random processes in genetics. Math Proc Camb Philos Soc 54(1):60–71. Cambridge University Press, Cambridge

Moran P et al (1962) The statistical processes of evolutionary theory. Clarendon, Oxford

Muthukrishna M, Shulman BW, Vasilescu V, Henrich J (2014) Sociality influences cultural complexity. Proc R Soc B Biol Sci 281: 20132511

Nakahashi W (2013) Cultural evolution and learning strategies in hominids. In: Akazawa T, Nishiaki Y, Aoki K (eds) Dynamics of learning in Neanderthals and modern humans. Replacement of Neanderthals by modern humans series, vol 1. Springer, Tokyo, pp 245–254

Neff H (1992) Ceramics and evolution. Archaeol Method Theory 4:141–193

Nickel M, Tresp V, Kriegel HP (2011) A three-way model for collective learning on multi-relational data. In: Getoor L, Scheffer T (eds) Proceedings of the 28th international conference on machine learning (ICML-11), Bellevue. ACM, New York, pp 809–816

Nishiaki Y, Aoki K, Akazawa T (2013) Introduction. In: Akazawa T, Nishiaki Y, Aoki K (eds) Dynamics of learning in Neanderthals and modern humans. Replacement of Neanderthals by modern humans series, vol 1. Springer, Tokyo, pp 1–3

O'Brien M, Lyman R, Mesoudi A, VanPool T (2010) Cultural traits as units of analysis. Philos Trans R Soc B Biol Sci 365: 3797–3806

O'Brien MJ, Shennan S (2010) Innovation in cultural systems: contributions from evolutionary anthropology. MIT, Cambridge

Otter R (1948) The number of trees. Ann Math 49:583–599

Premo L (2012) Local extinctions, connectedness, and cultural evolution in structured populations. Adv Complex Syst 15:1150002

Rotman JJ (1995) An introduction to the theory of groups. Graduate texts in mathematics, vol 148. Springer, New York

Schank RC, Abelson RP (1977) Scripts, plans, goals, and understanding: an inquiry into human knowledge structures. Artificial intelligence series. Erlbaum Associates, Hillsdale

Schiffer MB, Skibo JM (1987) Theory and experiment in the study of technological change. Curr Anthropol 28:595–622

Shennan S (2000) Population, culture history, and the dynamics of culture change. Curr Anthropol 41:811–835

Shennan S (2001) Demography and cultural innovation: a model and its implications for the emergence of modern human culture. Camb Archaeol J 11:5–16

Smilkov D, Kocarev L (2012) Influence of the network topology on epidemic spreading. Phys Rev E 85:016114

Sterelny K (2012) The evolved apprentice. MIT, Cambridge

Stout D (2002) Skill and cognition in stone tool production: an ethnographic case study from Irian Jaya. Curr Anthropol 43:693–722

Stout D (2011) Stone toolmaking and the evolution of human culture and cognition. Philos Trans R Soc B Biol Sci 366:1050–1059

Straus LG (2005) A mosaic of change: the middle–upper paleolithic transition as viewed from New Mexico and Iberia. Quat Int 137:47–67

Terashima H (2013) The evolutionary development of learning and teaching strategies in human societies. In: Akazawa T, Nishiaki Y, Aoki K (eds) Dynamics of learning in Neanderthals and modern humans. Replacement of Neanderthals by modern humans series, vol 1. Springer, Tokyo, pp 141–150

Tostevin GB (2012) Seeing lithics: a middle-range theory for testing for cultural transmission in the pleistocene. Oxbow Books, Oxford

Villa P, Roebroeks W (2014) Neandertal demise: an archaeological analysis of the modern human superiority complex. PLoS One 9:e96424

Wimsatt WC (2007) Re-engineering philosophy for limited beings: piecewise approximations to reality. Harvard University Press, Cambridge

Wimsatt WC, Griesemer JR (2007) Reproducing entrenchments to scaffold culture: the central role of development in cultural evolution. In: Sansom R, Brandon RN (eds) Integrating evolution and development: from theory to practice. Bradford Books/MIT, Cambridge, pp 227–323

Inferring Learning Strategies from Cultural Frequency Data

7

Anne Kandler and Adam Powell

Abstract

Social learning has been identified as one of the fundamentals of culture and therefore the understanding of why and how individuals use social information presents one of the big questions in cultural evolution. To date much of the theoretical work on social learning has been done in isolation of data. Evolutionary models often provide important insight into which social learning strategies are expected to have evolved but cannot tell us which strategies human populations actually use. In this chapter we explore how much information about the underlying learning strategies can be extracted by analysing the temporal occurrence or usage patterns of different cultural variants in a population. We review the previous methodology that has attempted to infer the underlying social learning processes from such data, showing that they may apply statistical methods with insufficient power to draw reliable inferences. We then introduce a generative inference framework that allows robust inferences on the social learning processes that underlie cultural frequency data. Using developments in population genetics—in the form of generative simulation modelling and approximate Bayesian computation—as our model, we demonstrate the strength of this method with an example based on simulated data.

Keywords

Social learning • Cultural evolution • Generative inference • Approximate Bayesian computation

7.1 Introduction

Much research over the last decades has been devoted to investigating the dynamics of social learning, one of the fundamentals of culture. Social learning, defined as learning facilitated by observations of, or interactions with, another individual or its products (Heyes 1994; Hoppitt and Laland 2013), occurs in various forms and is widespread across the animal kingdom (Hoppitt and Laland 2013). While it can facilitate the spread of adaptive information, accumulated over generations, throughout a population, it also allows the transmission of outdated, misleading or inappropriate information especially in changing environmental conditions (e.g. Giraldeau et al. 2002). So understanding why and how individuals use social information is one of the big challenges in cultural evolution (Rendell et al. 2010) and the focus of a highly active, interdisciplinary debate. However, much of the modelling work to date has been largely theoretical, in isolation of real data. While models often provide important insights into which social learning strategies are expected to have evolved in changing environmental conditions, they are often unable to make robust inferences about which strategies human populations actuallyuse based on

A. Kandler (✉)
Department of Mathematics, City University London, London, UK
e-mail: anne.kandler.1@city.ac.uk

A. Powell
Palaeogenetics Group, Institute of Anthropology, University of Mainz, Mainz, Germany

A. Mesoudi and K. Aoki (eds.) *Learning Strategies and Cultural Evolution during the Palaeolithic*, Replacement of Neanderthals by Modern Humans Series, DOI 10.1007/978-4-431-55363-2_7,

observed cultural datasets. Much available empirical data is in the form of frequency distributions of a number of different variants of a cultural trait in the population at a certain point in time (e.g. Bentley et al. 2004) or of time-series that describe the dynamics of the frequency change of cultural variants over time (e.g. Henrich 2001), and often comprise a limited or chronologically-sparse sample from the whole population. Attempting to infer the underlying evolutionary processes from such population-level data poses a classical inverse problem, but despite widespread interest from researchers across archaeology, biological anthropology, animal behaviour and psychology much of the currently available methodology may be limited in its ability to reliably identify which learning strategies (or combinations thereof) could have produced the observed data. While a large number of possible learning strategies have been identified, such as direct-biased social learning, conformist social learning or copy-when-uncertain (e.g. Boyd and Richerson 1985; Laland 2004; Morgan et al. 2012), the inferential frameworks utilised often lack the necessary statistical power to distinguish between the, often very similar, variant frequency distributions generated under those learning strategies. This is especially true when the empirical data is sparse, and it is important to recognise when frequency data lack sufficient information. In such situations inference has to be based on other lines of evidence.

In this chapter we review some of these theoretical inferential approaches, focusing on their application to human learning strategies, and then present a non-equilibrium generative framework which is aimed at robustly addressing the current mismatch between theory and data. This generalizable framework models the frequency change of different variants of a cultural trait present in a population under the influence of various parameterised asocial and social learning strategies, and coupled with a powerful statistical technique called approximate Bayesian computation (ABC) it allows us to delineate the conditions under which such frequency data carry reliable signatures of the underlying learning strategies which generated them. The suggested approach provides a comprehensive and statistically powerful basis for determining possible mixture of learning strategies that are consistent with even sparse observed cultural frequency data, and therefore may result in a substantial reduction of the set of evolutionary hypotheses that could have produced the data.

7.2 Previous Research

Research to establish the presence of particular learning strategies in human populations is mainly centred around experimental, laboratory-based and theoretical, modelling-based approaches. Laboratory-based experiments have focused on uncovering the variety and subtlety of human social learning strategies (Rendell et al. 2011). 'Microsocieties' (e.g. Coultas 2004; Baum et al. 2004; McElreath et al. 2008; Morgan et al. 2012) and diffusion chain experiments (e.g. Mesoudi and O'Brien 2008; Caldwell and Millen 2008; Kirby et al. 2008) have proven to provide powerful frameworks for revealing the existence of specific learning strategies in human populations. For a comprehensive review of this experimental literature see (Rendell et al. 2011). In this chapter we focus on theoretical approaches and consider in the following the merits and limitations of the use of adoption curves (Sect. 7.2.1), power-law distributions (Sect. 7.2.2), model selection frameworks (Sect. 7.2.3) and formal population genetic and game-theoretic modelling (Sect. 7.2.4) when investigating the learning strategies employed by human populations.

7.2.1 Adoption Curve

Adoption curves detail the cumulative number of individuals that have acquired a specific cultural variant at any given point in time and therefore describe the temporal dynamic of the spread of novel cultural variants through the population. Many researchers suggested that the shape of adoption curves can be used to infer information about the underlying learning strategies employed by human populations; in particular about the presence or absence of social learning. Here mathematical models predicted accelerating adoption curves in situations when social information is used. The acceleration is caused by the frequency-dependent nature of social learning: the more individuals that have acquired the novel cultural variant the higher the probability becomes of others learning it socially, and is often described by a sigmoidal (or *S*-shaped) adoption curve (Cavalli-Sforza and Feldman 1981; Boyd and Richerson 1985; Laland et al. 1996). In contrast, asocial learning is predicted to result in decelerating, often *r*-shaped, adoption curves (Boyd and Richerson 1985; Laland et al. 1996). Henrich (2001) argued that asocial learning alone cannot produce *S*-shaped adoption patterns, and therefore that *S*-shaped dynamics must imply the presence of social learning strategies, in various forms, in the population. This reasoning would suggest that the presence or absence of social learning can be inferred directly from the temporal adoption dynamic of a cultural variant, and in fact many episodes of the diffusion of innovations through human populations exhibit an *S*-shaped adoption pattern (Rogers 2003). While it is undoubtedly true that social learning results in accelerating *S*-shaped adoption curves it has been demonstrated that the diagnostic power of the shape of the adoption curve is greatly reduced when allowing for individual heterogeneity or complexity in the adoption process (Reader 2004) or population structure (Franz and Nunn 2009). For example, Rogers (2003) assumed that the adoption time (defined as the time it takes an individual to adopt the cultural trait) is normally distributed and showed that in this situation asocial learning alone can produce S-shaped adoption curves. Therefore, individual differences in the propensity to adopt a cultural trait are likely to influence the adoption dynamic and researchers have to be cautious when attempting to infer the presence or absence of social learning from adoption curve data (Cavalli-Sforza and Feldman 1981). Furthermore, if there is a number of discrete steps required to adopt a cultural variant, the cumulative diffusion curve may be S-shaped, even if individuals learn the tasks asocially (Kendal et al. 2007; Hoppitt et al. 2010b). These ambiguities in discriminating between asocial and social learning by analysing the shape of the adoption curve demonstrate that this approach might have limited power when used as an inference tool.

Adoption curves have also been the subject of intense research in marketing science. Bass (1969) proposed that the population of adopters can be divided into innovators and imitators and that the shape of the adoption curve will vary as a function of their relative importance. The Bass model (and its numerous generalisations) has been widely used in forecasting, especially the forecast of new products' sales (e.g. Mahajan et al. 1995). Alternatively to this social learning hypothesis, threshold models have been proposed to explain observed temporal adoption dynamics. Here individuals adopt an innovation as soon as the utility of the innovation exceeds some threshold e.g. when considering costly innovations as soon as individuals can afford to adopt the innovation (e.g. Van der Bulte and Stremersch 2004). It has been shown that both explanations, the social learning explanation described by the Bass model and the heterogeneity explanation, can lead to nearly the same adoption behaviour (Van der Bulte and Stremersch 2004; Kandler and Steele 2010). This again points to the limited power of inferences based on temporal adoption curves.

7.2.2 Power-Law Distribution

An important strand of cultural evolution has investigated the utility of models of unbiased social learning or neutral copying, applied mainly in evolutionary archaeology (e.g. Neiman 1995; Shennan and Wilkinson 2001; Bentley and Shennan 2003). The underlying common model is derived through direct analogy to the neutral model of population genetics, where changes in selectively neutral allele frequencies are due simply to the stochastic processes of drift and mutation (e.g. Kimura and Crow 1964). Various empirically calculated diversity statistics of the different archaeological types or variants in an assemblage can be compared with theoretically derived diversity estimators in order to determine whether they are consistent, or not, with the hypothesis of unbiased social learning (Neiman 1995; Kohler et al. 2004). A further extension of this approach relies on examining the full variant frequency distribution of a dataset, rather than just summary statistics, and comparing it to the right-skewed power-law distribution expected to arise under unbiased social learning. This expectation is again derived from the equivalent model in population genetics, where it was demonstrated that at mutation-drift equilibrium in a finite population the allele frequency distribution is of the power-law form $P(v) \sim v^{\alpha}$ (Kimura and Crow 1964), where $P(v)$ is the proportion of alleles in the population at frequency v. Such distributions were found across a number of different cultural datasets (e.g. Hahn and Bentley 2003; Herzog et al. 2004; Bentley et al. 2004), and thus the underlying processes in these various cultural systems were deemed to be consistent with an unbiased social learning model. Mesoudi and Lycett (2009) showed that introducing learning biases into the basic neutral model can drive the equilibrium variant frequencies away from a power-law, but relied on a relatively crude testing procedure to identify significant deviation. Clauset and colleagues (2009) provided a very comprehensive set of statistical methods for confirming whether empirical data sets are power-law distributed, demonstrating that the methods adopted, and thus the strength of the conclusions drawn, in the previous literature were necessarily limited.

However, from an information-theoretic viewpoint (e.g. Frank 2009) a power-law distributed dataset is not a firm diagnostic of any particular underlying generative process. The seeming ubiquity of power-law distributions across such a wide range of natural and cultural phenomena, from the magnitude and frequency of earthquakes (Gutenberg and Richter 1944) to word usage frequencies in languages (Zipf 1929), can be explained purely as an inevitable consequence of any large-scale aggregation process that preserves information solely about the geometric mean. In other words, the power-law is a limiting distribution with a large attractor basin, such that invoking a unique mechanism to explain power-law distributed data may be redundant since an infinite number of different mechanisms will generate indistinguishable distributions. Recent work has drawn on this idea by extending the framework outlined by Mesoudi and Lycett (2009), in order to approximately quantify the limits of the power-law attractor basin by testing for when significant deviation from power-law distributed data occurs (Powell et al. in prep, see also Clauset et al. 2009). This work begins with an unbiased copying simulation model (e.g. Bentley et al. 2004) and introduces incrementally increasing levels of conformist or anti-conformist social learning. While even very low levels of anti-conformity bias produces clear deviation from a power-law, surprisingly high levels of conformist social learning can occur over many generations without significant deviation from a power-law distribution. This work demonstrates that there is very limited statistical power with such an approach, and shows the clear limitations of using power-law distributed data as indicative of an underlying unbiased social learning process. Additionally, this power-law approach relies on the assumption that innovation-drift equilibrium is achieved and it is not immediately obvious that such equilibria can be reliably achieved or maintained in many cultural systems, e.g. due to changes in population size or structure.

A closely-related area of work (e.g. Bentley et al. 2007; Acerbi and Bentley 2014) draws on another facet of the neutral model, the rate of turnover within the n most popular variants in a population, as a further basis to test for divergence from unbiased transmission. Under the unbiased copying assumption this turnover can be well-described in functional form, against which empirical datasets can be examined within a robust model comparison framework (Acerbi and Bentley 2014). This approach appears promising but still relies on the innovation-drift equilibrium assumption and may be difficult to apply it to sparse archaeological data where the total number of observed cultural variants is low (such that a 'top list' of size sufficient to accurately calculate turnover cannot be defined).

7.2.3 Model Selection Framework

Stressing the existence of a large number of possible social learning strategies, McElreath and co-workers (McElreath et al. 2008) developed a model selection framework based on the Akaike information criterion (AIC) (Akaike 1974; Burnham and Anderson 2002), which allows the quantification of the relative support different learning strategies receive from observed data. To do so they designed an experiment where each individual controls a virtual farm and has to decide which of the two possible crop varieties to plant. Thereby each individual can access information about the pay-off of each crop variety and the frequencies of the different choices in the social group. This set-up produced time-series of decisions and records of which kind of information is used to make this decision. Now in order to use this data to infer the learning strategies employed by the individuals McElreath et al. (2008) developed a probabilistic model which defined the conditional probability of an individual choosing any behavioural option at any point in time given the asocial and social information the individual accessed. Based on these formulations they devised a maximum likelihood framework to fit the considered learning strategies to the observed data. Subsequently the different learning strategies can be compared by evaluating the relative support of the strategies from the data using the AIC. Based on the results of this analysis one can then conclude which learning strategy possesses the highest relative explanatory power among a set of competing and plausible strategies. McElreath et al. (2008) concluded that a so-called 'hierarchical compare means/frequency dependent social learning' strategy receives overwhelming support from the experimental data. This strategy means that individuals compare the mean value of pay-offs of both crop choices and if there is only a small difference they use frequency-dependent social learning to choose which crop to adopt.

This way of approaching the inference problem provides significant advantages over classical hypothesis testing as it can determine which of the strategies, consistent with the data, has the most explanatory power and therefore describes the most likely (however not certain) way learning happened. The ideas developed later in this chapter are in a very similar spirit but we do not restrict the choice of the modelling framework by the availability of an analytical representation of the corresponding likelihood function.

Further, aimed at inferring social learning strategies in animal populations Franz and Nunn (2009) and Hoppitt et al. (2010a) introduced network-based diffusion analysis (NBDA). NBDA explores the existence of social learning in small groups by monitoring whether information flow follows the pattern of association between individuals. The strength of this approach lies in the combination of different kinds of information about the spread process and the social network and an information-theoretic inference framework. NBDA fits simulation models of asocial and social learning to the observed data using maximum-likelihood estimation and infers the underlying learning strategies using model selection based on the AIC (Franz and Nunn 2009). It overcomes a number of the problems associated with the analysis of the adoption curve described in Sect. 7.2.1 but successful applications of this approach require either a controlled experimental set-up or the detailed knowledge of the underlying social network. For a comprehensive review of statistical inference techniques in the animal social learning literature see (Hoppitt and Laland 2013).

7.2.4 Formal Modelling

There exists a large body of modelling work focusing on the question of which learning strategies human populations are expected to use especially in spatially and temporally changing environments. It is assumed that cultural variants differ in the benefit they convey to their adopters in a particular environmental situation and population-genetic and game-theoretic analyses revealed the evolutionary stable learning strategies for the assumed environmental fluctuations, i.e. the strategies that would have evolved under natural selection. These equilibria indicated that asocial learning is favoured by natural selection when environmental changes occur on relatively short intervals while social learning is favoured when the environment is relatively constant (e.g. Aoki et al. 2005; Boyd and Richerson 1985, 1988; Feldman et al. 1996; Rogers 1988). Other approaches such as the simulation approach of the 'social learning strategies tournament' (Rendell et al. 2010, 2011) or the n-variant competition framework (Kandler and Laland 2013) suggested that social learning could be the dominant learning strategy even in rapidly changing environmental conditions. This difference is partly caused by the fact that in these more recent studies more than two cultural variants, offering different levels of benefit in a particular environment, are considered. The potential accumulation of social information then allows social learners to adjust their adopted variant after an environmental change by switching to an available, more adaptive variant.

Further, social learning can occur in a large number of different ways (e.g. Cavalli-Sforza and Feldman 1981; Boyd and Richerson 1985; Laland 2004), which raises the question of how social information should be used to adapt efficiently to changing environmental conditions. In this context special attention has been devoted to conformist social learning, in which common variants are disproportionately adopted (Boyd and Richerson 1985). Henrich and Boyd (1998) studied the conditions under which conformity can evolve and found that selection favours conformist learning as long as the environment does not change too rapidly and the evolution of social learning is more strongly influenced by environmental heterogeneity than the evolution of conformity. Contrary to those results Nakahashi (2007), Wakano and Aoki (2007), Kendal et al. (2009) and Kandler and Laland (2013) demonstrated a negative relationship between environmental stability and the reliance on conformist social leaning: the reliance on conformity tends to be larger in frequently changing environments. Wakano and Aoki (2007) pointed out that the Henrich and Boyd model results in the same conclusion when the time of evolution is sufficiently long. Additionally the models mentioned above differ in their assumptions, e.g. Henrich and Boyd (1998) and Wakano and Aoki (2007) used mixed learning strategies in a spatial and non-spatial setting, respectively whereas Nakahashi (2007), Kendal et al. (2009) and Kandler and Laland (2013) assumed pure strategies where individuals use either social or asocial learning. Eriksson et al. (2007) suggested that allowing for more than two cultural variant (and additionally that individuals have complete knowledge about all present variants) will disadvantage the evolution of conformist social learning. Further, Efferson et al. (2008) noted that the evolutionary advantage of the conformist strategy depends on the accuracy of asocial learning. McElreath et al. (2011) pointed to the importance of spatial heterogeneity in evolutionary modelling and claimed that non-spatial frameworks may underestimate the role of conformity. Similarly, Nakahashi et al. (2012) argued that the focus on (unrealistic) situations characterized by: (i) only two cultural variants present, (ii) only temporally varying environments and (iii) error-free social learning has obscured conditions favouring the evolution of conformity.

In summary, much theoretical work has been devoted to the analysis of evolutionary stable strategies in different environmental conditions. This approach revealed a large number of important insights into the evolutionary process: It allows for an elegant characterisation of the long-term outcomes of evolution through natural selection as well as of the associated evolutionary trajectories and for the derivation of analytical conditions for different evolutionary trajectories. In other words, it produces predictions of which learning strategies are expected to be present in the population given the environmental circumstances. However, in many real-world situations learning strategies cannot be observed directly and therefore the model predictions reviewed above are difficult to verify with the available data. Further, the analysis of stable evolutionary equilibria implicitly assumes that sufficient time has passed so that the equilibrium could be reached and that individual fitness (with respect to the considered cultural trait) is the quantity maximised by evolution (Maynard Smith 1978). But how do we know that the considered cultural system is at equilibrium? Are those stable evolutionary equilibria really an appropriate description of which learning strategies human populations use?

7.3 Generative Inference

In order to address the inverse problem of inferring underlying learning strategies that could have given rise to observed data, there exist two major types of modelling approaches: discriminative and generative models. Discriminative inference proceeds by fitting a, usually limited, number of model parameters conditional on a single instance of empirical data, generally assuming some correlational relationship between these observations and the implicit parameters, which may have no explicit or causal bearing on the true underlying mechanism. This kind of approach includes many standard and familiar statistical modelling procedures, such as regression or clustering techniques, in which parameter values can be estimated or structurally different models be compared. On the other hand, the generative approach proceeds by building a fully specified probabilistic model, in which the hypothesised causal mechanisms are explicitly defined. This model is used to repeatedly simulate pseudo-datasets under known parameter values such that their expected distribution can then be statistically compared with the observed data. This comparison allows certain hypothesised mechanisms to be rejected as inconsistent with the empirical data, or the estimation of the causal model parameters that provide the best fit to the data.

The idea of generative inference in socio-cultural evolution really first emerged with the introduction of 'Sugarscape', one of the earliest agent-based models designed to investigate social processes (Epstein and Axtell 1996). By repeatedly simulating a population of autonomous agents moving and interacting on a heterogeneous resource landscape the model demonstrated that large-scale social phenomena or patterns, analogous to those observed in the real world, could emerge from the aggregation of micro-level interactions. 'Sugarscape' and subsequent work (see also Epstein 2007) demonstrated the viability of this inference approach, by investigating the effects of hypothesised and explicitly defined individual-level causal mechanisms on population-level outcomes, that could then be compared with observed data in order to test models and estimate parameter values. While one of the major advantages of this approach is that expected outcomes of much more realistic, and potentially non-equilibrium, systems can easily be generated, the principle limitation has been the lack of a robust statistical methodology capable of comparing the outcomes from these complex models to empirical data. Much of the work discussed in Sect. 7.2.2 and some in Sect. 7.2.1 can be considered as taking this generative approach, but the inferences drawn are limited by the lack of power in the statistical methodology employed. However, a relatively recent major statistical advance in the form of approximate Bayesian computation (driven mainly by statistical geneticists, e.g. Beaumont et al. 2002) has started to transform the field of population genetics, by allowing inference on increasingly detailed generative genetic models of human evolutionary history using observed genomic data. Its recent successes should serve as a model for the field of cultural evolution, and we are strongly advocating this approach here.

7.3.1 Application in Population Genetics

A significant area of population genetics has recently emerged that relies on the generative inference approach outlined above. This work is usually in the form of explicit simulation modelling, in which simulated genetic data sets are generated under increasingly realistic genetic (McVean and Cardin 2005; Marjoram and Wall 2006) and/or demographic (e.g. Currat et al. 2004) models, and then statistically compared to observed data in order to infer population history (Veeramah et al. 2012; Eriksson et al. 2012) or past selection pressures (Itan et al. 2009; Wilde et al. 2014).

But from its origins in the early twentieth century (Fisher 1930; Wright 1931) up until fairly recently, population genetics had been a largely *prospective* discipline (Ewens 2004), concerned with predicting how allele frequencies in a population will change over time. These mainly analytical models were used to identify theoretical equilibria between the processes of mutation, drift and selection and the expected levels of genetic variation thus maintained in a population, as well as predicting the long-term fate of new alleles. However, an alternative way of viewing evolution is based on the idea of genealogies, or lineages, where the ancestral relationships of a contemporary sample of individuals are considered *retrospectively*. One major advantage of this approach is that the ancestry of non-sampled individuals can be disregarded, as, by definition, it has no bearing on the evolution of the sampled individuals. While in genetics these ideas date back at least to the 1940s (e.g. Malécot 1948), it was only the development of coalescent theory in the 1980s (Kingman 1982) that allowed the ancestral process to be robustly embedded within a population. The coalescent is a statistical model that describes the stochastic process by which lineages reach common ancestors, or 'coalesce', backwards in time within a population of defined size or structure, generating a dated genealogy rooted by the most recent common ancestor of the whole sample. Many classical population genetics results were re-derived, often more elegantly, using this retrospective approach, but the real strength of the coalescent is that it allows extremely efficient simulation of genealogies under a well-defined population history without having to assume any sort of equilibrium. In conjunction with a mutation model, these genealogies can then be used to rapidly generate simulated genetic data to match each observed sample. A number of extensions to the basic coalescent model have allowed the incorporation of evolutionary processes such as recombination (Hudson 1983; Marjoram and Wall 2006), as well as complex or geographically realistic demographic histories (e.g. Currat et al. 2004; Eriksson et al. 2012).

The coalescent approach is used (almost exclusively) for inference based on neutral genetic loci, in order to avoid the serious confounding effects of natural selection. Disentangling the effects of demography and selection can be very difficult, especially in human populations, given that many genetic adaptations in humans are thought to have been driven by cultural processes, e.g. adaptations due to dietary changes (see Laland et al. 2010). It is difficult to extend the coalescent to such complex cases, so instead genetic data can be simulated by modelling the whole population forward in time, and explicitly defining selection pressures that may vary in time and/or space or apply just to certain cultural groups (e.g. Itan et al. 2009; Wilde et al. 2014). This *forward simulation* approach can be far more computationally demanding, but still the underlying principle is the same: genetic data is generated under an explicitly defined evolutionary model which does not need to assume equilibrium. Either way, these simulated genetic samples can then be compared to observed samples, in order to estimate the parameters of the evolutionary causal mechanisms, such as selection strength, changes in population size or migration, or test between different models of population history.

This simulation, or generative inference, approach has only become fully viable since the development of an extremely powerful statistical methodology known as approximate Bayesian computation (ABC) (Beaumont et al. 2002). While it is generally preferable to base statistical inferences on a likelihood function, in most applied population genetic models their explicit specification and calculation can be challenging, if not impossible. For those (often simplistic) models in which the likelihood can be specified, approaches such as Markov chain Monte Carlo (MCMC) can be implemented in order to evaluate the likelihood. Such methods are, in general, extremely computationally demanding and it can be difficult to ensure that the chain converges on the target distribution efficiently. In contrast, ABC allows statistically robust inferences to be made on complex models without the consideration of explicit likelihood functions. The general idea is to approximate the joint posterior distribution of the model parameters $P\left(\theta \middle| D\right)$ directly, given observed data D. This is done by repeatedly simulating data D^* under a generative model with parameter values drawn from their prior distributions $P(\theta)$, and retaining just those parameter sets that generate data sufficiently 'close' to the real data. These retained parameter sets form a sample from the distribution $P\left(\theta \middle| d\left(D, D^*\right) \leq \varepsilon\right)$, where $d(.,.)$ is the distance between the observed and simulated data and ε is tolerance level that determines the level of approximation. As $\varepsilon \to 0$ the retained sample asymptotically approaches the true posterior distribution $P\left(\theta \middle| D\right)$. In practice, the data are usually reduced in dimension by using a set of summary statistics S, such that we are approximating the posterior $P\left(\theta \middle| S\right)$ with $P\left(\theta \middle| d\left(S, S^*\right) \leq \varepsilon\right)$. Parameter estimation is then performed via the joint or marginal posteriors for each parameter, such that modal values and credible intervals can be obtained. Many subsequent developments of this basic algorithm have allowed successful applications of ABC to a wide variety of inference problems both in population genetics and elsewhere (e.g. Toni et al. 2009, and see Sect. 7.3.2.2).

Recent debate in population genetics has reinforced the necessity of a rigorous model-based approach to inference (Nielsen and Beaumont 2009; Beaumont et al. 2010; Templeton 2009). This allows explicit parameterized models to be created, tested and fitted to data, in a way that other approaches, such as phylogeography, cannot. While a number of other non-generative model-based methods are producing exciting findings (e.g. Green et al. 2010; Hellenthal et al. 2014), an increasing number of researchers are utilising the generative approach, often coupled with ABC, to answer diverse questions about human evolutionary history using both standard population genetic (Veeramah et al. 2012; Eriksson et al. 2012) as well as gene-culture coevolutionary (Itan et al. 2009; Rasteiro et al. 2012) models.

7.3.2 Application to Cultural Evolution

In this section we turn back to our initial question of which learning strategies human populations use especially in changing environments. Based on the framework developed by Kandler and Laland (2013) we demonstrate in the following how a cultural evolutionary generative inference procedure can be constructed and how it can be used to explore which learning hypotheses are consistent with the empirical evidence. Thereby we assume that time series data detailing the usage or occurrence frequencies of different cultural variants are available. Similarly to the genetic applications the inference procedure consists of two steps. The first step comprises the development of a non-equilibrium generative model capturing the main cultural and demographic dynamics of the considered system and describing the frequency changes of different cultural variants present in a population under the assumed learning hypothesis. The focus on non-equilibrium situations is motivated by observations of e.g. drastic environmental changes or temporal variations in the total size of the population of cultural variants. The model is designed to establish a causal relationship between the various asocial and social learning strategies and temporal frequency changes of the present cultural variants. Consequently initializing the model with the frequencies observed at the beginning of the time series produces patterns of frequency change expected under a specific learning hypothesis at any future point in time. The second step uses ABC to derive conclusions about which mixtures of learning strategies are consistent with the observable frequency data and which are not. The aim of this framework, then, is to allow researchers to 'reverse engineer' which learning strategies are likely to have been used in current or past populations, given knowledge of how frequencies have changed over time, independent of any optimality or equilibrium assumptions. It should be noted that the generative model we introduce below is simply an example to demonstrate the utility of this approach, and that a number of the models reviewed in Sect. 7.2 could feasibly be adapted to use within this framework. The crucial point is that the generative model is appropriately chosen and adjusted to the considered application.

7.3.2.1 Generative Model

In Kandler and Laland (2013) we considered k different variants of a cultural trait that differ in the benefit they

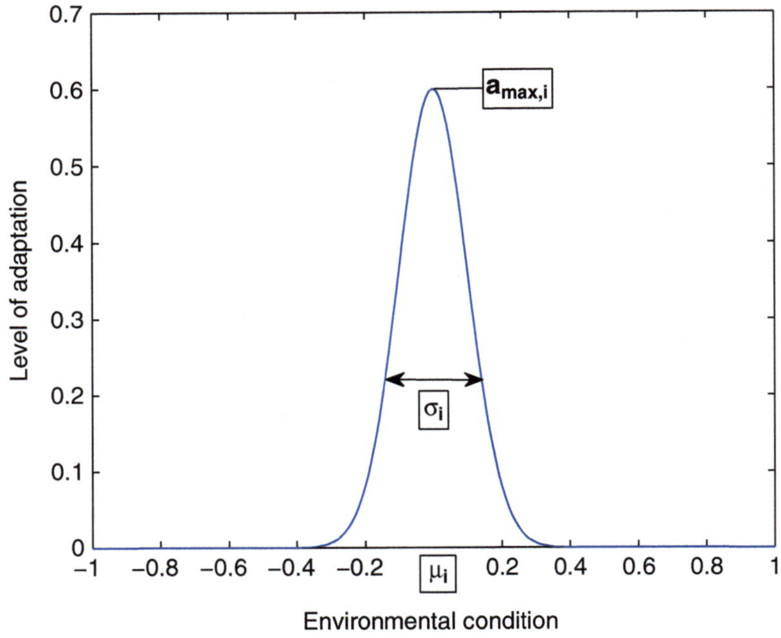

Fig. 7.1 Example of an adaptation function

convey to their adopters in a particular environment. Using a competition framework we modelled the frequency changes of those k variants of a cultural trait in the face of asocial and social learning in form of direct biased and conformist social learning. We assumed that a population is experiencing changing environmental conditions, expressed by the function $e(t)$ with $e(t) \in [-1, 1]$. Those changing conditions affect the adaptation levels of the different cultural variants and we characterised each variant i by its adaptation function $a_i(e(t))$. This function indicates the benefit the variant conveys to its adopters in condition $e(t)$ and is determined by three parameters: the environmental condition μ_i to which the variant is best adapted to, the maximum level of adaptation $a_{max,i}$ and the width σ_i (see Fig. 7.1 for an example).

Now asocial and social learning strategies cause the cultural variants to change in frequency and we described those changes using a differential equation framework of the form

$$\frac{\partial u_i}{\partial t} = -\nu u_i(t) + \xi \text{ (asocial learning)}$$
$$+ (1-\xi) \text{ (social leaning)}, \; i = 1, \ldots, k$$
$$\frac{dK}{dt} = (\lambda - \nu) \times K(t)\,(1 - K(t)) \tag{7.1}$$

The variable u_i describes the frequency of variant i at time t in the population, or in other words the fraction of the population that has adopted variant i. Further, K denotes the total population size, λ and ν the birth and death rates, respectively and ξ the reliance of the population on asocial learning. It follows from the equation describing the temporal dynamic of the population size in model (7.1) that $K(t) \leq 1, \; \forall t$ and further it holds $\sum_{i=1}^{k} u_i(t) \leq K(t)$. Consequently, the temporal change in frequency of variant i (denoted by du_i/dt) is determined by

- the loss of adopters of variant i (described by νu_i) and
- the cumulative influences of social and asocial learning (described by the terms ξ (asocial learning) + $(1-\xi)$ (social learning)).

Social learning is based on social cues (and therefore can only lead to learning of variants which are already present in the considered location), however the nature of the cues can be ambiguous. We considered two different social learning strategies: direct biased social learning and conformist social learning. Direct biased social learning is based on adaptation information (meaning the adaptation levels a_i of the cultural variants in the current environment $e(t)$) and defined as the adoption of variants found to be efficacious by individual assessment (Boyd and Richerson 1985). This is incorporated in model (7.1) by assuming the following dynamic: The higher the adaptation level of a cultural variant the stronger is its transmission supported by direct biased social learning. Contrary, conformist social learning is based solely on frequency information and leads to a disproportional adoption of variants whose frequencies are above a commonness threshold (Boyd and Richerson 1985). It holds: the higher the variant's frequency the stronger it is supported by the conformist bias. The reliance on adaptation information and frequency information is modulated by the parameter b where $b = 0$ indicates no conformist tendencies. Asocial learning on the other hand is based on environmental cues, specifically on the inference about the benefit of specific variants in observed environmental conditions. Naturally, this inference is error-prone but can lead to the introduction of new variants into a specific location. We note that the considered cultural system consist of at most k different variants; variants can possess a frequency of zero at times and be reintroduced into the system by the process of asocial learning. In order to explore the temporal dynamic of the learning process the birth and death rates, λ and ν, are kept constant. A detailed mathematical description of the considered model can be found in the Appendix. We note that the suggested framework can be easily extended to a spatially explicit model by allowing the adopters of the different cultural variants to move in space. These spatial movements can be described by a simple diffusion process and the appendix provides the description of a spatially explicit model.

Summarizing, system (7.1) provides a framework that generates temporal frequency patterns of different cultural variants possessing different levels of adaptation to the experienced environment conditioned on the assumed mixture of learning strategies, i.e. it assumes that a fraction ξ of the population engages in asocial learning and consequently a fraction $(1 - \xi)$ engages in social learning whereby the strength of conformist learning is determined by the parameter b. The adaptation level a_i of the present cultural variants is determined by experienced environmental conditions $e(t)$ and the specific adaptation functions as given in Fig. 7.1. Importantly, the environment $e(t)$ can vary in a deterministic or random manner (for more details see Kandler and Laland 2013). Now a systematic variation of the model parameters (ξ, b) in system (7.1) produces insights into the interplay between learning strategies and patterns of frequency changes (see (Kandler and Laland 2013) for a detailed analysis). In particular when initializing the developed generative model with the frequency data observed at the first time point we obtain expectations about possible frequency patterns at later time points conditioned on the assumed learning hypothesis (ξ, b) and adaptation levels a_i of the variants. In the following we aim to establish statistically which of the possible mixtures of learning strategies are consistent with the observed frequency data.

7.3.2.2 Statistical Inference

As already mentioned above, in order to explore which mixtures of learning strategies are consistent with the observed data we would ideally determine the likelihood function of system (7.1) indicating the likelihood of the assumed parameter values $\theta = [a_i, \xi, b]$ given the observed data D. In this framework the data consists of frequency information about different cultural variants at different points in time. Based on the likelihood function we could derive the region of parameter spacewhich is likely to have produced the observed frequency patterns and therefore conclude which mixtures of the learning strategies could adequately explain the data. However, in our case the likelihood function cannot be determined easily and so we use ABC in the form of a Sequential Monte Carlo (SMC ABC) algorithm (Toni et al. 2009; Beaumont 2010; Del Moral et al. 2012) to approximate the posterior distribution $P\left(\theta \middle| D\right)$. The key idea of the SMC ABC algorithm is to decompose this problem into a series of simpler sub-problems as efficient sampling from approximate posterior distribution $P\left(\theta \middle| d\left(D, D^*\right) \le \varepsilon\right)$ can prove to be difficult. The algorithm starts by choosing a sequence of tolerance levels $\{\varepsilon_0, \varepsilon_1, \ldots, \varepsilon_n\}$ with $\varepsilon_0 < \varepsilon_1 \cdots < \varepsilon_n \ge 0$. In the first step a sample is drawn from $P\left(\theta \middle| d\left(D, D^*\right) \le \varepsilon_0\right)$ with ε_0 large and subsequently from an increasingly constrained sequence of distributions $P\left(\theta \middle| d\left(D, D^*\right) \le \varepsilon_i\right)$, $i = 1, 2, \ldots, n$. In other words the tolerance level is sequentially decreased until the desired level $\varepsilon = \varepsilon_n$ is reached. In this context, Del Moral et al. (2012) introduced a procedure which adaptively determines the best tolerance level ε_i in each step.

The SMC ABC algorithm produces distributions of the model parameters that indicate the ranges of the parameter space which are able to produce frequency data within a given tolerance margin ε of the observed data. Consequently, the mixture of learning strategies that are consistent with the data can be deduced. As suggested in Kandler and Laland (2013) the widths of the marginal posterior distributions will allow an estimation of the amount of information contained in the data on each parameter. If the ranges are narrow then only a small region of the parameter space is consistent with the data and therefore we conclude that a large number of mixtures of learning strategies are *not* able to produce the observed frequency change. In this way the suggested approach can help reduce the possible range of strategies that could have produced the observed frequency data. On the other hand, if the posterior distributions of the model parameters are wide a large region of the parameter space is consistent with the data. This might be indicative of either an inadequate description of the cultural system by the generative model or the fact that the observed frequency data do not carry a strong signature of the underlying learning strategies: many different mixtures of learning strategies can result in very similar frequency changes and therefore not much information about the underlying learning processes can be gained from analysing frequency data. Even though this appears as a negative result it is crucial to understand when frequency data is *not* informative about the underlying evolutionary processes, as in this situation researchers have to base inference on other lines of evidence.

7.3.2.3 Example

In order to demonstrate the applicability of the suggested framework we consider the following theoretical example. We use frequency data generated by the previously described generative model (7.1) in the time interval [0,100] and sample n data points as shown in Fig. 7.2. In more detail, we assume three cultural variants to be present initially (with initial fractions of adopters $u_1(0) = 0.2$, $u_2(0) = 0.1$ and $u_3(0) = 0.07$) and the environment to stay constant until time $t = 50$. Then a shock occurs causing the environment to change, the present variants to decrease in frequency and two new variants to be introduced. The model parameters are chosen as follows

$\xi = 0.1$, $b = 0$, $a_1 = 0.3$, $a_2 = 0.45$, $a_3 = 0.4$ for the time interval [0,50] and

$\xi = 0.1$, $b = 0$, $a_1 = 0.35$, $a_2 = 0.1$, $a_3 = 0.07$, $a_4 = 0.6$, $a_5 = 0.45$ for the time interval [50,100].

Using this approach we have full knowledge of the system (i.e. the adaptation level of the variants, learning processes and environmental fluctuations) that generated the data, and this enables us to explore how well our proposed inference framework recovers the true parameters from disturbed data (see coloured squares in Fig. 7.2). The data is disturbed by adding Gaussian noise distributed as $\mathcal{N}\left(0, \sigma^2\right)$ with $\sigma^2 = 0.03^2$.

Now we apply the suggested inference framework to the noisy frequency data and aim to estimate the adaptation levels of variants present in the time intervals [0,50] and [50,100], respectively, as well as the fraction of asocial learners ξ and the strength of conformist learning strategy b. Figure 7.3 shows the resulting posterior distribution for ξ and b for the interval [0,50] (top row) and the interval [50,100] (bottom row) for $n = 11$ data points. These distributions describe the range of the parameter space of ξ and b where the squared sum of the differences between the theoretical and observed frequencies is smaller than ε. Thereby ε is chosen to be the squared sum of the differences of disturbed and undisturbed data as shown in Fig. 7.2 (Toni et al. 2009). It is obvious that in both cases the posterior distributions are tightly clustered around the true value $\overline{\xi} = 0.1$ and $\overline{b} = 0$ and therefore the underlying learning processes can be recovered relatively accurately from the noisy data. For sake of shortness we omitted the posterior distributions for the adaptation level a_i, however note that those are also tightly clustered around the true values. Figure 7.3c, f show the joint distributions of the parameter ξ and b in the different time intervals. The light areas indicate areas with high probability whereas the darker areas indicate less likely parameter regions. The joint distributions provide valuable information about the interactions of the investigated processes, in our case asocial learning and conformist social learning. In other words, they provide an indication of which parameter constellation give rise to similar population-level frequency data. It is obvious that large deviations from the true value in both parameters do not produce frequency patterns consistent with the observed data (see the triangular shapes of the light areas in Fig. 7.3c, f). We conclude that the developed inference framework is able to recover the true model parameter on the base of $n = 11$ data points.

This set-up also allows us to explore the question of how many data points are needed to obtain reliable inference results. Figure 7.4 shows the posterior distributions for ξ and b for the time interval [0,50] if only $n = 3$ data points at $t = 0;25;50$\$ (top row) and $n = 2$ data points at $t = 0;50$ (bottom row) are available. Not surprisingly the ranges of consistent parameter values are broadened and therefore more mixtures of learning processes are consistent with the sparse data. Especially if only two data points are available we cannot make conclusive inferences about the role of conformist social learning. The range [0,0.08] of the parameter space for b is consistent with the data which means that we cannot exclude the presence of a weak or medium-strength conformist bias in the population. Nevertheless based on only two data points, we can firstly reject the hypothesis of the existence of a strong conformist tendency and secondly establish that the fraction of asocial learner is between 4 and 18 %. Further, the joint posterior distribution (Fig. 7.4f) shows that a stronger tendency for conformist social learning is accompanied by a larger fraction of asocial learning in the population (see shape of the light region). In other words, increasing the strength of conformist social learning b and the fraction of asocial learning ξ in the populations simultaneously can lead to very similar frequencies of the three present cultural variants at $t = 50$.

7.4 Conclusion

Model-based inference approaches have already started to transform the field of population genetics, by allowing inference on increasingly detailed generative genetic models of human evolutionary history using observed genomic data. Acknowledging that fundamental questions in cultural evolution and population genetics are of similar nature, namely which underlying evolutionary mechanisms could have given rise to observed frequency distributions, we explored in this chapter the applicability of this inference technique to cultural data. The key to success of such a framework is the development of a generative model capturing the main demographic and cultural dynamics of the considered cultural system. This means that properties like population structure, changes in population size or patterns of environmental change are input to the model and need to be appropriately chosen. Here we considered a competition framework modelling frequency changes of k different variants of a cultural

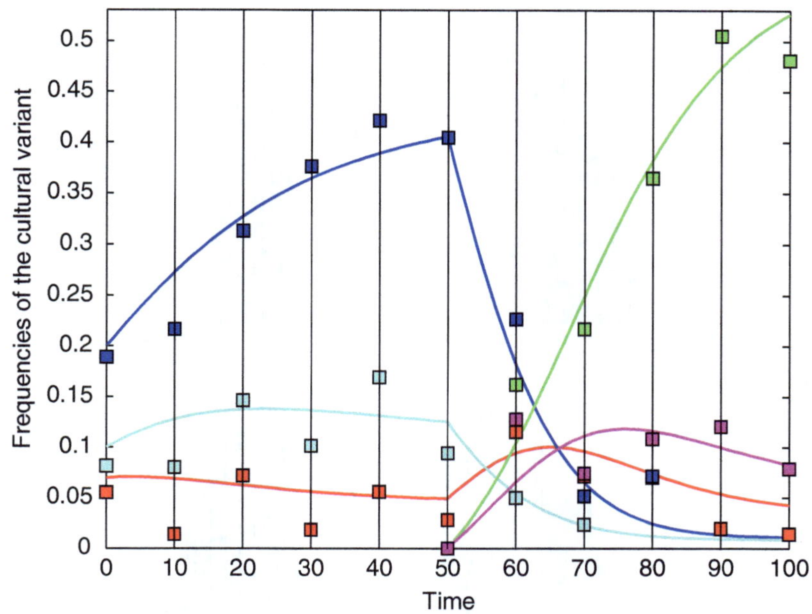

Fig. 7.2 Time course of the frequencies of the different variants

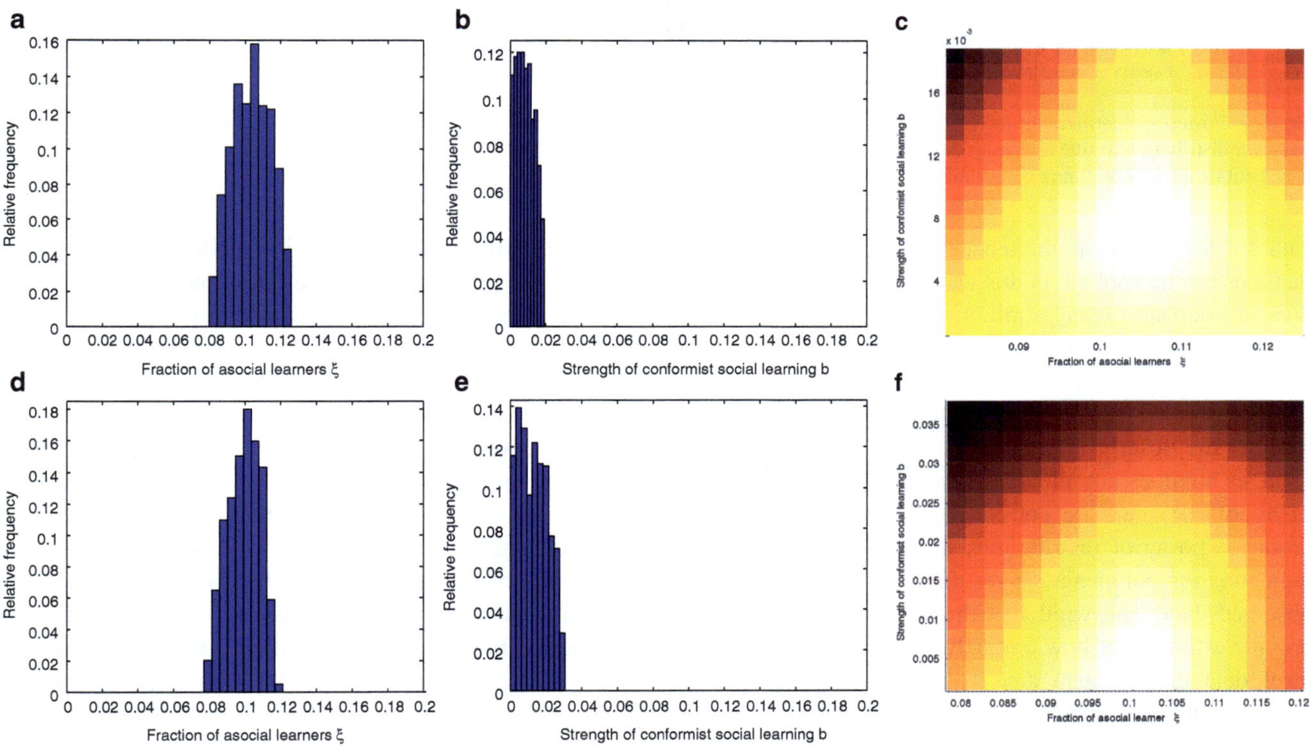

Fig. 7.3 Posterior distributions for ξ, b and the corresponding joint posterior distribution for the interval [0,50] (**a–c**), posterior distributions for ξ, b and the corresponding joint posterior distribution for the interval [50,10] (**d–f**). In all cases $n = 11$ data points are sampled and the true parameter values are $\overline{\xi} = 0.1$ and $\overline{b} = 0$

trait in the face of asocial and social learning in the form of direct biased and conformist learning in a spatially and temporally variable environment. We note that this chosen framework is only an example of a generative model; other approaches such as simulation or network-based frameworks can be used in exactly the same way (see Crema et al. 2014 for an application of a simulation based inference framework applied to archaeological data). The crucial point is that the generative model firstly produces data to which observed data can be compared to and secondly establishes a causal relationship between the processes under investigation (in our case the employed learning process) and the generated

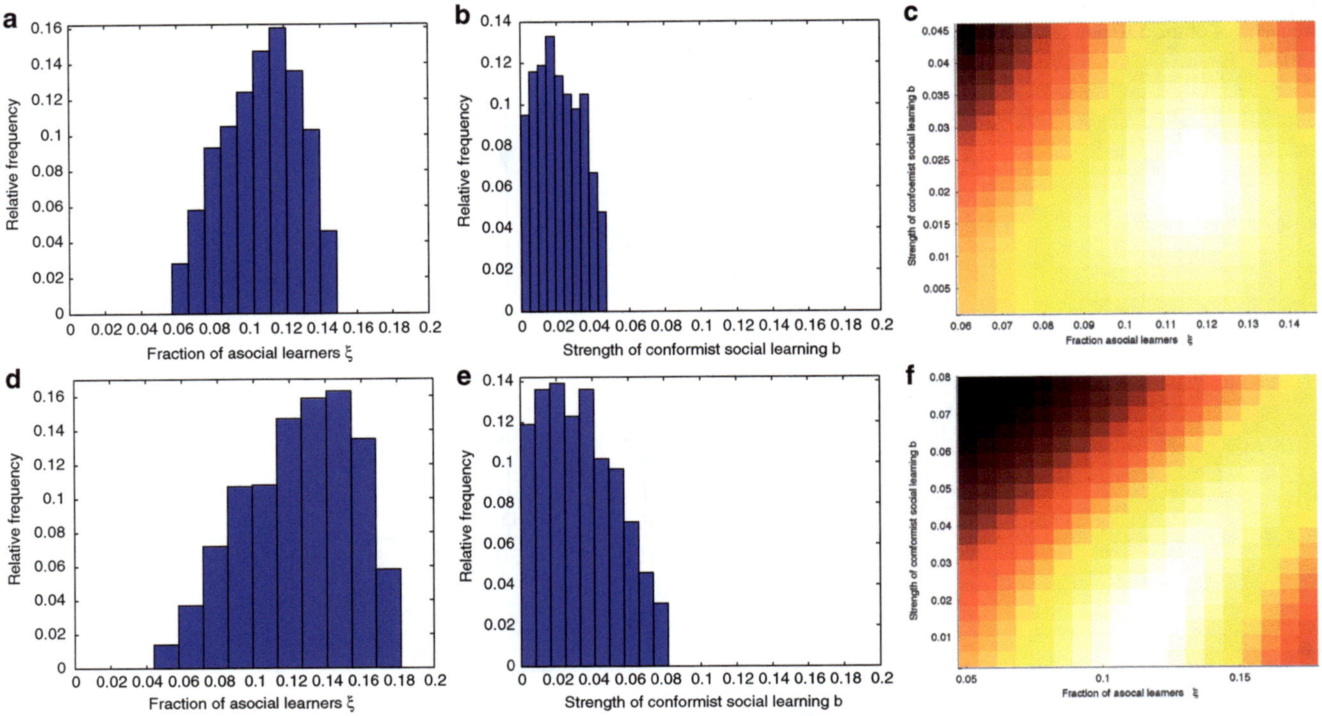

Fig. 7.4 Posterior distributions for ξ, b and the corresponding joint posterior distribution for the interval [0,50] when $n = 3$ (**a–c**), posterior distributions for ξ, b and the corresponding joint posterior distribution for the interval [0,50] when $n = 2$ (**d–f**). The true parameter values are $\overline{\xi} = 0.1$ and $\overline{b} = 0$

data so that statistical inference procedures such as ABC methods can be applied. In this way the suggested procedure does not depend on any equilibrium assumption and infers conclusions from a comparison of theoretical and observed patterns of frequency change.

While conventional hypothesis testing is able to establish whether the observed data are consistent with a single hypothesis but does not allow for conclusions about the range of hypotheses that are consistent, the ABC methodology produces posterior distributions for the model parameters indicating the parameter ranges that are consistent with the observed data. The widths of those posterior distributions provide estimates of how uniquely the observed population-level data can be generated from a specifically defined mixture of learning strategies. In other words, this framework should help to address the long-standing question of how much information about the underlying evolutionary processes is really contained in observed population-level data. While narrow posterior distributions for the model parameter describing the strength of the different learning processes indicate that the observed patterns of frequency change can only be produced by a small number of hypothesis wide distributions do not allow for informative conclusions about underlying learning processes based on the considered data *and* the generative model. We expect the developed inference framework to be valuable in helping to narrow down the range of possible hypotheses that could have produced observed data and therefore to be instructive especially in the face of sparse data.

Additionally, careful application of an ABC model selection framework can also potentially allow robust comparison between structurally different hypotheses. Under the Bayesian paradigm competing models can be evaluated on the basis of Bayes factors. While implicitly penalising over-parameterized complex models these factors serve as a summary of the evidence provided by the data in favour of a specific cultural hypothesis out of the possibilities considered. Early applications of this idea in connection with ABC (e.g. Pritchard et al. 1999) used the notion that Bayes factors can be empirically approximated by calculating the posterior probabilities of the competing models, i, $P\left(\theta^i \middle| D\right)$. In practice, this is achieved by assigning each competing hypothesis equal prior probability and then, given a fixed tolerance or error margin ε, approximating model posterior probabilities by their relative acceptance rates in the rejection algorithm (Toni and Stumpf 2010). In our example application we would distinguish between two models: our original model (7.1) accounting for asocial learning and social learning in the form of direct biased and conformist learning anda less complex model which does not allow for a conformist tendency in the population (This is achieved by setting $b = 0$ in model (7.1).). As expected we find that the less complex model has a higher Bayes factor as the original data were generated under the assumption of no conformity in the population.

In the application presented in Sect. 7.3.2.3 we compared theoretical and observed frequency data D directly in the SMC ABC algorithm, using the squared distance between the data points. But in most real applications using the data directly might not be possible, and then the ABC framework relies on reducing the full data D to a set of summary statistics S such that the approximated posterior distribution $P\left(\theta \middle| S \leq \varepsilon\right)$ is adequately representative of $P\left(\theta \middle| D \leq \varepsilon\right)$. In statistical terms this requires S to be 'sufficient' for D, such that no information is lost in the dimension reduction. In practice this is almost impossible to guarantee, so choosing S can be difficult. A number of different heuristic schemes to guide the choice of S have been proposed, such that S 'approaches sufficiency' (e.g. Joyce and Marjoram 2008) or choice is otherwise optimized (e.g. Nunes and Balding 2010) or automated (e.g. Fearnhead and Prangle 2012). In addition, the model selection approach described above can become problematic when the Bayes factors are approximated on the basis of summary statistics S rather than the full data D (e.g. Robert et al. 2011), as even if S is sufficient for each model it may not be sufficient to compare between competing models. This is currently a topic of intense research in both population genetics and statistics in general, and regular improvements allow these limitation to be somewhat mitigated in careful analyses. Notwithstanding these limitations, the ABC family of techniques has in just a short time permitted a revolutionary change in inferential power using increasingly complex and realistic models, especially those in population genetics. However, we stress that the accuracy of the obtained inference depends crucially on the appropriate description of the considered cultural or genetic system by the generative model. Naturally this points to the importance of the understanding of the nature and complexity of the learning strategies employed by human populations.

In order to infer which learning strategies human populations really use we need to base inference on available empirical evidence. Large individual-level datasets on social learning are relatively rare, outside of experimental conditions. However, two recent examples have drawn on detailed historical records of the board game Go (Beheim et al. 2014) or contemporary ethnographic data from Fiji (Henrich and Broesch 2011) to directly infer the relative importance of different learning-biases across and how this may impact population-level cultural dynamics. While both studies clearly demonstrate the advantages of such high-resolution data, in most pre-modern contexts (including the palaeolithic period) we are necessarily limited to aggregate population-level data. The framework we have outlined here successfully demonstrates that we can often still use such sparse data to make inferences on the relative weights of various underlying individual-level learning processes. Importantly, the framework also indicates when the available population-level data do not carry a strong signature about underlying learning processes and therefore when the analysis of this data will not provide meaningful information. The ABC inference framework offers an alternative way of studying cultural phenomena, independent of any equilibrium or optimality assumptions and is highly appropriate for application to typical archaeological datasets (see also Crema et al. 2014). In particular, this approach provides a useful bridge between the important on-going theoretical work in cultural evolution and the continually growing empirical data on human evolutionary history.

Acknowledgement We thank Kenichi Aoki and Jeremy Kendal for their constructive comments which helped improving this chapter.

A.1 Appendix

In this chapter we assume that asocial and social learning strategies cause the cultural variants to change in frequency and describe those changes using a reaction–diffusion framework of the form

$$\begin{aligned} \frac{\partial u_i}{\partial t}(t,x) &= d\Delta u_i(t,x) - \nu u_i(t,x) + \xi\,(\text{asocial learning}) \\ &\quad + (1-\xi)\,(\text{social learning}),\ i = 1,\ldots,k \\ \frac{\partial K}{\partial t}(t,x) &= d\Delta K(t,x) + (\lambda - \nu) \\ &\quad \times K(t,x)\,(1 - K(t,x)). \end{aligned} \tag{7.2}$$

Thereby the variable u_i describes the frequency of variant i at time t in the population, or in other words the fraction of the population that has adopted variant i. The variable K denotes the population size at location x and in the following we assume the population size to be the same for all locations x. It follows from the second equation in model (7.2) that $K(t,x) \leq 1,\ \forall t$ and further, it holds $\sum_{i=1}^{k} u_i(t,x) \leq K(t,x)$. The diffusion coefficient d describes the scale of spatial interactions, λ and ν the birth and death rates, respectively and ξ the reliance of the population on asocial learning. For sake of simplicity we stated the non-spatial version in the main text. All dynamics describes below hold in a similar way for this model.

In more detail, learning in various forms can increase or decrease the frequency of variant i. Asocial learning is based on the judgement about the benefit of specific variants in observed environmental conditions and consequently has two error sources: misjudgement of the current environmental condition and misjudgement of the adaptation levels of the different variants. Despite the conceptual differences both error sources lead to the same outcome in the modelling framework: a variant i is chosen for which holds $\mu_i \neq e$. Therefore the inaccuracy of asocial learning is modelled by assuming that asocial learning is based on $\overline{e} = e + \omega$ with $\omega \sim \mathcal{N}\left(0, \sigma_\omega^2\right)$. However, besides being error-prone asocial learning can lead to the introduction of new variants and its dynamic is modelled by (for sake of shortness we write $\overline{e} = \overline{e}(t, x)$)

$$P_i\left(a_i\left(\overline{e}\right)\right)\left(K(t,x) - \sum_{j=1}^{k} u_j\right) + \sum_{j=1, j\neq i}^{k}\left(P_{ji}\left(a_i\left(\overline{e}\right), a_j\left(\overline{e}\right)\right) u_j(t,x) - P_{ij}\left(a_i\left(\overline{e}\right), a_j\left(\overline{e}\right)\right) u_i(t,x)\right) \tag{7.3}$$

The parameter P_i describes the rate at which the fraction of the population which has not yet adopted a variants (described by the difference between the population size $K(t,x)$ at time t and the sum of the fractions of the population which have adopted one of the k variants, $K(t,x) - \sum_{j=1}^{k} u_j(t,x)$) learns variant i asocially. P_i depends on the adaption level $a_i(\bar{e})$ meaning that asocial learning is not completely random: the higher the adaptation level in the estimated environment $\bar{e}$ the higher is the adoption rate. Further, we allowed for the switching of variants which describes the process that individuals who already have adopted a cultural variant can switch to adopting a different variant. The coefficient P_{ij} models the rate at which the fraction of the population which has adopted variant i switches to variant j due to the evaluation of environmental cues. Again it holds the larger the difference $a_j\left(\overline{e}\right) - a_i\left(\overline{e}\right)$ between the estimated adaption levels the higher is the switching rate. Contrary to asocial learning, social learning is based on social cues and therefore can only lead to learning of variants which are already present in the considered location. In the considered framework we only considered two social learning strategies: direct biased social learning and conformist social learning. Direct biased social learning is modelled by (for sake of shortness we write $e = e(t,x)$)

$$r_i\left(a_i(e)\right) u_i(t,x)\left(1 - \frac{u_i(t,x)}{K(t,x) - \sum_{j=1, j\neq i}^{k} u_j(t,x)}\right) + \sum_{j=1, j\neq i}^{k}\left(c_{ji}\left(a_i(e), a_j(e)\right) - c_{ij}\left(a_i(e), a_j(e)\right)\right) u_i(t,x) u_j(t,x). \tag{7.4}$$

Similarly to the dynamic of asocial learning the first term

$$r_i\left(a_i(e)\right) u_i(t,x)\left(1 - \frac{u_i(t,x)}{K(t,x) - \sum_{j=1, j\neq i}^{k} u_j(t,x)}\right)$$

models the adoption of variant i by the population which has not adopted any variants yet. However contrary to asocial learning, this term is frequency-dependent. It is a logistic growth process with adoption rate (or intrinsic rate of increase) r_i and broadly speaking describes cultural reproduction. Per definition, the population size $K(t,x)$ at location x is the upper limit of the total fraction of adopters in the population at this location (given by $\sum_{j=1}^{k} u_j(t,x)$), regardless of the adopted variant. Consequently, the upper limit for the fraction of the population that has adopted variants i is given by $K(t,x) = \sum_{j=1}^{k} u_j(t,x)$ (i.e. we assume that our cultural variants compete for a common pool of adopters). The adoption rate r_i is assumed to be proportional to the adaptation level a_i in the currently experienced environmental condition e. It holds: The higher the adaptation level the higher is the adoption rate. The second term

$$\sum_{j=1, j\neq i}^{k}\left(c_{ji}\left(a_i(e), a_j(e)\right) - c_{ij} a_i(e), a_j(e)\right) u_i(t,x) u_j(t,x)$$

describes the switching dynamic between the fractions of the population which has already adopted a variant. Again we assumed that individuals who have already adopted a variant have the chance to switch to another variant and therefore the different cultural variants compete with each other for use. These interactions between the variants are described by the terms $c_{ij}(a_i(e), a_j(e)) u_i(t,x) u_j(t,x)$ which model the switch process from variant i to variant j. The strength of this process is determined by the rate c_{ij} and it holds: The higher the difference $a_j(e) - a_i(e)$ of the adaptation levels of both variants the higher is the switching rate. In order to include conformist social learning we allowed these model parameters to be frequency-dependent. We assumed

$$\tilde{r}_i = (1-b)\, r_i\, (a_i(e)) + b\, (u_i\,(t,x) - c_b K\,(t,x)) \text{ and}$$
$$\tilde{c}_{ij} = \left[(1-b)\, c_{ij}\, (a_i(e), a_j(e)) + b\, (u_j\,(t,x) - c_b K\,(t,x))\right]^+$$

where b controls the reliance on adaptation information and frequency information, respectively. For $b = 0$ we obtain direct biased learning while $b > 0$ supports variants with a frequency higher than the commonness threshold $c_b K(t,x)$. In this case the difference $(u_i\,(t,x) - c_b K\,(t,x))$ is positive and the adoption rate $\tilde{r}_i$ is increased. Contrary if $(u_i\,(t,x) - c_b K\,(t,x))$ is negative (and therefore variant i has a relatively small frequency) the adoption rate $\tilde{r}_i$ is decreased. A similar dynamic applies to the switching rate $\tilde{c}_{ij}$. If the frequency of variant j (the target of the switch process) exceeds the commonness threshold $c_b K(t,x)$ then the rate $\tilde{c}_{ij}$ with which variant i is substituted by variant j is increased. The symbol $[.]^+$ denotes the positive part of any real number (e.g. $[3.4]^+ = 3.4$ but $[-3.4]^+ = 0$) ensures that there is no reversal of the switch direction.

We note that when considering a single cultural variant the dynamic of asocial learning (7.3) results in r-shaped adoption curve while the dynamic of asocial learning (7.4) results in a S-shaped curve whereby the existence of a conformist tendency ($b > 0$) produces long tails at the beginning and an accelerated adoption behaviour when the commonness threshold is exceeded. System (7.2) can be solved using the Finite-Element method and we obtain the time course of the frequencies u_i. of each cultural variant that are expected under the assumed learning hypothesis and environmental change.

References

Acerbi A, Bentley RA (2014) Biases in cultural transmission shape the turnover of popular traits. Evol Hum Behav 35:228–236

Akaike H (1974) A new look at the statistical model identification. IEEE Trans Autom Control 19:716–723

Aoki K, Wakano JY, Feldman MW (2005) The emergence of social learning in a temporally changing environment: a theoretical model. Curr Anthropol 46:334–340

Bass FM (1969) A new product growth model for consumer durables. Manag Sci 15:215–227

Baum WM, Richerson PJ, Efferson CM, Paciotti BM (2004) Cultural evolution in laboratory microsocieties including traditions of rule giving and rule following. Evol Hum Behav 25: 305–326

Beaumont MA (2010) Approximate Bayesian computation in evolution and ecology. Annu Rev Ecol Evol Syst 41:379–406

Beaumont MA, Zhang W, Balding DJ (2002) Approximate Bayesian computation in population genetics. Genetics 162(4):2025–2035

Beaumont MA, Nielsen R, Robert C, Hey J, Gaggiotti O, Knowles L, Estoup A, Panchal M, Corander J, Hickerson M, Sisson SA, Fagundes N, Chikhi L, Beerli P, Vitalis R, Cornuet J-M, Huelsenbeck J, Foll M, Yang Z, Rousset F, Balding D, Excoffier L (2010) In defence of model-based inference in phylogeography. Mol Ecol 19:436–446

Beheim BA, Thigpen C, McElreath R (2014) Strategic social learning and the population dynamics of human behavior: the game of Go. Evol Hum Behav 35:351–357. doi:10.1016/j.evolhumbehav.2014.04.001

Bentley RA, Shennan SJ (2003) Cultural transmission and stochastic network growth. Am Antiq 68:459–485

Bentley RA, Hahn MW, Shennan SJ (2004) Random drift and culture change. Proc R Soc B 271:1443–1450

Bentley RA, Lipo CP, Herzog HA, Hahn MW (2007) Regular rates of popular culture change reflect random copying. Evol Hum Behav 28:151–158

Boyd R, Richerson PJ (1985) Culture and the evolutionary process. The University of Chicago Press, Chicago

Boyd R, Richerson PJ (1988) An evolutionary model of social learning: the effects of spatial and temporal variation. In: Zentall T, Galef BG Jr (eds) Social learning. Erlbaum, Hillsdale, pp 29–48

Burnham K, Anderson D (2002) Model selection and multi-model inference: a practical information-theoretic approach. Springer, Berlin

Caldwell CA, Millen AE (2008) Experimental models for testing hypotheses about cumulative cultural evolution. Evol Hum Behav 29:165–171

Cavalli-Sforza L, Feldman MW (1981) Cultural transmission and evolution: a quantitative approach. Princeton University Press, Princeton

Clauset A, Shalizi CR, Newman MEJ (2009) Power-law distributions in empirical data. Soc Ind Appl Math Rev 51(4):661–703

Coultas JC (2004) When in Rome ... an evolutionary perspective on conformity. Group Process Intergroup Relations 7(4):317–331

Crema ER, Edinborough K, Kerig T, Shennan SJ (2014) An approximate Bayesian computation approach for inferring patterns of cultural evolutionary change. J Archaeol Sci 50:160–170

Currat M, Ray N, Excoffier L (2004) Splatche: a program to simulate genetic diversity taking into account environmental heterogeneity. Mol Ecol Notes 4(1):139–142

Del Moral P, Doucet A, Jasra A (2012) An adaptive sequential Monte Carlo method for approximate Bayesian computation. Stat Comput 22(5):1009–1020

Efferson C, Lalive R, Richerson PJ, McElreath R, Lubell M (2008) Conformists and mavericks: the empirics of frequency-dependent cultural transmission. Evol Hum Behav 29:56–64

Epstein JM (2007) Generative social science: studies in agent-based computational modeling. Princeton University Press, Princeton

Epstein JM, Axtell RL (1996) Growing artificial societies: social science from the bottom up. MIT and Brookings Institution, Washington, DC

Eriksson K, Enquist M, Ghirlanda S (2007) Critical points in current theory of conformist social learning. J Evol Psychol 5(1–4): 67–87

Eriksson A, Betti L, Friend AD, Lycett SJ, Singarayer JS, von Cramon-Taubadel N, Valdes PJ, Balloux F, Manica A (2012) Late Pleistocene climate change and the global expansion of anatomically modern humans. Proc Natl Acad Sci U S A 109:16089–16094

Ewens WJ (2004) Mathematical population genetics, 2nd edn. Springer, New York

Fearnhead P, Prangle D (2012) Constructing summary statistics for approximate Bayesian computation: semi-automatic approximate Bayesian computation. J R Stat Soc Ser B 74:419–474

Feldman MW, Aoki K, Kumm J (1996) Individual versus social learning: evolutionary analysis in a fluctuating environment. Anthropol Sci 104:209–232

Fisher RA (1930) The genetical theory of natural selection. Clarendon, Oxford

Frank SA (2009) The common patterns of nature. J Evol Biol 22:1563–1585

Franz M, Nunn CL (2009) Network-based diffusion analysis: a new method for detecting social learning. Proc R Soc B 276(1663):1829–1836

Giraldeau L-A, Valone TJ, Templeton JJ (2002) Potential disadvantages of using socially acquired information. Philos Trans R Soc Lond B 357:1559–1566. doi:10.1098/rstb.2002.1065

Green RE, Krause J, Briggs AW et al (2010) A draft sequence of the Neandertal genome. Science 328:710–722

Gutenberg B, Richter CF (1944) Frequency of earthquakes in California. Bull Seismol Soc Am 34:185–188

Hahn MW, Bentley RA (2003) Drift as a mechanism for cultural change: an example from baby names. Proc R Soc B 270:120–123

Hellenthal G, Busby GBJ, Band G, Wilson JF, Capelli C, Falush D, Myers S (2014) A genetic atlas of human admixture history. Science 343(6172):747–751

Henrich J (2001) Cultural transmission and the diffusion of innovations: adoption dynamic indicate that biased cultural transmission is the predominant force in behavioural change. Am Anthropol 103(4):992–1013

Henrich J, Boyd R (1998) The evolution of conformist transmission and the emergence of between-group differences. Evol Hum Behav 19:215–241

Henrich J, Broesch J (2011) On the nature of cultural transmission networks: evidence from Fijian villages for adaptive learning biases. Philos Trans R Soc B 366:1139–1148

Herzog HA, Bentley RA, Hahn MW (2004) Random drift and large shifts in popularity of dog breeds. Proc R Soc B 271:1443–1450

Heyes CM (1994) Social learning in animals: categories and mechanisms. Biol Rev 69:207–231

Hoppitt WJE, Laland KN (2013) Social learning: an introduction to mechanisms, methods, and models. Princeton University Press, Princeton

Hoppitt WJE, Boogert NJ, Laland KN (2010a) Detecting social transmission in networks. J Theor Biol 263(4):544–555

Hoppitt WJE, Kandler A, Kendal JR, Laland KN (2010b) The effect of task structure on diffusion dynamics: implications for diffusion curve and network-based analyses. Learn Behav 38(3):243–251

Hudson RR (1983) Properties of a neutral allele model with intragenic recombination. Theor Popul Biol 23:183–201

Itan Y, Powell A, Beaumont MA, Burger J, Thomas MG (2009) The origins of lactase persistence in Europe. PLoS Comput Biol 5:e1000491

Joyce P, Marjoram P (2008) Approximately sufficient statistics and Bayesian computation. Stat Appl Genet Mol Biol 7:26

Kandler A, Laland KN (2013) Tradeoffs between the strength of conformity and number of conformists in variable environments. J Theor Biol 332:191–202

Kandler A, Steele J (2010) Social learning, economic inequality and innovation diffusion. In: O'Brian MJ, Shennan S (eds) Innovation in cultural systems. The MIT Press, Cambridge, MA, pp 193–214

Kendal JR, Kendal RL, Laland KN (2007) Quantifying and modelling social learning processes in monkey populations. Int J Psychol Psychol Ther 7(2):123–138

Kendal JR, Giraldeau L-A, Laland KN (2009) The evolution of social learning rules: payoff-biased and frequency-dependent biased transmission. J Theor Biol 260:210–219

Kimura M, Crow JF (1964) The number of alleles that can be maintained in a finite population. Genetics 49:725–738

Kingman JFC (1982) The coalescent. Stochastic Process Appl 13:235–248

Kirby S, Cornish H, Smith K (2008) Cumulative cultural evolution in the laboratory: an experimental approach to the origins of structure in human language. Proc Natl Acad Sci U S A 105(31):10681–10686

Kohler TA, Van Buskirk S, Ruscavage-Barz S (2004) Vessels and villages: evidence for conformist transmission in early village aggregations on the Pajarito Plateau, New Mexico. J Anthropol Archaeol 23:100–118

Laland KN (2004) Social learning strategies. Learn Behav 32(1):4–14

Laland KN, Richerson PJ, Boyd R (1996) Developing a theory of animal social learning. In: Heyes CM, Galef BG Jr (eds) Social learning in animals: the roots of culture. Academic Press, London, pp 129–154

Laland KN, Odling-Smee FJ, Myles S (2010) How culture shaped the human genome: bringing genetics and the human sciences together. Nat Rev Genet 11:137–148

Mahajan V, Muller E, Bass FM (1995) Diffusion of new products: empirical generalizations and managerial uses. Mark Sci 14(3):79–88

Malécot G (1948) The mathematics of heredity (trans: Yermanos DM 1969). WH Freeman, San Francisco

Marjoram P, Wall JD (2006) Fast "coalescent" simulation. BMC Genet 7:16

Maynard Smith J (1978) Optimization theory in evolution. Annu Rev Ecol Syst 9:31–56

McElreath R, Bell AV, Efferson C, Lubell M, Richerson PJ, Waring T (2008) Beyond existence and aiming outside the laboratory: estimating frequency-dependent and pay-off-biased social learning strategies. Philos Trans R Soc B 363:3515–3528

McElreath R, Fasolo B, Wallin A (2011) The evolutionary rationality of social learning. In: Hertwig R, Hoffrage U (eds) Simple heuristics in a social world. Oxford University Press, New York

McVean GAT, Cardin NJ (2005) Approximating the coalescent with recombination. Philos Trans R Soc B 360:1387–1393

Mesoudi A, Lycett SJ (2009) Random copying, frequency-dependent copying and culture change. Evol Hum Behav 30(1):41–48

Mesoudi A, O'Brien MJ (2008) The cultural transmission of great basin projectile-point technology I: an experimental simulation. Am Antiq 73(1):3–28

Morgan TJH, Rendell L, Ehn M, Hoppitt WJE, Laland KN (2012) The evolutionary basis of human social learning. Proc R Soc B 279:653–662

Nakahashi W (2007) The evolution of conformist transmission in social learning when the environment changes periodically. Theor Popul Biol 72:52–66

Nakahashi W, Wakano JY, Henrich J (2012) Adaptive social learning strategies in temporally and spatially varying environments. Hum Nat 23:386–418

Neiman FD (1995) Stylistic variation in evolutionary perspective: inferences from decorative diversity and interassemblage distance in Illinois woodland ceramic assemblages. Am Antiq 60:7–36

Nielsen R, Beaumont MA (2009) Statistical inferences in phylogeography. Mol Ecol 18(6):1034–1047

Nunes MA, Balding DJ (2010) On optimal selection of summary statistics for approximate Bayesian computation. Stat Appl Genet Mol Biol 9:34

Powell A, Shennan SJ, Thomas MG (in prep) The power of power-laws in cultural evolution

Pritchard JK, Seielstad MT, Perez-Lezaun A, Feldman MW (1999) Population growth of human Y chromosomes: a study of Y chromosome microsatellites. Mol Biol Evol 16(12): 1791–1798

Rasteiro R, Bouttier P-A, Sousa V, Chikhi L (2012) Investigating sex-biased migration during the Neolithic transition in Europe. Proc R Soc B 279:2409–2416

Reader SM (2004) Distinguishing social and asocial learning using diffusion dynamics. Anim Learn Behav 32(1):90–104

Rendell L, Boyd R, Cowden D, Enquist M, Eriksson K, Feldman MW, Fogarty L, Ghirlanda S, Lillicrap T, Laland KN (2010) Why copy others? Insights from the social learning strategies tournament. Science 328(5975):208–213

Rendell L, Fogarty L, Hoppitt WJE, Morgan TJH, Webster MM, Laland KN (2011) Cognitive culture: theoretical and empirical insights into social learning strategies. Trends Cogn Sci 15(2):68–76

Robert CP, Corneut J-M, Marin J-M, Pillai NS (2011) Lack of confidence in approximate Bayesian computation model choice. Proc Natl Acad Sci U S A 108:15112–15117

Rogers AR (1988) Does biology constrain culture? Am Anthropol 90:819–831

Rogers EM (2003) Diffusion of innovations, 5th edn. Free Press, New York

Shennan SJ, Wilkinson JR (2001) Ceramic style change and neutral evolution: a case study from Neolithic Europe. Am Antiq 66:577–594

Templeton AR (2009) Statistical hypothesis testing in intraspecific phylogeography: nested clade phylogeographical analysis vs. approximate Bayesian computation. Mol Ecol 18:319–331

Toni T, Stumpf MPF (2010) Simulation-based model selection for dynamical systems in systems and population biology. Bioinformatics 26(1):104–110

Toni T, Welch D, Strelkowa N, Ipsen A, Stumpf MPH (2009) Approximate Bayesian computation scheme for parameter inference and model selection in dynamical systems. J R Soc Interface 6:187–202

Van der Bulte C, Stremersch S (2004) Social contagion and income heterogeneity in new product diffusion: a meta-analytic test. Mark Sci 23(4):530–544

Veeramah KR, Wegmann D, Woerner A, Mendez F, Watkins J, Destro-Bisol G, Soodyall H, Louie L, Hammer MF (2012) An early divergence of KhoeSan ancestors from those of other modern humans is supported by an ABC-based analysis of autosomal resequencing data. Mol Biol Evol 29:617–630

Wakano JY, Aoki K (2007) Do social learning and conformist bias coevolve? Henrich and Boyd revisited. Theor Popul Biol 72: 504–512

Wilde S, Timpson A, Kirsanow K, Kaiser E, Kayser M, Unterlaender M, Hollfelder N, Potekhina ID, Schier W, Thomas MG, Burger J (2014) Direct evidence for positive selection of skin, hair, and eye pigmentation in Europeans during the last 5,000y. Proc Natl Acad Sci U S A 111(13):4832–4837

Wright S (1931) Evolution in Mendelian populations. Genetics 16:97–159

Zipf GK (1929) Relative frequency as a determinant of phonetic change. Harv Stud Class Philol 15:1–95

Simulating Geographical Variation in Material Culture: Were Early Modern Humans in Europe Ethnically Structured?

8

Mirna Kovacevic, Stephen Shennan, Marian Vanhaeren, Francesco d'Errico, and Mark G. Thomas

Abstract

A high degree of structuring is seen in the spatial distribution of symbolic artefact types associated with the Aurignacian culture in Upper Palaeolithic Europe, particularly the degree of sharing of ornament types across archaeological sites. Multivariate analyses of these distributions have been interpreted as indicating ethno-linguistic differentiation (Vanhaeren and d'Errico 2006), although simpler explanations such as isolation-by-distance have not been formally discounted. In this study we have developed a spatiotemporally explicit cultural transmission simulation model that generates expectations of a range of spatial statistics describing the distribution of shared ornament types. We compare these simulated spatial statistics to those observed from archaeological data for Aurignacian Europe—using Approximate Bayesian Computation—in order to test and compare a range of hypotheses concerning group interaction dynamics for the period. Among the set of hypotheses examined, we include ones where material culture does or does not drive group interaction dynamics.

Keywords

Simulation modelling • Culture evolution • Palaeolithic • Demography • Ethnicity

8.1 Introduction

"The purpose of models is not to fit the data but to sharpen the questions." – Samuel Karlin, 11th R A Fisher Memorial Lecture, Royal Society April 1983

Over the last 30 years, there has been a movement from qualitative towards quantitative approaches to the study of

M. Kovacevic (✉)
CoMPLEX (Centre for Mathematics and Physics in the Life Sciences and Experimental Biology), University College London, Physics Building, Gower Street, London WC1E 6BT, UK

Research Department of Genetics, Evolution and Environment, University College London, Darwin Building, Gower Street, London WC1E 6BT, UK
e-mail: m.kovacevic@ucl.ac.uk

S. Shennan
Institute of Archaeology, University College London, 31-34 Gordon Square, London WC1H 0PY, UK

M. Vanhaeren
CNRS (Centre National de la Recherche Scientifique), UMR (Unité Mixte de Recherche) 5199, PACEA (De la Préhistoire à l'Actuel: Culture, Environement et Anthropologie), Université Bordeaux 1, Avenue des Facultés, 33405 Talence, France

F. d'Errico
CNRS (Centre National de la Recherche Scientifique), UMR (Unité Mixte de Recherche) 5199, PACEA (De la Préhistoire à l'Actuel: Culture, Environement et Anthropologie), Université Bordeaux 1, Avenue des Facultés, 33405 Talence, France

Department of Archaeology, History, Cultural Studies and Religion, University of Bergen, Øysteinsgate 3, 7805, 5020 Bergen, Norway

M.G. Thomas
Research Department of Genetics, Evolution and Environment, University College London, Darwin Building, Gower Street, London WC1E 6BT, UK

A. Mesoudi and K. Aoki (eds.) *Learning Strategies and Cultural Evolution during the Palaeolithic*, Replacement of Neanderthals by Modern Humans Series, DOI 10.1007/978-4-431-55363-2_8,

archaeological material culture, and a progression towards modelling approaches to understanding past processes. Archaeologists are now more widely postulating explicit hypotheses to explain the material culture records recovered from archaeological sites, and developing various methods to test these hypotheses. As a result, the fields of archaeology, anthropology and the social sciences in general have become increasingly systematic and multidisciplinary. In archaeology, there has been an increase in the application of computer simulation modelling and statistical techniques to study the relationship between cultural and demographic processes (Clark and Hagemeister 2007; Powell et al. 2009; Costopoulos and Lake 2010; Gerbault et al. 2014) in order to address longstanding archaeological and anthropological questions that are difficult to address through interpretation of archaeological data alone.

The evolution and spread of cultures have been studied using computational modelling methods, with particular focus on processes of cultural innovation and the transmission and accumulation of cultural traits (Neiman 1995; Shennan 2001; Henrich 2004; Powell et al. 2009). The formation of cultural boundaries has also been studied using a group of computational methods labelled agent based modelling (ABM). ABM has been applied throughout the social sciences to investigate how large-scale effects emerge as a result of interactions between agents in the system (Premo 2007; Powell et al. 2009) and for studies of hominin dispersal (Mithen and Reed 2002; Nikitas and Nikita 2005; Hughes et al. 2007). In particular, Robert Axelrod has used ABM methods to investigate the persistence of cultural heterogeneity as a result of interactions between individual agents that are dependent on the extent of cultural similarity between those agents (Axelrod 1997).

Simulation modelling of this kind is a powerful approach that allows the incorporation of stochasticity (variation in demographic and cultural processes arising from random events) into the models. Simulation modelling, and computational modelling in general, also allows researchers to account for sample sizes and the spatial distribution of sample sites, effectively incorporating sampling error and some archaeological bias in inferences on the past (Shennan et al. 2013; Gerbault et al. 2014). The use of modelling in archaeology has resulted in a better understanding of behaviours of agents within the complex systems modelled, as well as helping to refine the questions that are asked and hypotheses that are postulated. With such methods, archaeologists are able to develop robust frameworks that allow a qualitative comparison of alternative modelled scenarios with each other and with observed material culture records, in effect creating virtual experiments to test the effect of varying parameter values on the similarity between simulated and observed material culture data.

In addition to simulation modelling, statistical modelling methods are widely used to describe distributions of, and relationships between, archaeological variables; for example, regression modelling is used to infer correlations between variables of interest. As in many other disciplines, Bayesian methods in archaeology have surged in popularity in recent years. In brief, Bayesian inference is a branch of statistics that uses particular datasets to infer the probability that a proposed hypothesis, or a parameter value of that hypothesis, is true. In contrast to frequentist statistics, where the hypothesis is fixed and variation in outcomes (data) is explored, in Bayesian inference the data becomes fixed and some space of possible explanations (hypotheses) is explored. This means that Bayesian approaches are naturally well suited to archaeological inference since observed data from the past is fixed but only one of a number of possible outcomes of a set of stochastic processes of interest. In Bayesian approaches, various models with set numbers of parameters are proposed, and the posterior probability distributions of these parameters are inferred using information from prior probability distributions of the parameters and information provided by the observed data.

In archaeology, Bayesian methods are primarily associated with dating; for example, to integrate stratigraphic information with radiocarbon date estimates in order to calibrate the probability density distributions (Buck 2001). Other branches of Bayesian methods have not been extensively implemented in archaeological studies. Of particular interest in this paper is a family of Bayesian methods called Approximate Bayesian Computation (ABC) (Tavare et al. 1997; Fu and Li 1997; Beaumont et al. 2002; Bertorelle et al. 2010).

In ABC techniques, a large number of datasets are simulated under a model assuming different, randomly chosen, parameter values from within prior ranges, and appropriate summary statistics are used to measure the extent to which the simulated datasets emulate the observed data. Parameter values under which the model generates datasets closest to the observed data are retained and form a sample of the posterior probability distributions of the parameters. This approach allows the researcher to postulate a number of hypotheses and, provided that they are sufficiently well defined to allow data to be simulated, test which of these hypotheses are more likely given the observed data. An important advantage of ABC over traditional Bayesian approaches is that it is not necessary to formulate an exact function to calculate the probability of the data given some conditions (the likelihood function). The ABC framework and algorithms are further discussed in Appendix 1: Bayesian Inference and Approximate Bayesian Computation (ABC), Appendix 2: Approximate Bayesian Computation (ABC) Algorithm and Appendix 3: Summary Statistics.

In this paper we present a case study in which a spatiotemporally explicit cultural transmission simulation framework has been developed and integrated with observed material culture data (Upper Palaeolithic bead types identified as personal ornaments), using ABC, in order to aid the interpretation of quantitative data analyses on the observed material culture data (Vanhaeren and d'Errico 2006).

8.2 Case Study: Applying Simulation Modelling and ABC Methods

8.2.1 Introduction

The transition from the Middle Palaeolithic to the Upper Palaeolithic period in Europe occurred as early as approximately 44,000 years ago (Kuhn et al. 2001; Bar-Yosef 2002; Mellars 2005; Higham et al. 2012; Banks et al. 2013). This transition is widely seen as marking the appearance of modern human behaviour in Europe, as evidenced in the Upper Palaeolithic material culture by increased and consistent symbolic activity, and other technological and cultural advances (Powell et al. 2009). These changes in behavioural patterns appear in the archaeological record in the form of abstract and figurative art, the use of personal ornaments, systematically produced microlithic stone tools, bone, ivory and antler artefacts, and increasingly complex hunting technologies. The initial appearance of such items in the European territory dates to the beginnings of the Upper Palaeolithic transition and is thought to coincide with the appearance of AMH in Europe (Kuhn et al. 2001; Zilhão 2007).

The earliest evidence of anatomically modern humans in Europe remains a subject of debate, but is estimated to date to between approximately 45 Ka (Benazzi et al. 2011; Higham et al. 2011) and 40 Ka (Zilhão et al. 2007; Trinkaus and Zilhão 2012). Due to the lack of reliably dated Neanderthal fossils younger than approximately 40 Ka (Pinhasi et al. 2011), archaeological findings dating to 40 Ka or later are assumed to be mostly the result of activities of anatomically modern human populations. Little is known about the migration routes of the first anatomically modern human populations inhabiting Europe at the onset of the Upper Palaeolithic, the extent of biological, cultural and linguistic diversity among them, and the nature and extent of their interactions with the local Neanderthals (but see, for example, Prufer et al. (2014)).

Personal ornaments are considered to be among the first material objects used to communicate social and ethnic identity within and across cultural boundaries (Kuhn et al. 2001). In relation to ethnic identity, personal ornaments can therefore be considered to be the most diagnostic components of material culture surviving in the archaeological record. It has been argued that personal ornaments and beadwork can be used as a proxy for ethno-linguistic identity (Vanhaeren and d'Errico 2006), and that they offer archaeological advantages over other components of the material record for inferring ethno-linguistic structuring, including their exclusively symbolic function, and the frequency and wide assortment in which they occur at archaeological sites associated with the Upper Palaeolithic (Kuhn et al. 2001; Vanhaeren and d'Errico 2006).

In their study, Vanhaeren and d'Errico (2006) considered bead types, identified as personal ornaments, from European Aurignacian sites. Seriation and correspondence analyses of the data identified geographically non-randomly distributed clusters of sites sharing bead types. Seriation analysis is a relative dating method used to chronologically order artefacts recovered from different sites and belonging to the same culture. It is based on the relative chronological order of artefacts and is often applied when absolute dates are not available. Correspondence analysis is related to principal components analysis and is a method used to identify dimensions of variation in categorical data and rank them by the amount of variance explained. The authors argued that the observed variation in spatial distributions was not due to changes over time in personal ornament preference or local availability of raw materials, but rather represented cultural differences among the human groups using Aurignacian technologies. They further argued that the identified trends may have reflected ethno-linguistic diversity among Aurignacian populations.

While this is an interesting interpretation, simpler explanations of these results have not been formally discounted. There are many factors that could cause spatial variation or geographical structuring of material culture, including ethnicity, chronology, local availability of raw materials, environmental influences or simply isolation-by-distance and identity by descent. It is also important to distinguish between spatial variation and ethnic structuring, the latter referring to the ability of individuals, or groups of individuals, to consciously identify with a specific social group *"based on a particular locality or origin"* (Shennan 1989). Considering this definition, it is clear that drawing conclusions about ethnic identity and structure for prehistoric populations is difficult since there are no data in the material record relating to individual's conscious identification; the challenges of invoking ethnic structuring and reconstructing patterns of ethnicity through analysis of material culture data have been discussed by several authors (Shennan 1989, 2002; Jones 1997). With this in mind, invoking ethnic structure for the Upper Palaeolithic in Europe is a challenging task given the paucity of material culture and other data for the period.

However, ethnic identity and structuring are universals in the modern world and are therefore frequently assumed for peoples in the past. Identifying the earliest appearance of ethnicity is an issue of general importance for the history of human evolution that has implications for the emergence of languages, and may inform on the evolutionary dynamics of human populations, as well as the role of identity construction in people today.

The current study therefore aims to test whether the distribution of artefact types reported by Vanhaeren and d'Errico (2006) can be explained by a model of cultural identity-by-descent with modification and isolation-by-distance, or whether it is necessary to invoke cultural group interaction processes that would be expected if material culture was symbolically marking ethnic group identity. For example, an interaction between two culturally similar populations may result in sharing of cultural traits between the two, causing them to become overall more culturally similar, while an interaction between two culturally different populations might result in the two undergoing conflict, dependent on the extent of the cultural difference between them, and possibly the imposition of one culture on another. An analogous distinction is that between the existence of inter-group differences arising through cultural mutation and drift (the null model), of which actors are not consciously aware, versus that null model plus the intentional adherence to behavioral norms that imply identity and actively shape interaction processes, and through that, the spread, loss and mixing of culturally inherited traits.

In this study, spatiotemporally explicit cultural transmission simulation models that generate simulated material culture data under each of the scenarios described above have been developed and explored through simulation. The archaeological dataset published by Vanhaeren and d'Errico (2006) is used to assess the validity of each model. The underlying principle here is that conditions under which the simulated data is very similar to the observed archaeological data—as reflected in a range of spatial statistics describing the distribution of artefact types—are more likely to be true than conditions under which the simulated data is unlike the observed data. This assessment of the goodness-of-fit between the simulated and observed data is quantified using ABC.

8.2.2 Simulation Modelling

Each simulation is initialised at the onset of the Aurignacian period, approximately 42 Ka, and simulated forward in time to the end of the Aurignacian period, approximately 29 Ka (Higham et al. 2012). Each simulation spans a total of 13,000 years, or 520 generations assuming a 25 year generation time (Tremblay and Vezina 2000; Thomas et al. 2006). Since this may be an overestimate of the length of the Aurignacian period (Zilhão and Pettitt 2006), data is also collected when each simulation reaches 10,000 years, or 400 generations, though these results are not presented here. Each simulation includes a 1,000 year, or 40 generation, burn-in period at the start of the simulation during which no simulated data is collected, in order to allow for possible inaccuracies in initial locations of simulated groups.

8.2.2.1 Simulation World

The geographic region considered in this study is the range of latitudes and longitudes corresponding to the European territory. The longitude, ϕ, ranges from $-11°$ to $30°$, which, relative to modern day country boundaries, is approximately the area from the western Irish boundary to the western Russian boundary at the Urals. The latitude, λ, ranges from $35°$ to $60°$, which is approximately the area from the northern boundary of Africa to the northern boundary of Scotland. Although it would be possible to incorporate changes in sea levels through time by using available bathymetry data, dramatic geostatic rebound for northern latitudes makes it difficult to accurately estimate coastlines for northern Europe. For this reason, modern coastlines are currently used in simulations.

Within the defined region, each geographic location is assigned a local carrying capacity. The carrying capacity of a location determines the habitability, and therefore potential population density, of that location; a zero carrying capacity corresponds to an uninhabitable region, for example sea or ice covered land. In order to estimate these local carrying capacities for geographic locations in the modelled domain, two distinct environmental scenarios have been considered; each simulation is conditioned on only one of these two environmental scenarios.

The first is a simple scenario in which Europe is assumed to be a flat space. This corresponds to a distribution with all locations within the modelled domain having equal relative carrying capacity values.

In the second of the environmental scenarios, instead of treating Europe as a flat space, we have taken information on estimated population densities during the Aurignacian from Bocquet-Appel et al. (2005) to inform on carrying capacities for the modelled domain, shown in Fig. 8.1. Bocquet-Appel et al. used databases of archaeological sites corresponding to the Upper Palaeolithic period, together with simulated climatic variables and ethnography of hunter-gatherers, to estimate the distribution of hunter-gatherer populations in Upper Palaeolithic Europe.

The original estimate of population density, shown in Fig. 8.1 for the Aurignacian period, was not made available, so the distribution used in this study is approximated based on the original figure. Since we are concerned with distributions rather than exact numbers estimated in the original study (Bocquet-Appel et al. 2005), this estimate is normalised to give relative distributions. The normalised distribution is used in simulations as the relative carrying capacity value for each location.

In both scenarios, the potential, or target, population density for each location is calculated as the product of the relative carrying capacity value at that location and the G_{max} parameter, which specifies the total maximum number of groups that the modelled domain can sustain. We treat this as an unknown parameter and explore a range of values.

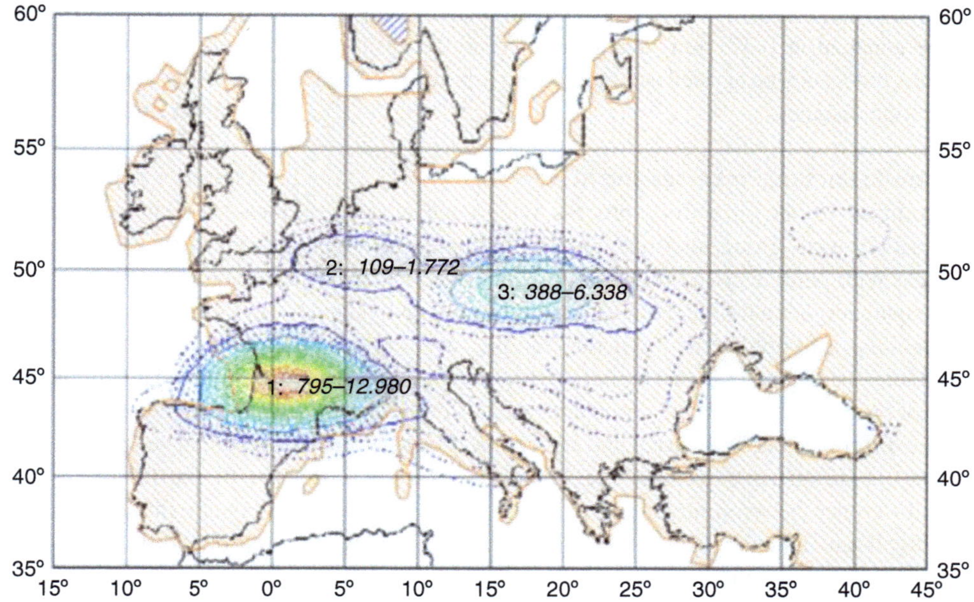

Fig. 8.1 Estimate of the regional distribution of the metapopulation of hunter-gatherers during the Aurignacian period of the Upper Palaeolithic in Europe, superimposed on the IOS3 project maps. The boundaries (*in black*) of the accretion zones, with the corresponding numbers, account for roughly 90 % of the distribution of the local population (Image and edited caption from (Bocquet-Appel et al. 2005)

8.2.2.2 Demographic Processes

Each simulation is initialised with a fixed number of groups, G_O, placed in randomly chosen habitable locations in the modelled domain; all attributes and processes are defined at the level of the group, rather than individuals in that group, and groups are assumed to be the same size. Groups migrate locally and undergo fission/extinction processes. These demographic processes are analogous to an isolation-by-distance model in population genetics (Wright 1943; Slatkin 1993).

Migratory Processes

At each generation groups are subjected to migratory processes modelled as parameterised Gaussian random walks. The distance each group traverses in a migration process is picked from a normal distribution with mean μ_{mig} and standard deviation $\sigma_{mig} = d_{mig}$. Positive and negative values picked from the distribution correspond to movement in opposite directions, namely East and West and North and South, respectively. The mean of the distribution is therefore set to $\mu_{mig} = 0$ to ensure that movement in opposite directions is equally likely. Parameter d_{mig} corresponds to the standard deviation, or width, of the normal distribution and specifies the range of values that the migration distance is most likely to take in each of the East-West and North-South directions. We treat d_{mig} as an unknown parameter and explore a range of values.

The distance travelled by each group at each generation in the East-West and North-South directions is picked independently from the above-described normal distribution. The distance, d, and direction, θ, that define each group's movement are given by:

$$d = \sqrt{\left(\Delta x\right)^2 + (\Delta y)^2}, \text{ and} \tag{8.1}$$

$$\theta = \arctan 2\,(\Delta x,\ \Delta y), \tag{8.2}$$

where *arctan2* corresponds to a variant of the *arctan* function that takes into account the sign of both vectors in question and distinguishes diametrically opposite directions, therefore specifying unique angle values in the range (0, 2π).

The new proposed position of each group is then calculated based on the group's current location, the distance, d, and the direction, θ, of movement. If the longitude and latitude of the group's current positions are $\phi_{current}$ and $\lambda_{current}$, respectively, and the same for the group's new locations are ϕ_{new} and λ_{new}, respectively, then:

$$\lambda_{new} = \sin^{-1}\left(\sin(\lambda_{current}) \cdot \cos\left(\frac{d}{R}\right) + \cos(\lambda_{current}) \cdot \sin\left(\frac{d}{R}\right) \cdot \theta\right), \tag{8.3}$$

$$\phi_{new} = \phi_{current} + \arctan 2\left(\cos\left(\frac{d}{R}\right) + \cos(\lambda_{current}) \cdot \sin\left(\frac{d}{R}\right) \cdot \theta,\ \cos\left(\frac{d}{R}\right) - \sin(\lambda_{current}) \cdot \sin(\lambda_{new})\right). \tag{8.4}$$

These formulae are introduced to allow for the curvature of the Earth when calculating the new group positions. Although the curvature of the Earth has little effect in the current framework, as the migration distance d is small relative to the radius of the Earth (denoted R and assumed to be constant at 6,371 km) using these formulae ensures that the model can be applied accurately with arbitrarily large migration distances.

Fission/Extinction Processes and KDE

In addition to the migratory process undergone at each generation, each group also undergoes a fission/extinction process with parameterised probability. The probability that a group undergoes a fission/extinction process is given by the probability of fission/extinction parameter, $p_{f/e}$; we treat this parameter as an unknown and explore a range of values. The type of process that a selected group undergoes is determined by the difference between target and current local population density at the group's location (i.e. fission or extinction). The difference between target and current local population density is an indicator of potential for growth; a positive value indicates that the location is below carrying capacity (i.e. the target local population density is greater than the current local population density—the location is under-populated and so there is potential for growth) and therefore results in a fission event, while a negative value indicates that the location is above carrying capacity (i.e. the target local population density is smaller than the current local population density—the location is over-populated and so there is no potential for growth) and therefore results in an extinction event. The population density at the current generation is estimated from the group locations using kernel density estimation (Wand and Jones 1995).

An extinction event results in the group being deleted from the simulation, while a fission event results in a replication such that two groups, the parent and offspring, are present in the next generation. The offspring group retains the cultural traits of the parent group (i.e. the offspring group is an exact replica of the parent group, except for any mutation events), analogous to identity-by-descent in population genetics. In subsequent generations, the parent and offspring groups migrate and undergo fission/extinction processes independently, and their respective cultures also evolve independently.

8.2.2.3 Cultural Processes: Modelling Ethnic Diversity

Axelrod's Model of Cultural Dissemination

The models developed in this study simulate innovation in culture and so require the concept of culture to be mathematically defined. For this purpose we have used an adapted version of Axelrod's definition (Axelrod 1997) in which the culture of an agent (an individual or a group of individuals) is defined to be a set of attributes that are subject to social influence. In Axelrod's definition, the culture of an agent consists of some number of these attributes, referred to as cultural *features*, and each can assume one of a predefined number of values, referred to as *traits*, thus, each agent is monomorphic for each cultural feature. In this definition, the culture of an agent is then described as a list of digits, with the position of a digit corresponding to the feature and the value of a digit specifying the current trait for that feature. In Axelrod's definition, the trait—or value that a feature takes—is assigned at the start of the simulation and is only influenced by social interactions (i.e. it does not undergo any mutation processes).

In Axelrod's formulation, social interactions are constrained to occur only between agents that are immediate neighbours. The simulations occur on a square lattice with agents arrayed at discrete points over the lattice. Most agents therefore have four immediate neighbours, with those on the edge of the lattice having three and those in the corners having two immediate neighbours. Also in Axelrod's model, the probability of an interaction between two agents is proportional to the cultural similarity between them. This similarity is quantified as the proportion of their features that have the same trait. The interaction then consists of an agent, and an immediate neighbour to that agent, being chosen at random. A single feature on which the chosen agent's culture and the neighbour's culture differ is selected at random, and the value of this feature (trait) in the chosen agent's culture is set to the value of the same feature in the neighbour's culture.

This formulation is a good basis; however, it is very limited in diversity of cultural features and traits and is inadequate to capture the high dimensionality of the observed data used in the current study (Vanhaeren and d'Errico 2006). In addition, over long chronological periods such as those simulated here, it is necessary to consider the effect of cultural mutation and drift processes. This definition must therefore be modified so that it can be applied to the current problem.

Observed and Simulated Datasets

The observed dataset (Vanhaeren and d'Errico 2006) consists of 157 distinct bead types recorded at 98 Aurignacian sites in Europe and the Near East, with records specifying presence/absence of distinct bead types in sites only. These distinct bead types are divided between 11 features according to different raw materials, with *"62 representing ornaments made of shells, 31 of teeth, 30 of ivory, 11 of stone, 11 of bone, 7 of deer antler, and one each of belemnite, nummulite, ammonite, sea urchin and amber"* (Vanhaeren and d'Errico 2006).

In the models developed in this study, we have adapted Axelrod's definition of culture described above so that each agent, in our case a group, is polymorphic for each cultural feature. To allow for this, each group carries a parameterised number of items, or beads, in its cultural repertoire, specified by the N_{items} parameter, treated as an unknown and chosen at the onset of each simulation from a pre-defined range of values. These items are then divided between the 11 features probabilistically (using a multinomial function), such that the probability of an item being assigned to a particular feature is proportional to the ratio of unique items observed in that feature and the total number of unique items observed (39.5 % shells, 19.7 % teeth, 19.1 % ivory, 7 % stone, 7 % bone, 4.5 % deer antler, and 0.6 % each for belemnite, nummulite, ammonite, sea urchin and amber). Within each feature, each item can then take one of a number of unique possible values, corresponding to the number of distinct bead types for that feature in the observed ornament data (Vanhaeren and d'Errico 2006).

Mutation and Drift

The culture of each group undergoes mutation and drift processes at each generation, such that the culture of each group at the next generation will be the combined result of mutation and drift processes acting on the culture of that group at the current generation. In addition to testing various environments as described above, two different models of cultural variation have been considered; in each simulation data is simulated under only one of these two cultural variation models.

In the first, mutation is modelled according to the bounded stepwise model often used to model mutations at microsatellite loci in population genetics (Kimura and Ohta 1978; Valdes et al. 1993), and occurs at each generation for each item in each group's culture with probability proportional to the p_{mut} parameter. We treat this parameter, which specifies the probability of mutation, as an unknown and explore a range of possible values. Under this stepwise mutation model, a cultural trait in a particular feature at the current generation is constrained to mutate to one of the cultural traits on either side of it, within that feature, at the next generation—mutation therefore changes the frequency with which each trait occurs in the next generation. In this case, we assume that cultural traits are ordered in such a way that adjacent traits are more similar than traits that are further apart in the sequence. Since cultural traits considered in this study are discrete and fixed (i.e. one trait cannot morph into another trait), this stepwise mutation model corresponds to a group being more likely to add an item to its cultural repertoire that is morphologically similar to one that is already present in its cultural repertoire than one that is very different. Similarly to population genetics, cultural mutation has the effect of increasing diversity.

In the second, mutation is discrete within the bounds of each feature. Similarly to the stepwise mutation model, in this bounded discrete model mutation occurs at each generation for each item in each group's culture with probability proportional to the p_{mut} parameter. This parameter again specifies the probability of mutation; it is treated as an unknown and a range of possible values are explored. Under this mutation model, however, a cultural trait in a particular feature at the current generation is permitted to mutate to any of the other cultural traits within that feature with equal probability at the next generation. The mutation process again changes the frequency with which each trait occurs in the next generation and has the effect of increasing diversity.

Drift has the opposite effect and decreases the amount of diversity in each group's culture. It is modelled based on genetic drift, where allele frequencies change as a result of random differences in reproduction; in finite populations drift corresponds to the intergeneration sampling error (see, for example, Tishkoff and Verrelli (2003)). The drift process is modelled by using a multinomial function to sample the traits of each cultural feature independently. This implementation takes into account frequencies of cultural traits in the current generation, such that, for a particular group, cultural traits that are at higher frequencies in the group's culture at the current generation are more likely to be present in the group's culture at the next generation.

Depositing Cultures in Sites

Locations of sites in the model are defined to correspond to the locations of the archaeological sites in the observed data (Vanhaeren and d'Errico 2006). A group will deposit its culture at a site when within a specified geographic distance of that site. This catchment distance is initially set to be equal for all sites, with the further constraint that if two groups are within the catchment distance then the group closest to the site will be the one to deposit its culture there.

The distance measure used to calculate the distance between group locations and archaeological sites is the geodesic distance, which is the aerial path between two points, also called the as-the-crow-flies, great-circle or orthodromic distance. To account for curvature of the Earth, geographic distances are calculated using the Haversine Formula (Sinnott 1984). This calculates the great-circle distance between two points on a sphere given their respective longitudes and latitudes. If the longitude and latitude of the points are ϕ_1 and λ_1 for point one and ϕ_2 and λ_2 for point two, respectively, and:

$$\Delta\phi = \frac{\phi_1 - \phi_2}{2}, \tag{8.5}$$

$$\Delta\lambda = \frac{\lambda_1 - \lambda_2}{2}, \tag{8.6}$$

then the distance, D, between the two points is calculated as:

$$D = 2R\sin^{-1}\left(\left(\sin^2(\Delta\phi) + \cos(\phi_1)\cdot\cos(\phi_2)\cdot\sin^2(\Delta\lambda)\right)^{\frac{1}{2}}\right), \tag{8.7}$$

where R is the radius of the Earth, assumed to be constant at 6,371 km.

Each site is assigned an item capacity, which corresponds to the number of items recovered from that site as reported in the original study (Vanhaeren and d'Errico 2006). When a group comes within the specified distance for a particular site, a number of unique items, equal to the item capacity for that site, are selected probabilistically (using a multinomial function so that trait frequencies are taken into account) from the group's entire culture to be deposited at the site—this corresponds to one copy of each cultural trait that is selected being deposited at the site. The original dataset contains presence/absence records of distinct bead types only, so specifying the item capacity and 'uniqueness' of items deposited at each site in simulations should theoretically minimise archaeological and sampling bias. A group arriving at a site at which a deposit has previously been made will deposit its culture at that site and overwrite the existing deposit only if it comes within closer proximity to that site than the last group that deposited its culture there.

Simulated material culture data deposited at the locations of the archaeological sites in the observed data (Vanhaeren and d'Errico 2006) are collected at the end of each simulation, which corresponds to the end of the Aurignacian period.

Cultural Interactions

A pair of groups will interact if they are within a parameterised geographical distance, d_{int}, of each other; we treat this parameter as an unknown and explore a range of values. As above, the distance between groups is calculated as the geodesic distance and the Haversine Formula (Sinnott 1984) is used to account for curvature of the Earth. Two cultural interaction processes are modelled in this framework—conflict and sharing—and a pair of interacting groups will undergo one of these two processes.

The outcome of a conflict process is the replacement of the culture of one group by that of the other. To model this, we assign one in each pair of interacting groups as a winning, and one as a losing group, and replace the culture of the losing group entirely by that of the winning group. The decision on which is assigned to be the winning group, and which the losing group, is made at random due to the assumption that the aspects of material culture we are considering (personal ornamentation) do not have an effect on, and are not a proxy for, group fitness. Additionally, since groups are modelled such that they are the same size, group size cannot be used as a proxy for group fitness. The conflict interaction process is analogous to a group imposing its culture on a group that they have defeated, or, alternatively, assimilating the defeated group into their own, followed by a fission process.

The other interaction process considered is sharing of cultures between interacting groups. Sharing is modelled by permutation, whereby the cultures of the two interacting groups are pooled, permuted and then divided between the two. This is analogous to culturally similar groups swapping cultural traits.

8.2.2.4 Null and Culture-Dependent Interaction Models

The Null Model and Culture-Dependent Interaction Model are both models of cultural identity-by-descent with modification and isolation-by-distance, and are made up of the demographic and cultural processes described above.

The difference between the two models lies in the method of deciding which type of interaction will occur between two interacting groups. In the Null Model, the type of interaction is decided at random; groups are equally as likely to share material culture as they are to undergo conflict. The Null Model is therefore a scenario in which group interactions are independent of similarities or differences in groups' ornamental material cultures. Conversely, in the Culture-Dependent Interaction Model, the type of interaction is decided probabilistically and depends on the extent of cultural similarity between the two interacting groups; groups that are relatively culturally similar are more likely to share cultures while those that are relatively culturally different are more likely to undergo conflict. The main aim of this study is to test which of these two models best explains the observed spatial distribution of ornament types in the archaeological record; the latter is intended to represent the effects of ethnic structuring on the spatial distribution of material culture.

Measures of Cultural Similarity

The extent of cultural similarity between a pair of interacting groups is quantified differently depending on which of the models of cultural variation described above is considered. In simulations that follow the stepwise mutation model, the extent of cultural similarity is quantified using a measure akin to the $(\delta\mu)^2$ measure used to quantify the genetic similarity between populations using microsatellite data (Goldstein et al. 1995a, b). We define this cultural $(\delta\mu)^2$ measure as:

$$(\delta\mu)^2 = \sum_i \sum_j (i-j)^2 x_i y_j - \frac{1}{2}\left[\sum_i \sum_j (i-j)^2 x_i x_j + \sum_i \sum_j (i-j)^2 y_i y_j\right], \quad (8.8)$$

where x_i and y_j are frequencies of traits i and j in (interacting) groups x and y respectively. This measure therefore quantifies the cultural similarity between the two interacting groups by taking into account the frequencies with which all traits occur in their respective cultural repertoires—this measure does not discriminate between differences in cultural features. The calculated value of $(\delta\mu)^2$ is normalised by the maximum $(\delta\mu)^2$ recorded up to that generation of that simulation, giving a measure of cultural similarity that is relative to the maximum measured cultural similarity.

In simulations that follow the non-stepwise mutation model, the extent of cultural similarity between a pair of interacting groups is quantified using a measure akin to the F_{ST} measure used to quantify the genetic similarity between populations using allele frequency data (Wright 1978; Cavalli-Sforza et al. 1996). We define the cultural F_{ST} measure as:

$$F_{ST} = \frac{H_T - \overline{H_S}}{H_T}. \quad (8.9)$$

In this definition, H_T is the amount of variation in all traits in the whole population (considered to be the two interacting groups) and is defined as:

$$H_T = 1 - \sum_i \overline{p_i}^2, \quad (8.10)$$

where $\overline{p_i}$ is the average frequency of trait i calculated over the two interacting groups. H_S is the amount of variation between traits within each group (calculated separately for each of the two interacting groups); $\overline{H_S}$ is the average of H_S calculated over the two interacting groups. H_S is defined as:

$$H_S = 1 - \sum_i p_i^2, \quad (8.11)$$

where p_i is the frequency of trait i. Similarly to the $(\delta\mu)^2$ measure discussed above, the F_{ST} measure takes into account the frequencies with which all traits occur in the cultural repertoires of the two interacting groups—as with the $(\delta\mu)^2$ measure, this measure does not discriminate between differences in cultural features. It is therefore an estimate of the proportion of the total variation in a set of traits that is the result of between-group differences (Bell et al. 2009). Similarly again to $(\delta\mu)^2$, the calculated value of F_{ST} is normalised by the maximum F_{ST} recorded up to that generation of that simulation, giving a measure of cultural similarity that is relative to the maximum measured cultural similarity.

The relative values of $(\delta\mu)^2$ (bounded stepwise mutation model) and F_{ST} (bounded discrete mutation model) can take values between 0 and 1 and are treated as probabilities to decide which of the interaction processes described above occurs between the two interacting groups; a value of 0 indicates that the two groups have identical cultural repertoires and are therefore more likely to share cultures, while a value of 1 indicates complete cultural difference and indicates that the two groups are more likely to undergo conflict.

8.2.2.5 Models, Model Parameters and Prior Ranges

Given that one of two environmental scenarios and one of two models of cultural variation are considered for each simulation in both the Null Model and the Culture-Dependent Interaction Model, data is simulated under eight distinct scenarios. These are summarised in Table 8.1. The acronym and text colour associated with each model correspond to those used in Fig. 8.2 for that model.

In total there are six parameters that govern the processes considered in the Null and Culture-Dependent Interaction Models. Both models have 4 key processes: migration, fission/extinction, cultural mutation and cultural interaction, governed by 4 parameters: d_{mig}, $p_{f/e}$, p_{mut} and d_{int}, respectively. In addition to these, there are two further parameters in both models, namely the maximum number of groups, G_{max}, and the number of items in each group's culture, N_{items}.

There is little information in the archaeological record relating to the precise values that these parameters may take. Each parameter is therefore constrained to a uniform prior range, with the value of each parameter in each simulation randomly assigned from this uniform prior. Prior ranges for each parameter are listed in Table 8.2.

8.2.3 Analysis

Once a large number of simulations have been performed under the models described above, the objective of the data analysis is to quantify the extent of similarity between observed and simulated material culture data. To do this, ABC techniques are used to compare the differences in goodness-of-fit between the observed data (Vanhaeren and d'Errico 2006) and data simulated by different proposed models. To be able to compare the observed and simulated datasets, robust statistics that sufficiently describe the full properties of the data, referred to as summary statistics, are used. Summary statistics used in this study are discussed in detail

Table 8.1 Summary of combinations of environmental scenarios and cultural variation models under which data is simulated in both the Null Model and Culture-Dependent Interaction (CDI) Model

Cultural variation model \ Environmental scenario	Flat space (FLAT)	Bocquet-Appel et al. (2005) distribution (B-A)
Bounded stepwise mutation model (SW)	Null model - - - - - CDI model ———	Null model - - - - - CDI model ———
Bounded discrete mutation model (DIS)	Null model - - - - - CDI model ———	Null model - - - - - CDI model ———

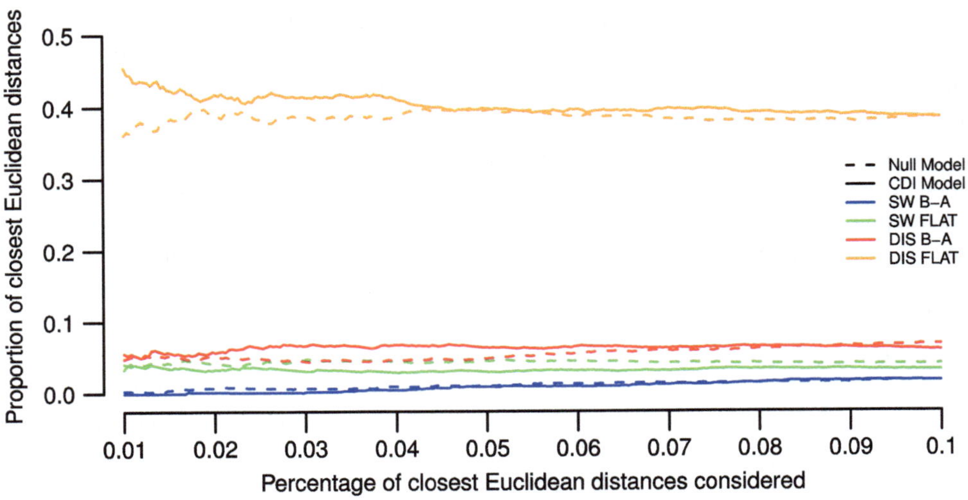

Fig. 8.2 Relative marginal likelihood estimates (y-axis) of each Null Model (*dashed lines*) and Culture-Dependent Interaction Model (*solid lines*) for each percentage (x-axis) of closest simulations, taking into consideration 2,680,000 simulations (335,000 simulations for each Null Model and Culture-Dependent Interaction Model)

Table 8.2 Model parameters and their prior ranges, for both the Null Model and the Culture-Dependent Interaction Model

		Prior range	
Parameter		Null Model	Culture-Dependent Interaction Model
Migration distance (km)	d_{mig}	[0,100]	[0,100]
Probability of fission/extinction	$p_{f/e}$	[0, 1]	[0, 1]
Probability of cultural mutation	p_{mut}	[0, 0.2]	[0, 0.2]
Number of items	N_{items}	[500, 1,500]	[500, 1,500]
Maximum number of groups	G_{max}	[50, 1,000]	[50, 1,000]
Interaction radius (km)	d_{int}	[0, 50]	[0, 50]

in Appendix 3: Summary Statistics. By comparing summary statistics calculated for each simulated dataset to those for the observed data, this method allows us to accept those simulations with summary statistics sufficiently close to the target summary statistics—these are the best simulations, that is, those generating data most similar to the observed data.

Another useful feature of this approach is the ability to formally compare the performance of different models using Bayes Factors (Kass and Raftery 1995). In short, a Bayes Factor is a summary of the evidence provided by the data in favour of one model over another; this is further discussed in Appendix 4: Bayes Factors for Model Comparison. What we are estimating in this study are the relative marginal likelihoods of each proposed model given the data. More explicitly, given models M_1 and M_2 that we want to compare, their respective relative marginal likelihoods l_1 and l_2 are defined as:

$$l_1 = \frac{N_1}{N}, \text{ and} \tag{8.12}$$

$$l_2 = \frac{N_2}{N}, \tag{8.13}$$

where N_1 and N_2 are the number of simulation that come from models M_1 and M_2, respectively, and N $(=N_1 + N_2)$ is the total number of simulations considered; the relative marginal likelihood of each model is defined to be the proportion of total number of simulations considered that come from that model. This is therefore a measure of which model explains the observed data better, given that N simulations are considered.

This form of model comparison is independent of the number of parameters for each model, and instead estimates the likelihood of the model considering all possible parameter values. In cases where models with different numbers of parameters are compared, this method automatically and correctly penalises model complexity; for models with a large number of parameters there is a larger parameter space to explore and so it is more difficult to find those parameter sets that generate data similar to the observed data. Models with more parameters are therefore penalised for the increased complexity compared to simpler models, resulting in a comparison weighted by model complexity. Such an approach prevents us from overfitting—from invoking parameters to explain aspects of the data that are in fact due to randomness. However, in this particular study the number of parameters is equal in all models.

8.2.4 Results

Results shown are from 335,000 simulations for each of the Null and Culture-Dependent Interaction Models. For the relative marginal likelihood estimation, results for the eight models are considered together; a total of 2,680,000 simulations are therefore taken into account. In this analysis, we estimate the relative marginal likelihood of each model, taking into account the extent of similarity between simulated and observed data (Vanhaeren and d'Errico 2006).

Figure 8.2 is a plot of the estimated relative marginal likelihood of each version of the Null Model (dashed lines) and Culture-Dependent Interaction Models (solid lines) at different thresholds. It shows what proportion (y-axis) of the best simulations—those generating data most similar to the observed data—are coming from each model for each percentage (x-axis) of closest simulations. The colours refer to the combination of the environmental scenario and the model of cultural variation considered, as detailed in Table 8.1. Since the plot depicts proportions, for any particular percentage of closest simulations (i.e. for any particular value on the x-axis), the sum of the proportions of the closest Euclidean distances coming from each model (i.e. the sum of the values on the y-axis) will always be 1. The relative marginal likelihood estimates of each version of Null Model and Culture-Dependent Interaction Model for 0.1 % of closest simulations (i.e. Fig. 8.2, x = 0.1) are also detailed in Table 8.3.

Figure 8.2 indicates that, for all scenarios modelled (scenario here referring to a pairwise combination of an environmental scenario and a model of cultural variation as explained in Models, Model Parameters and Prior Ranges), there is little difference in how well the Null Model and Culture-Dependent Interaction Model perform. The best fits of simulated to observed data are generated by data simulated under the scenario that combines the bounded discrete mutation model and the environmental scenario in which Europe is assumed to have a flat distribution of carrying capacities (represented by orange lines in Fig. 8.2), with approximately 38.4 % of the best 0.1 % of simulations coming from each the Null Model (dashed orange line) and Culture-Dependent Interaction Model (solid orange line).

Table 8.3 Relative marginal likelihood estimate of each Null Model and Culture-Dependent Interaction (CDI) Model for 0.1 % of closest simulations

Environmental scenario / Cultural variation model	Flat space (FLAT)	Bocquet-Appel et al. (2005) distribution (B-A)
Bounded stepwise mutation model (SW)	Null model: 4.0% CDI model: 3.2%	Null model: 1.6% CDI model: 1.6%
Bounded discrete mutation model (DIS)	Null model: 38.4% CDI model: 38.4%	Null model: 6.8% CDI model: 6.0%

8.2.5 Discussion and Extensions of Simulated Model

This study does not support the hypothesis that Aurignacian populations in Early Upper Palaeolithic Europe were ethnically structured in a manner related to ornamental material culture. The spatially explicit simulation models and ABC analysis presented here, conditioned on the data presented by Vanhaeren and d'Errico (2006), indicate that there is little difference between the simple scenario of cultural identity-by-descent with modification and isolation-by-distance, and the more complex one that, in addition, invokes cultural group interaction processes that would be expected if material culture was symbolically marking ethnic group identity.

Prior to discussing the results presented above, it is important to note that any scenario considered will only be *relatively* better or worse than any other scenario considered; it is not possible to rate how good a scenario is *absolutely*.

Considering the results firstly in view of the two environmental scenarios used to condition the demography of the simulation space, we see that there is no improvement in the fit of simulated to observed data when conditioning simulations on the distribution from the Bocquet-Appel et al. study (Bocquet-Appel et al. 2005) rather than the scenario in which Europe is assumed to have a flat distribution of carrying capacities (i.e. Europe is assumed to be a flat space). Indeed, for each of the two mutation models considered, simulations in which the demography is conditioned on the latter environmental scenario generate a better fit to the observed data. Since this latter scenario is clearly not realistic, this result implies that *both* environmental scenarios used to condition the demography of the simulation space in this study are unrealistic; this is further discussed as a caveat of the current modelling framework below, along with suggestions for possible improvements.

Analysing the results now in view of the two mutation models considered, we see that, regardless of the assumed environmental scenario, data simulated under the bounded discrete mutation model generate a better fit to observed data than that simulated under the bounded stepwise mutation model. Although this result requires further investigation, it could be speculated that this suggests that, in the context at least of group interactions, there is little scaling of item similarity in material culture repertoires; little or no scaling of item similarity implies that a particular item would have been treated as either the same as or different to items already in the repertoire.

The fact that the best fits of simulated to observed data are generated by data simulated under the scenario that combines the bounded discrete mutation model and the environmental scenario in which Europe is assumed to be a flat space, and that these are a far better fit than any of the other scenarios considered, implies that both the assumed mutation model and the assumed environmental model strongly drive the fit of simulated to observed data. Although we are cautious about interpreting the following, it is interesting to note that the assumed mutation model makes a bigger difference than the assumed environmental scenario to the fit of simulated to observed data, suggesting that continuously scaled cultural similarities were not important in distinguishing inter-group identity.

This study is a work in progress and there are several caveats, discussed below, which should be taken into consideration when interpreting our results, but these methods offer the opportunity to formally investigate whether observed material culture distributions are better explained under the assumption that ethnic structuring exists and that identities reflected in ornamental material culture influence how people interact.

It should be noted that, for the ABC approach adopted here, the number of simulations per model is relatively small and may not be enough to adequately explore the parameter space considered; for this reason, the number of simulations performed under each combination of environmental scenario and cultural variation model should be systematically increased.

The culture transmission process used in this framework assumes neutrality in that bead types are not assumed to differentially affect group fitness. A number of authors have been unable to reject neutrality using cultural transmission models (Neiman 1995; Steele et al. 2010); however, this may be due to the lack of statistical methods available to test for deviations from neutrality. Tests for deviations from neutrality have only been carried out on post-Palaeolithic datasets and have not been applied in a Palaeolithic context. However, there is certainly no a priori reason why use of different bead types should differentially affect group fitness.

As detailed in the description of the framework above, each group in the simulation deposits its material culture at the locations of the archaeological sites in the observed data (Vanhaeren and d'Errico 2006) and overwrites any existing deposits in the site if it comes within closer proximity to that site than the last group that deposited its culture there. The simulated material culture data is therefore a collection of items selected from different groups' material cultures (each of which is the result of mutation, drift and cultural interaction processes) and deposited at different points throughout the time period of interest; the process of a group depositing its culture is only dependent on the geographic distance between the group and the location of the site and deposits are made with equal probability throughout the simulation. Each site is considered to be single occupancy—only the material culture of the last group that deposited at a particular site is considered. Assuming that each site is single occupancy may be misleading since the observed data (Vanhaeren and d'Errico 2006) cannot be chronologically resolved and some sites may feature multiple layers that were deposited thousands of years apart within the period of interest. To address the inconsistency of this assumption with the cumulative aspect of the archaeological record, the depositing process could be modified such that, instead of overwriting previous deposits at a particular site, a group depositing its culture at that site would simply add its entire culture, including information on the frequency of each trait, to the existing deposits. At the end of the simulation, a number of unique items, equal to that recovered from the site as reported in the original study (Vanhaeren and d'Errico 2006), could then be selected probabilistically (using a multinomial function so that trait frequencies are taken into account) for each site, such that the probability of an item being selected is proportional to the frequency with which it is occurs in that site.

The two environmental scenarios used to condition the demography of the simulation space in this study are not realistic. In the first scenario Europe is assumed have a flat distribution of carrying capacities; this is clearly a simplistic and unrealistic view since topographic and climatic variation within the geographic region considered during the time period of interest would have had an impact on differences in habitability, and therefore the carrying capacity values, of different geographic locations at different points in time throughout the time period of interest. In the second scenario, information on estimated population densities is taken from the Bocquet-Appel et al. (2005) study to inform on carrying capacities. The reported geographic distribution and relative estimates of Upper Palaeolithic population size are an indicative starting point; however, the study itself could be considered somewhat controversial since the millennial scale climatic variation observed during the time periods that are considered is not taken into account. The geographic region during the time period of interest in the current study is characterised by a number of rapid climatic changes (Banks et al. 2008) and it is therefore unrealistic to consider the environment, and the resulting potential population densities, static for the entire duration of a simulation.

Since these environmental scenarios are unrealistic, future work could consider how the results are affected when simulations are dependent on modelled environments that take into account the climatic variability across the geographic region considered during the time period of interest. This could be achieved by using simulated Palaeoclimate data to inform on the relative carrying capacity values, and therefore potential population densities, of locations in the region of interest. Since Palaeoclimate data are available at different time points throughout the time period of interest, this approach would allow us to take into account the observed climatic variability by updating the carrying capacities in the modelled domain throughout the simulation. On way of doing this would be to use Palaeoclimate data (Banks et al. 2008; Singarayer and Valdes 2010) to approximate Net Primary Productivity values for each location in the region of interest, following the precedent set by Eriksson et al. (2012). Net Primary Productivity provides a proxy for food availability and has been shown to be a predictor of demographic patterns in ecological studies (Binford 2001; Luck 2007) it is therefore an informative proxy for carrying capacity values, and thus potential population densities.

Group migrations could be conditioned on topographic roughness by using Topographic Roughness Index values calculated at the required resolution for the geographic area of interest using high-resolution (3 arc-sec or 90 m) elevation data (Jarvis et al. 2008). In this case, the value of the Topographic Roughness Index at a particular location would effectively scale the distance that a group can travel at that location; at locations with low values of the index (low topographic roughness) migrations would be relatively easier, while at locations with high values of the index (high topographic roughness) migrations would be relatively more difficult.

Additionally, migratory processes could be modelled as parameterised Lévy random walks, instead of as parameterised Gaussian random walks as presented above. Lévy walks are a type of random walk in which movement distances follow power-law distributions, and studies (Brown et al. 2007; Raichlen et al. 2014) looking at foraging patterns in human hunter-gatherer populations have suggested that Lévy walks are the optimal movement pattern when foraging for heterogeneously located resources (with little or no prior knowledge of resource distribution patterns). With this in mind, migratory processes in this study could be modelled as parameterised Lévy random walks, with the distance that each group traverses in a migration process selected from parameterised power-law distributions.

More generally, we have to face up to the degree of archaeological resolution we have available. Just as we cannot assume constant climatic conditions during the course of the time period considered, with climatic fluctuations that occurred during the approximately 13,000 years of the Aurignacian inevitably affecting regional population densities, we cannot necessarily assume that the aggregate data set we are dealing with represents interaction processes acting uniformly over that period; it might represent a spurious averaging of a variety of different processes. However, this is not an argument against modelling approaches; such approaches are the *only* way we can get an insight into the accumulated outcomes of iterated processes going on for hundreds or thousands of years. It is instead an argument for improving the archaeology of the time period, as well as for further comparison. The results presented here would gain further significance if they could be compared with those from the subsequent Gravettian and later cultural periods of the same region. Similarly, we may gain further insight into group interaction dynamics during the Aurignacian by comparing the results of the bead analysis (Vanhaeren and d'Errico 2006) with patterns derived from similarities and differences between lithic assemblages at the same sites.

8.3 General Discussion and Conclusions

Many fields, including archaeology, are becoming increasingly systematic and interdisciplinary through integration of traditional methods with techniques developed in other fields. Simulation modelling involves the use of theory developed for problems in physical and biological sciences and allows archaeologists to propose and test explicit hypotheses in order to address longstanding archaeological and anthropological questions. Our paper has demonstrated a novel and rigorous approach to a topic of major interest, namely the role of social structuring in archaic humans. As pointed out above, the appearance of personal ornaments has long been considered a distinctive feature differentiating Neanderthals and anatomically modern humans. While that may be the case, our results show that we have to be careful of making the further inferential step of assuming that this reflects ethno-linguistic structuring specific to anatomically modern humans. As Kuhn (2013, p. 208) points out, apparently complex large-scale phenomena can arise *"as a function of simple transmission rules operating on bounded social networks"*, thus other, and simpler, processes accounting for the observed patterning need to be considered and rejected. Simulation modelling within the Bayesian ABC framework provides a means of doing this.

As far as we are aware, the approach reported here has not been attempted when considering archaeological evidence for ethnic structuring. We fully accept that there are strengths and weaknesses to this approach, just as there are with other approaches, and these should be considered when interpreting and comparing these results to those of others. Given that there is, to the best of our knowledge, little or no representation in the literature of explicit simulation modelling approaches to questions of ethnic structuring, while interpretative approaches are well represented, we believe that this study begins to fill an important gap in the literature.

A simulation modelling approach is considerably more complex and laborious to implement compared to the interpretation of descriptive statistics or patterns in data alone. It is, however, a formal scientific approach that proposes a model with an explicit prediction of the distribution of material culture data, and tests this formally by comparing the simulated data to the observed data for validation. Taking this approach necessitates reduced models and these, by definition, will never fully describe the complexity of the true processes that shaped the material culture data. However, the model building and testing process is not a closed one; the previous section has already indicated various ways in which aspects of the current model could potentially be improved. None-the-less, the approach adopted here is explicit and transparent and therefore less likely to be influenced by the subjective biases that guide interpretation (Gerbault et al. 2014).

Acknowledgments The authors thank Pascale Gerbault, Kevin Bryson, Bill Croft, Mark Maslin and Cynthia Beall for helpful comments and discussions. The authors also wish to thank Ken Aoki and an anonymous reviewer for constructive comments on an earlier version of this manuscript. The authors acknowledge the use of the UCL *Legion* High Performance Computing Facility, and associated support services, in the completion of this work. Mirna Kovacevic is funded by EPSRC through UCL CoMPLEX. Francesco d'Errico acknowledges the European Research Council (FP7/2007/2013, TRACSYMBOLS 249587).

Appendices

Appendix 1: Bayesian Inference and Approximate Bayesian Computation (ABC)

Bayesian inference is a branch of statistics that uses observations of particular datasets to infer the probability that a proposed hypothesis, or a parameter of that hypothesis, is true. To do this, various models with set numbers of parameters are proposed, and the posterior probability distributions of these parameters are inferred using information from prior probability distributions of the parameters and information provided by the observed data, through implementing Bayes theorem. Bayes theorem states that, given parameter (or set of parameters) θ and observed dataset D, the posterior distribution of θ, denoted $P(\theta|D)$, is proportional to the product of the probability of observing dataset D given model with parameter θ, denoted $P(D|\theta)$, and the likelihood of θ, denoted $\pi(\theta)$, which is the distribution of θ prior to any observations being made. Mathematically, this can be written as:

$$P\left(\theta\middle|D\right) \propto P\left(D\middle|\theta\right) \cdot \pi\left(\theta\right). \qquad (8.14)$$

Since the explicit form of the likelihood $P(D|\theta)$ is difficult to compute in many complex problems, a family of Bayesian methods, referred to as Approximate Bayesian Computation (ABC), which do not require the likelihood function to be theoretically specified, are used (Tavare et al. 1997; Fu and Li 1997; Beaumont et al. 2002; Bertorelle et al. 2010).

In ABC techniques, a large number of datasets are simulated under a model assuming different, randomly chosen, parameter values from within prior ranges, and appropriate summary statistics are used to measure the extent to which the simulated datasets emulate the observed data. Parameter values under which the model generates datasets closest to the observed data are retained in the posterior probability distributions of the parameters.

To be able to compare the observed and simulated datasets, robust statistics that sufficiently describe the full properties of the data are used. These are called summary statistics and those developed for the current framework are discussed in detail in Appendix 3: Summary Statistics. By comparing summary statistics calculated for each simulated dataset to those for the observed data, we are able to accept to the posterior those simulations with summary statistics sufficiently close to the summary statistics for the observed dataset, referred to as the target summary statistics. The similarity δ between observed data, S, and simulated data, S', is calculated as the sum of normalised Euclidean distances of individual summary statistics:

$$\delta\left(S, S_j'\right) = \sqrt{\sum\nolimits_{i=1}^{n} \frac{\left(s_i - s_{ij'}\right)^2}{\sigma\left(s_{i'}\right)^2}}, \qquad (8.15)$$

where s and s' are values of each of the summary statistics for the observed and simulated datasets, respectively, subscript i denotes the ith of n statistics, subscript j denotes the jth of N simulations and $\sigma(s_i')$ is the standard deviation of the ith statistics over all N simulations. In performing the data analysis, we regard the ε quantile of the distribution of distances between the observed and simulated data, $\delta(S, S_j')$, as the best simulations—those generating data most similar to the observed data.

Appendix 2: Approximate Bayesian Computation (ABC) Algorithm

Let M denote the chosen model and the set of parameters of M be $\theta = (\theta_1, \ldots, \theta_m)$. Let $S = (s_1, \ldots, s_n)$ and $S' = (s_1', \ldots, s_n')$ denote the values of the summary statistics for the observed and simulated datasets, respectively. Values $S = (s_1, \ldots, s_n)$ are referred to as the target values for each of the summary statistics. The ABC algorithm is applied as follows:

1. Define a set of summary statistics that capture relevant information contained in the observed dataset.
2. Compute summary statistics values $S = (s_1, \ldots, s_n)$ for the observed dataset—these are the target values.
3. Sample parameters $\theta^* = (\theta_1^*, \ldots, \theta_m^*)$ from an appropriate prior distribution.
4. Simulate data by using parameter θ^* set with model M.
5. Compute summary statistics values $S' = (s_1', \ldots, s_n')$ for the simulated data.
6. Compute $\delta(S, S')$, where δ is an appropriately chosen distance measure.
7. For a chosen tolerance ε, retain parameter set θ^* in the posterior distribution of θ if $\delta(S, S') < \varepsilon$.

8. Repeat steps 1–7 until the desired number of parameter values have been accepted to the posterior distribution.

In order for ABC methods to be effective, appropriate summary statistics that sufficiently describe the observed dataset need to be developed and appropriate choices for the distance measure, δ, and tolerance, ε, must be made.

Appendix 3: Summary Statistics

As explained previously, to be able to compare simulated and observed datasets using ABC methods, summary statistics that capture the information contained in the observed data must be developed. These should be robust statistics and should describe sufficiently the full properties of the observed dataset considered. For the current dataset, these are:

- shared information between bead types and sites, respectively
- mutual dependence between bead types and sites, respectively
- diversity in the number of occurrences of different bead types
- cultural diversity of sites as represented by the variation in the number of distinct bead types recovered from each sites
- spatial distribution of sites

For each of these statistics, we consider the values of the mean and variance in the data analysis.

Shared Information (SI)

Shared information, denoted *SI*, is a statistic that measures the extent of similarity between two variables. For measuring the shared information between bead types, *SI* is defined to be:

$$SI\left(t_i, t_j\right) = \frac{f_i f_j}{\overline{f}^2} \log \frac{r\left(t_i\right) + r\left(t_j\right)}{r\left(t_i, t_j\right)}, \tag{8.16}$$

where $r(t_i)$ and $r(t_j)$ denote the ratio of the number of occurrences of bead types *i* and *j* to the total number of sites, $r(t_i, t_j)$ is the ratio of the number of concurrent occurrence of bead types *i* and *j* to the total number of sites, f_i and f_j represent the number of sites in which bead types *i* and *j* occur, respectively, and $\overline{f}$ is the average number of times any bead type occurs over all sites. In this case, *SI* measures the similarity between pairwise bead types in terms of which sites the are present in. When two bead types never occur in the same site,

$$r\left(t_i\right) + r\left(t_j\right) = r\left(t_i, t_j\right), \text{ and} \tag{8.17}$$

$$SI\left(t_i, t_j\right) = 0. \tag{8.18}$$

A similar equation can be used to measure the shared information between sites:

$$SI\left(s_i, s_j\right) = \frac{g_i g_j}{\overline{g}^2} \log \frac{r\left(s_i\right) + r\left(s_j\right)}{r\left(s_i, s_j\right)}, \tag{8.19}$$

where $r(s_i)$ and $r(s_j)$ denote the ratio of the number of sites in which bead types *i* and *j* occur to the total number of bead types, $r(s_i, s_j)$ is the ratio of the number of sites that share bead types *i* and *j* to the total number of bead types, g_i and g_j represent the total number of bead types present in sites *i* and *j*, respectively, and $\overline{g}$ is the average number of bead types occurring per site. In this case, *SI* measures the extent of similarity between pairwise sites in terms of bead types present in those sites. Similarly to above, if two sites have no bead types in common,

$$r\left(s_i\right) + r\left(s_j\right) = r\left(s_i, s_j\right), \text{ and} \tag{8.20}$$

$$SI\left(s_i, s_j\right) = 0. \tag{8.21}$$

Mutual Information (MI)

The mutual information, *MI*, between two random variables *X* and *Y* is a measure of the mutual dependence between them. It is defined as:

$$MI\left(X;Y\right) = \sum_{y \in Y} \sum_{x \in X} p\left(x,y\right) \log \frac{p\left(x,y\right)}{p_1(x) + p_2(y)}, \tag{8.22}$$

where $p(x,y)$ denotes the joint probability of *x* and *y* (the probability of *x* and *y* occurring together), and $p_1(x)$ and $p_2(y)$ denote the marginal probabilities of *x* and *y* respectively (the probabilities of the specified values of *x* and *y* occurring).

For the observed dataset in this study, setting $X = t_i$ and $Y = t_j$, where t_i and t_j correspond to the number of occurrences of bead type *i* and *j* in all sites respectively, allows the mutual information between all pairs of bead types to be computed. Analogously, setting $X = s_i$ and $Y = s_j$, where s_i and s_j correspond to the total number of bead types present in sites *i* and *j* respectively, allows the mutual information between all pairs of sites to be computed.

In contrast to the *SI* statistic, which only examines the common presences between sites or bead types, the *MI* statistic examines both the common presences and common absences. It therefore represents the *dependence* between the pairwise vectors in question.

Mean Absolute Deviation (MAD)

The observed dataset shows large fluctuations both in the number of bead types recovered at individual sites, and

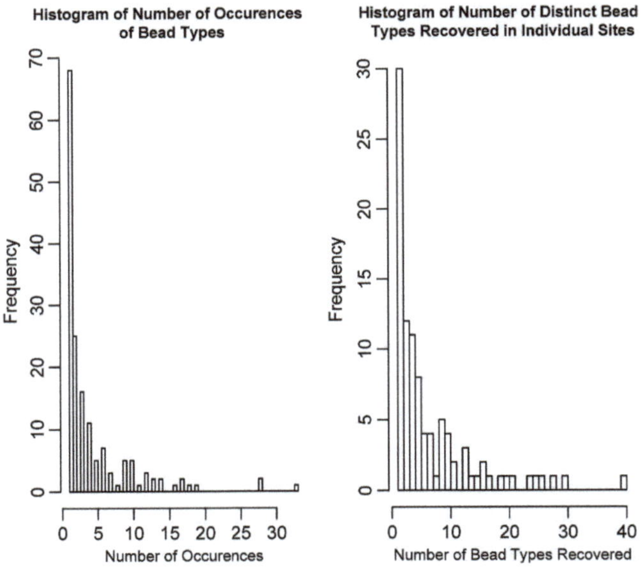

Fig. 8.3 Histograms of the number of occurrences of bead types (*left*) and number of distinct bead types recovered from individual sites (*right*) for the observed data

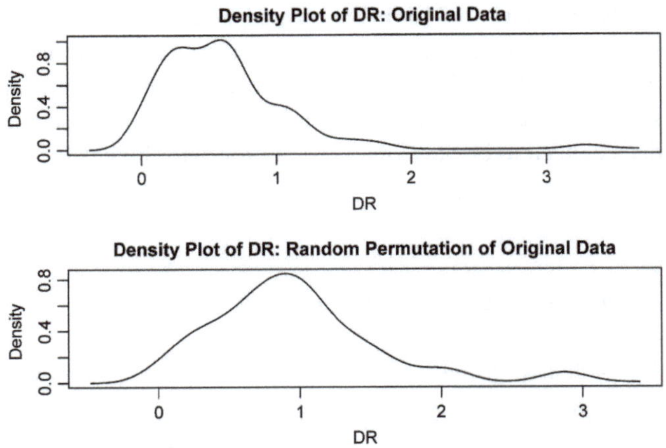

Fig. 8.4 Density plots of the *DR* statistic for the original observed data (*top*) and a random permutation of the same (*bottom*)

the number of times each particular bead type occurs, as shown in Fig. 8.3. Assuming that this is not the result of archaeological bias, these differences could be attributed to cultural wealth at sites, and the preference for particular bead types, respectively. To quantify this, the median absolute deviation statistic, *MAD*, is used. It is a measure of the variability of a random sample, and is defined to be:

$$MAD = median\left(\left|X_i - median_j\left(X_j\right)\right|\right). \quad (8.23)$$

Letting $X_i = T = \frac{f_i}{\overline{f}}$, where f_i represents the number of sites in which bead type *i* occurs and $\overline{f}$ is the average number of times any bead type occurs over all sites, the *MAD* statistic is a measure the variability in the number of occurrences of bead types. This can be thought of as a measure of variability in the popularity of, or preference for, bead types.

Letting $X_i = S = \frac{g_i}{\overline{g}}$, where g_i represents the total number of bead types present in site *i* and $\overline{g}$ is the average number of bead types occurring per site, the *MAD* statistic measures the variability in the number of beady types recovered. This can be thought of as a measure of variability in the cultural wealth recovered from sites.

Spatial Distribution of Sites (DR)

The extent to which sites share bead types may be a function of the distance between those sites. It is logical to expect that sites which are located near to each other share bead types more frequently than those which are far apart. The spatial distribution of sites can be explored by considering the average distance between sites sharing bead type *i*, $\overline{d}_i$, in relation to the average distance between all sites, $\overline{d}$, as follows:

$$DR_i = \frac{\overline{d}_i}{\overline{d}}. \quad (8.24)$$

DR therefore quantifies the spatial distribution of sites in terms of the shared bead types between them. Figure 8.4 shows density plots for the original observed dataset (top) and a random permutation of the same (bottom). The obvious shift to the right in the density plot of the permuted dataset implies that the distance between sites sharing a particular bead type is on average larger if bead types are randomly assigned to sites. For the original observed dataset this implies that sites which are located closer to one another on average share bead types more frequently with each other than with sites that are further away, as expected.

Appendix 4: Bayes Factors for Model Comparison

Another useful feature of the ABC approach is the ability to formally compare the performance of different models using Bayes Factors (Kass and Raftery 1995). A Bayes Factor is a summary of the evidence provided by the data in favour of one model over another. Given models M_0 and M_1, not necessarily with the same number of parameters, Bayes Factor *B* is given by:

$$B = \frac{P\left(M_1 \middle| D\right)}{P\left(M_0 \middle| D\right)} = \frac{P\left(D \middle| M_1\right)\pi\left(M_1\right)}{P\left(D \middle| M_0\right)\pi\left(M_0\right)}, \quad (8.25)$$

where $\pi(M_i)$ is the prior probability of model M_i, $P(D| M_i)$ is the probability of data D given model M_i and $P(M_i|D)$ is the posterior probability of the model, defined as:

$$P(M_i|D) = \frac{P\left(D\middle|M_i\right)\pi(M_i)}{P(D)}, \tag{8.26}$$

where $P(D)$ is the unconditional marginal likelihood of the data.

This form of model comparison is independent of the parameters for each model, and instead calculates the probability of the model considering all possible parameter values. This method automatically and correctly penalises model complexity; for models with a large number of parameters there is a larger parameter space to explore and so it is more difficult to find those parameter sets that generate data similar to the observed data. Therefore, models with more parameters are penalised for the increased complexity compared to simpler models, resulting in a comparison weighted by model complexity.

References

Axelrod R (1997) The dissemination of culture—a model with local convergence and global polarization. J Confl Resolut 41(2):203–226

Banks WE, d'Errico F, Peterson AT, Kageyama M, Sima A, Sanchez-Goni MF (2008) Neanderthal extinction by competitive exclusion. PLoS One 3(12):e3972. doi:10.1371/journal.pone.0003972

Banks WE, d'Errico F, Zilhao J (2013) Human-climate interaction during the Early Upper Paleolithic: testing the hypothesis of an adaptive shift between the Proto-Aurignacian and the Early Aurignacian. J Hum Evol 64(1):39–55

Bar-Yosef O (2002) The Upper Paleolithic revolution. Annu Rev Anthropol 31:363–393. doi:10.1146/Annurev.Anthro.31.040402.085416

Beaumont MA, Zhang W, Balding DJ (2002) Approximate Bayesian computation in population genetics. Genetics 162(4):2025–2035

Bell AV, Richerson PJ, McElreath R (2009) Culture rather than genes provides greater scope for the evolution of large-scale human prosociality. Proc Natl Acad Sci U S A 106(42):17671–17674. doi:10.1073/pnas.0903232106

Benazzi S, Douka K, Fornai C, Bauer CC, Kullmer O, Svoboda J, Pap I, Mallegni F, Bayle P, Coquerelle M, Condemi S, Ronchitelli A, Harvati K, Weber GW (2011) Early dispersal of modern humans in Europe and implications for Neanderthal behaviour. Nature 479(7374):525–528. doi:10.1038/nature10617

Bertorelle G, Benazzo A, Mona S (2010) ABC as a flexible framework to estimate demography over space and time: some cons, many pros. Mol Ecol 19(13):2609–2625. doi:10.1111/J.1365-294x.2010.04690.X

Binford LR (2001) Constructing frames of reference: an analytical method for archaeological theory building using hunter-gatherer and environmental data sets. University of California Press, Berkeley

Bocquet-Appel JP, Demars PY, Noiret L, Dobrowsky D (2005) Estimates of Upper Palaeolithic meta-population size in Europe from archaeological data. J Archaeol Sci 32(11):1656–1668. doi:10.1016/J.Jas.2005.05.006

Brown CT, Liebovitch LS, Glendon R (2007) Levy flights in dobe ju/'hoansi foraging patterns. Hum Ecol 35(1):129–138

Buck CE (2001) Applications of the Bayesian statistical paradigm. In: Brothwell DR, Pollard AM (eds) Handbook of archaeological sciences. Wiley, Chichester, pp 695–702

Cavalli-Sforza LL, Menozzi P, Piazza A (1996) The history and geography of human genes. Abridged paperback edn. Princeton University Press, Princeton/Chichester

Clark JT, Hagemeister EM (2007) Digital discovery: exploring new frontiers in human heritage; CAA 2006: computer applications and quantitative methods in archaeology. In: Proceedings of the 34th conference, Fargo, Apr 2006. Archaeolingua, Hungary

Costopoulos A, Lake M (2010) Simulating change: archaeology into the twenty-first century, Foundations of archaeological inquiry. University of Utah Press, Salt Lake City

Eriksson A, Betti L, Friend AD, Lycett SJ, Singarayer JS, von Cramon-Taubadel N, Valdes PJ, Balloux F, Manica A (2012) Late Pleistocene climate change and the global expansion of anatomically modern humans. Proc Natl Acad Sci U S A 109(40):16089–16094

Fu YX, Li WH (1997) Estimating the age of the common ancestor of a sample of DNA sequences. Mol Biol Evol 14(2):195–199

Gerbault P, Allaby RG, Boivin N, Rudzinski A, Grimaldi IM, Pires JC, Vigueira CC, Dobney K, Gremillion KJ, Barton L, Arroyo-Kalin M, Purugganan MD, de Casas RR, Bollongino R, Burger J, Fuller DQ, Bradley DG, Balding DJ, Richerson PJ, Gilbert MTP, Larson G, Thomas MG (2014) Storytelling and story testing in domestication. Proc Natl Acad Sci U S A 111(17):6159–6164

Goldstein DB, Linares AR, Cavallisforza LL, Feldman MW (1995a) An evaluation of genetic distances for use with microsatellite loci. Genetics 139(1):463–471

Goldstein DB, Ruiz Linares A, Cavalli-Sforza LL, Feldman MW (1995b) Genetic absolute dating based on microsatellites and the origin of modern humans. Proc Natl Acad Sci U S A 92(15):6723–6727

Henrich J (2004) Demography and cultural evolution: how adaptive cultural processes can produce maladaptive losses—the Tasmanian case. Am Antiq 69(2):197–214

Higham T, Compton T, Stringer C, Jacobi R, Shapiro B, Trinkaus E, Chandler B, Groning F, Collins C, Hillson S, O'Higgins P, FitzGerald C, Fagan M (2011) The earliest evidence for anatomically modern humans in northwestern Europe. Nature 479(7374):521–524. doi:10.1038/nature10484

Higham T, Basell L, Jacobi R, Wood R, Ramsey CB, Conard NJ (2012) Testing models for the beginnings of the Aurignacian and the advent of figurative art and music: the radiocarbon chronology of Geissenklosterle. J Hum Evol 62(6):664–676. doi:10.1016/j.jhevol.2012.03.003

Hughes JK, Haywood A, Mithen SJ, Sellwood BW, Valdes PJ (2007) Investigating early hominin dispersal patterns: developing a framework for climate data integration. J Hum Evol 53(5):465–474. doi:10.1016/J.Jhevol.2006.12.011

Jarvis A, Reuter HI, Nelson A, Guevara E (2008) Hole-filled SRTM for the globe version 4. International Centre for Tropical Agriculture (CIAT); data available from the CGIAR-CSI SRTM 90 m Daab ase. (http://srtm.csi.cgiar.org)

Jones S (1997) The archaeology of ethnicity: constructing identities in the past and present. Routledge, London/New York

Kass RE, Raftery AE (1995) Bayes factors. J Am Stat Assoc 90(430):773–795

Kimura M, Ohta T (1978) Stepwise mutation model and distribution of allelic frequencies in a finite population. Proc Natl Acad Sci U S A 75(6):2868–2872

Kuhn SL (2013) Questions of complexity and scale in explanations for cultural transitions in the Pleistocene: a case study from the

Early Upper Paleolithic. J Archaeol Method Theory 20(2):194–211. doi:10.1007/s10816-012-9146-3
Kuhn SL, Stiner MC, Reese DS, Gulec E (2001) Ornaments of the earliest Upper Paleolithic: new insights from the Levant. Proc Natl Acad Sci U S A 98(13):7641–7646. doi:10.1073/pnas.121590798
Luck GW (2007) The relationships between net primary productivity, human population density and species conservation. J Biogeogr 34(2):201–212. doi:10.1111/J.1365-2699.2006.01575.X
Mellars P (2005) The impossible coincidence. A single-species model for the origins of modern human behavior in Europe. Evol Anthropol 14(1):12–27. doi:10.1002/Evan.20037
Mithen S, Reed M (2002) Stepping out: a computer simulation of hominid dispersal from Africa. J Hum Evol 43(4):433–462. doi:10.1006/Jhev.2002.0584
Neiman FD (1995) Stylistic variation in evolutionary perspective—inferences from decorative diversity and interassemblage distance in Illinois Woodland ceramic assemblages. Am Antiq 60(1):7–36
Nikitas P, Nikita E (2005) A study of hominin dispersal out of Africa using computer simulations. J Hum Evol 49(5):602–617. doi:10.1016/J.Jhevol.2005.07.001
Pinhasi R, Higham TFG, Golovanova LV, Doronichev VB (2011) Revised age of late Neanderthal occupation and the end of the Middle Paleolithic in the northern Caucasus. Proc Natl Acad Sci USA 108(21):8611–8616
Powell A, Shennan S, Thomas MG (2009) Late Pleistocene demography and the appearance of modern human behavior. Science 324(5932):1298–1301. doi:10.1126/science.1170165
Premo LS (2007) Exploratory agent-based models: towards an experimental ethnoarchaeology. In: Clark JT, Hagemeister EM (eds) Digital discovery: exploring new frontiers in human heritage; CAA 2006: computer applications and quantitative methods in archaeology: proceedings of the 34th conference, Fargo, Apr 2006. Archaeolingua, Hungary, pp 22–29
Prufer K, Racimo F, Patterson N, Jay F, Sankararaman S, Sawyer S, Heinze A, Renaud G, Sudmant PH, de Filippo C, Li H, Mallick S, Dannemann M, Fu Q, Kircher M, Kuhlwilm M, Lachmann M, Meyer M, Ongyerth M, Siebauer M, Theunert C, Tandon A, Moorjani P, Pickrell J, Mullikin JC, Vohr SH, Green RE, Hellmann I, Johnson PL, Blanche H, Cann H, Kitzman JO, Shendure J, Eichler EE, Lein ES, Bakken TE, Golovanova LV, Doronichev VB, Shunkov MV, Derevianko AP, Viola B, Slatkin M, Reich D, Kelso J, Paabo S (2014) The complete genome sequence of a Neanderthal from the Altai Mountains. Nature 505(7481):43–49. doi:10.1038/nature12886
Raichlen DA, Wood BM, Gordon AD, Mabulla AZP, Marlowe FW, Pontzer H (2014) Evidence of Levy walk foraging patterns in human hunter-gatherers. Proc Natl Acad Sci U S A 111(2):728–733
Shennan S (1989) Archaeological approaches to cultural identity, vol 10, One world archaeology. Unwin Hyman, London/Boston
Shennan S (2001) Demography and cultural innovation: a model and its implications for the emergence of modern human culture. Camb Archaeol J 11(1):5–16
Shennan S (2002) Genes, memes, and human history: Darwinian archaeology and cultural evolution. Thames & Hudson, London
Shennan S, Downey SS, Timpson A, Edinborough K, Colledge S, Kerig T, Manning K, Thomas MG (2013) Regional population collapse followed initial agriculture booms in mid-Holocene Europe. Nat Commun 4:2486. doi:10.1038/ncomms3486
Singarayer JS, Valdes PJ (2010) High-latitude climate sensitivity to ice-sheet forcing over the last 120 kyr. Quat Sci Rev 29(1–2):43–55. doi:10.1016/J.Quascirev.2009.10.011
Sinnott RW (1984) Virtues of the Haversine. Sky Telescope 68(2):159
Slatkin M (1993) Isolation by distance in equilibrium and nonequilibrium populations. Evolution 47(1):264–279
Steele J, Glatz C, Kandler A (2010) Ceramic diversity, random copying, and tests for selectivity in ceramic production. J Archaeol Sci 37(6):1348–1358. doi:10.1016/J.Jas.2009.12.039
Tavare S, Balding DJ, Griffiths RC, Donnelly P (1997) Inferring coalescence times from DNA sequence data. Genetics 145(2):505–518
Thomas MG, Stumpf MP, Harke H (2006) Evidence for an apartheid-like social structure in early Anglo-Saxon England. Proc Biol Sci/R Soc 273(1601):2651–2657. doi:10.1098/rspb.2006.3627
Tishkoff SA, Verrelli BC (2003) Patterns of human genetic diversity: implications for human evolutionary history and disease. Annu Rev Genomics Hum Genet 4:293–340. doi:10.1146/annurev.genom.4.070802.110226
Tremblay M, Vezina H (2000) New estimates of intergenerational time intervals for the calculation of age and origins of mutations. Am J Hum Genet 66(2):651–658. doi:10.1086/302770
Trinkaus E, Zilhão J (2012) Paleoanthropological implications of the Peştera cu Oase and its contents. In: Trinkaus E, Constantin S, Zilhão J (eds) Life and death at the Peştera cu Oase: a setting for modern human emergence in Europe. Oxford University Press, Oxford, pp 881–911
Valdes AM, Slatkin M, Freimer NB (1993) Allele frequencies at microsatellite loci: the stepwise mutation model revisited. Genetics 133(3):737–749
Vanhaeren M, d'Errico F (2006) Aurignacian ethno-linguistic geography of Europe revealed by personal ornaments. J Archaeol Sci 33(8):1105–1128. doi:10.1016/J.Jas.2005.11.017
Wand MP, Jones MC (1995) Kernel smoothing. Monographs on statistics and applied probability, vol 60, 1st edn. Chapman & Hall, London/New York
Wright S (1943) Isolation by distance. Genetics 28(2):114–138
Wright S (1978) Evolution and the genetics of populations: a Treatise in four volumes, volume 4 variability within and among natural populations. The University of Chicago Press, Chicago
Zilhão J (2007) The emergence of ornaments and art: an archaeological perspective on the origins of "behavioral modernity". J Archaeol Res 15(1):1–54. doi:10.1007/S10814-006-9008-1
Zilhão J, Pettitt P (2006) On the new dates for Gorham's Cave and the late survival of Iberian Neanderthals. Before Farm 3:1–9
Zilhão J, Trinkaus E, Constantin S, Milota S, Gherase M, Sarcina L, Danciu A, Rougier H, Quilès J, Rodrigo R (2007) The Peştera cu Oase people, Europe's earliest modern humans. In: Mellars PB, Boyle K, Bar-Yosef O, Stringer C (eds) Rethinking the human revolution. McDonald Institute for Archaeological Research, Cambridge, pp 249–262

Transmission of Cultural Variants in the North American Paleolithic

9

Michael J. O'Brien, Briggs Buchanan, Matthew T. Boulanger, Alex Mesoudi, Mark Collard, Metin I. Eren, R. Alexander Bentley, and R. Lee Lyman

Abstract

North American fluted stone projectile points occur over a relatively short time span, ca. 13,300–11,900 calBP, referred to as the Early Paleoindian period. One long-standing topic in Paleoindian archaeology is whether variation in the points is the result of drift or adaptation to regional environments. Studies have returned apparently conflicting results, but closer inspection shows that the results are not in conflict. At one scale—the overall pattern of flake removal—there appears to have been an early continent-wide mode of point manufacture, but at another scale—projectile-point shape—there appears to have been regional adaptive differences. In terms of learning models, the Early Paleoindian period appears to have been characterized by a mix of indirect-bias learning at the continent-wide level and guided variation at the regional level, the latter a result of continued experimentation with hafting elements and other point characters to match the changing regional environments. Close examination of character-state changes allows a glimpse into how Paleoindian knappers negotiated the design landscape in terms of character-state optimality of their stone weaponry.

Keywords

Clovis • Cultural transmission • Fluted point • Guided variation • Paleolithic • Social learning

M.J. O'Brien (✉) • M.T. Boulanger • M.I. Eren • R.L. Lyman
Department of Anthropology, University of Missouri, Columbia, MO 65211, USA
e-mail: obrienm@missouri.edu

B. Buchanan
Department of Anthropology, University of Tulsa, Tulsa, OK 74104, USA

A. Mesoudi
Department of Anthropology and Centre for the Coevolution of Biology and Culture, Durham University, Durham DH1 3LE, UK

M. Collard
Human Evolutionary Studies Program and Department of Archaeology, Simon Fraser University, Burnaby V5A 1S6, Canada

Department of Archaeology, University of Aberdeen, Aberdeen AB24 3UF, UK

R.A. Bentley
Department of Archaeology and Anthropology, University of Bristol, Bristol BS8 1UU, UK

9.1 Introduction

Cultural-transmission theory has as its purpose the identification, description, and explanation of mechanisms that humans use to acquire, modify, and retransmit cultural information in particular instances, whether it be rules concerning eligible marriage partners or instructions for how to produce fishing nets or any of a countless number of other cultural features (Eerkens et al. 2014). As Mesoudi (2013:131) put it, "this surely places cultural transmission at the heart of pretty much every social science discipline." This certainly is the case in American archaeology and anthropology, where interest in the process and mechanisms of cultural transmission runs deep (e.g., Boas 1904; Kroeber 1923; Mason 1895; Sapir 1916; Tylor 1871). Franz Boas, the oft-identified "father" of American anthropology, for example, pointed out that "the theory of transmission has

A. Mesoudi and K. Aoki (eds.) *Learning Strategies and Cultural Evolution during the Palaeolithic*, Replacement of Neanderthals by Modern Humans Series, DOI 10.1007/978-4-431-55363-2_9,

induced investigators to trace the distribution and history of [cultural traits] with care so as to ascertain empirically whether they are spontaneous creations or whether they are borrowed and adapted" (1904:522). He later noted that "we must investigate the innumerable cases of transmission that happen under our very eyes and try to understand how transmission is brought about and what are the conditions that favor the grouping of certain new elements of an older culture" (Boas 1911:809). The many discussions of cultural transmission that have appeared from the 1980s on rarely mention this earlier work, making it sound as if our forebears ignored the issue, when a more appropriate way of phrasing it would be to say that common sense substituted for rigorous models of transmission (Lyman 2008; Lyman and O'Brien 1997, 2003).

That lack of rigor began to be eclipsed in the 1970s with the mathematical-modeling work of Luca Cavalli-Sforza, a population geneticist, and Marcus Feldman, a theoretical biologist (e.g., Cavalli-Sforza and Feldman 1973, 1981; Feldman and Cavalli-Sforza 1976). The innovative aspect of their approach, which they labeled "gene–culture coevolutionary theory," was that they not only modeled the differential transmission of genes between generations but also incorporated cultural information into the analysis, which allowed the evolution of the two systems to be mutually dependent (Laland and Brown 2011). Cavalli-Sforza and Feldman's work was followed by that of Robert Boyd and Peter Richerson, whose 1985 book, *Culture and the Evolutionary Process*, laid the foundation for what they labeled as "dual-inheritance theory," which for purposes here we view as synonymous with Cavalli-Sforza and Feldman's "gene–culture coevolutionary theory." Boyd and Richerson's (1985) discussion of individual (asocial) versus social learning, especially their attention to transmission biases, would have a significant effect on anthropological and archaeological thought.

There now exist many applications of cultural-transmission theory, both in anthropology and archaeology, that attempt to define these mechanisms mathematically and to model their effects over time (e.g., Aoki 2013; Aoki et al. 2011; Atkisson et al. 2012; Bentley and O'Brien 2011; Bentley and Shennan 2003; Bentley et al. 2004; Bettinger and Eerkens 1997, 1999; Derex et al. 2013; Eerkens and Lipo 2005, 2007; Henrich 2001, 2004, 2006, 2010; Henrich and Boyd 1998; Hoppitt et al. 2010; Kameda and Nakanishi 2002, 2003; Kandler and Shennan 2013; Kandler and Steele 2010; Kempe and Mesoudi 2014; Kempe et al. 2012; Kendal et al. 2009; Kobayashi and Aoki 2012; Kohler et al. 2004; Kuhn 2013; Lipo et al. 1997; McElreath et al. 2005; Mesoudi 2008, 2011a; Mesoudi and Lycett 2009; Mesoudi and O'Brien 2008a, b, c; Nakahashi 2013; Neiman 1995; Powell et al. 2009; Premo 2012, 2014; Premo and Scholnick 2011; Rendell et al. 2011a, b; Rendell et al. 2010; Schillinger et al. 2014; Sharon 2009; Shennan 2000; Steele et al. 2010). Our goal here is not to summarize this extensive body of work (see Laland 2004; Laland and Brown 2011; Mesoudi 2011b; various chapters in this volume) but rather to extract a few points that would appear to be of considerable interest to archaeologists interested in how cultural information is acquired and transmitted. We use as a basis for discussion several studies that have examined variation in North American projectile points that date ca. 13,300–11,900 calendar years before present [calBP], a time span referred to as the Early Paleoindian period. To align our contribution with others in this volume, we can easily refer to that period as the American "Paleolithic."

9.2 Learning Models

Cultural transmission involves learning, which can be usefully subdivided into two categories, social learning and individual learning (Cavalli-Sforza and Feldman 1981; Laland 2004; Mesoudi 2011b). Although the division is analytically useful, it obscures the fact that humans are neither purely social nor purely individual learners. Rather, certain conditions, perceived or real, dictate which is used in any particular situation (Aoki et al. 2012; Bentley et al. 2014; Enquist et al. 2008; O'Brien and Bentley 2011). Humans use social learning for a variety of adaptive reasons (Bentley and O'Brien 2011; Boyd and Richerson 1996; Ehn and Laland 2012; Enquist et al. 2011; Henrich and Broesch 2011; Kameda and Nakanishi 2002; Laland 2004; Mesoudi 2011b; Reader and Laland 2002; Rendell et al. 2010; Richerson and Boyd 2005; Tomasello et al. 1993). They learn their language, morals, technology, how to behave socially, what foods to eat, and most ideas from other people. This process is the basis for human culture, organizations, and technology (Whiten et al. 2011); thus the first published definition of human culture by an anthropologist reads "that complex whole which includes knowledge, belief, art, morals, law custom, and any other capabilities and habits *acquired by man as a member of society*" (Tylor 1871:1, emphasis added). Humans continue to "learn things from others, improve those things, transmit them to the next generation, where they are improved again, and so on," and this process continues to lead to the "rapid *cultural* evolution of superbly designed adaptations to particular environments" (Boyd and Richerson 2005:4, emphasis in original).

Much of the time, social learning is an effort to replicate another's behavior accurately without embellishment. It is a powerful adaptive strategy that allows others to risk failure first (Henrich 2001; Laland 2004)—that is, to let others filter behaviors and to pass along those that have the highest payoff (Rendell et al. 2011a). Copying others is itself a set of competing strategies in that one might preferentially copy someone based on that individual's skill level (copy those who are better at something than you are, copy good social learners, copy those who are successful, and so on), whereas others might base their decisions on social criteria (copy the majority, copy kin or friends, copy older individuals). The various factors that can affect one's choice of whom or what to copy are often referred to as "biases," which in Boyd and Richerson's program are unique evolutionary forces for the selective retention of cultural variants (Marwick 2005). Hence, the term "biased learning" is commonly used as a synonym for certain social-learning strategies (Boyd and Richerson 1985; Laland 2004). Of importance is the difference in the effects of copying based on selection for knowledge or a skill level as opposed to copying based on random social interaction (that is, the term "bias" is used here in a statistical sense to indicate some deviation from random or "unbiased" copying; it is not used in any normative sense, such as "gender bias" or "racial bias"). Our view mirrors that of Rendell et al. (2011a): Copying confers an adaptive plasticity on populations, which allows them to draw on deep knowledge bases in order to respond to changing environments rapidly. High-fidelity copying leads to an exponential increase in the retention of cultural knowledge.

We should insert a few caveats here with respect to copying. First, the term "copying" carries a connotation that someone simply looks over someone else's shoulder, views a product, and then replicates it. This behavior might work on homework or a classroom exam, but it does not apply in many situations. One cannot, for example, watch a homebuilder and his crew construct a house and expect to replicate the behaviors and create a successful product. If the observer starts with moderate skills, he might learn some tricks of the trade, but he can never hope to go away and build a house that mirrors what he saw being constructed. Second, nonrandom copying can take several forms, including (1) "indirect bias," where learners use such criteria as success or prestige as a basis for selecting a model, and (2) frequency-dependent copying, where learners perhaps copy the most-frequent variant, which is often labeled as "conformity." Third, cultural models of model-based transmission often implicitly assume that individuals can find a master teacher from whom to learn. Likewise, models of conformist transmission often implicitly assume that individuals can sense how popular a behavior is in the population. These assumptions are fine for small groups but unrealistic for large populations, where individuals have only local, imperfect knowledge of what models, and hence what behaviors, are optimal (Bentley and O'Brien 2011; Mesoudi and Lycett 2009). Fourth, acquisition costs could affect the ability to copy faithfully (Mesoudi 2011c). This applies to all modes of social learning.

As opposed to learning socially, one can learn individually, or asocially. This is a slow process, wherein an individual modifies existing behaviors through trial and error to suit his or her own needs[1] Perhaps a learner obtains the basic behavior from a parent or master and then begins to tinker with it with no influence from other people. He or she then passes the behavior on to a few others. Boyd and Richerson (1985) refer to this as "guided variation." The guided-variation model shows that, in the absence of selection for a particular trait, a population will move toward whichever trait is favored by people's individual-learning biases. This occurs even when the strength of guided variation is weak (Mesoudi 2011b).

This form of learning is called "unbiased" (Boyd and Richerson 1985; Henrich 2001) because at the *population* level it approximately replicates the distribution of behaviors from the previous generation. After acquiring a behavior or tool, an individual can obtain environmental information about the relative payoffs of alternative skills or tools. If the difference in payoffs is clear, the individual adopts the behavior indicated by the environmental information. If not, the individual sticks with the behavior acquired through unbiased cultural transmission (Henrich 2001). Thus, Boyd and Richerson's (1985) "guided variation" has two equally important components: unbiased transmission and environmental (individual) learning. Henrich (2001) uses the term "environmental learning model" to include both the individual-level learning process, which may occur many times per generation, and its transgenerational counterpart, guided variation (unbiased transmission and individual learning).

9.2.1 Learning Models in Archaeology

Archaeologists have taken advantage of these perspectives on learning to help explain certain patterns in the archaeological record (Mesoudi 2010). One example is Bettinger and Eerkens's (1997, 1999) study of Great Basin projectile points manufactured ca. 1,500–1,200 calBP, following the replacement of the atlatl (throwing stick) with the bow and arrow. Bettinger and Eerkens (1999) observed that specimens of two point types found in adjacent regions of the Great

[1]For an example of individual learning involving stone tools, see Eren et al. (2011a, b). In this example, it took the experimental knapper 18 months to master a Middle Paleolithic lithic technology called "Preferential Levallois," in which a stone nodule's upper surface is carefully shaped such that a large "predetermined" flake can be removed with specific, beneficial morphometric properties (Eren and Lycett 2012). Some researchers have cited this long learning time as evidence for the difficulty of learning the Levallois technique and the high skill necessary to master it (Bar-Yosef 2013; Bar-Yosef and van Peer 2009; Putt et al. 2014). While undoubtedly "Preferential Levallois" represents expert learning (Wynn and Coolidge 2010), it is reasonable to hypothesize that the 18-month-long Levallois learning period of the experimental knapper would have decreased significantly had the learning been social rather than predominantly individual.

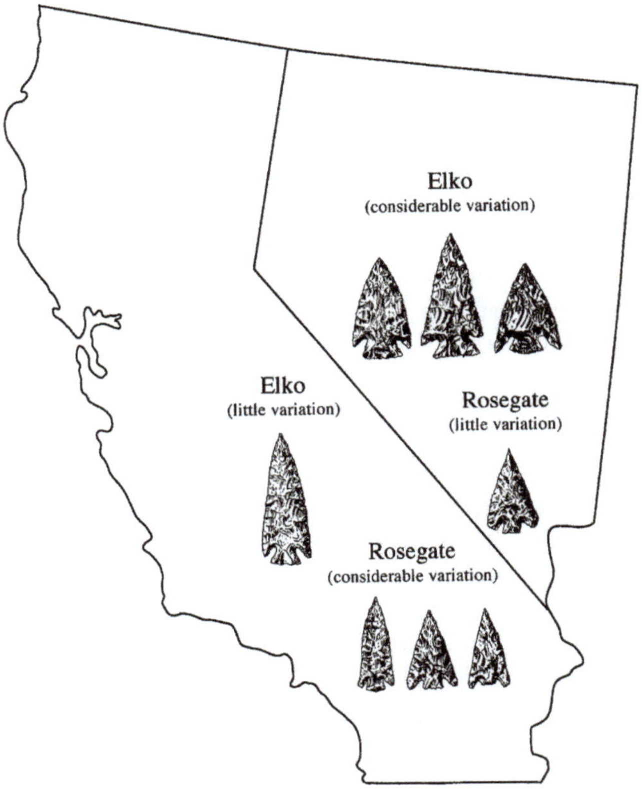

Fig. 9.1 Map of California and Nevada showing variation in the earlier Elko dart points and the later Rosegate arrow points from central Nevada and eastern California. Bettinger and Eerkens (1999) proposed that regional variation among Elko points from central Nevada was perhaps attributable to resharpening whereas those from eastern California were resharpened much less frequently. For Rosegate points, they attributed regional variation to different learning models—indirect bias in central Nevada and guided variation in eastern California

Basin—central Nevada and eastern California—differ in the degree to which attributes such as weight, width, and length correlate with each other (Fig. 9.1). The earlier, Elko points, which were used to tip darts, all have a similar base shape—the primary character used to place specimens in the type—but specimens from central Nevada vary considerably in weight and length, often being light and stubby, whereas those from eastern California are uniformly heavy and long relative to their width. Bettinger and Eerkens (1999) suggested that excessive resharpening drove the highly variable weight and length of Elko specimens from central Nevada. The lack of resharpening seen on specimens from eastern California is perhaps explained by the abundance of high-quality obsidian sources present. Elko points were simply discarded rather than resharpened.

Resharpening, however, cannot explain why the later, Rosegate points, which tipped arrows, are more variable in basal width in eastern California than in central Nevada. Bettinger and Eerkens (1999) attributed these differences to the manner in which prehistoric people of the two regions acquired and transmitted projectile-point technology. Specifically, the attributes of points found in eastern California were found to be poorly correlated with each other, which Bettinger and Eerkens argued was because point designs in that region originally spread as a result of guided variation. Hence, each attribute was subject to separate individual trial-and-error experimentation, causing them to vary independently. In contrast, projectile points of the same material and from around the same period found in central Nevada featured uniform designs with highly correlated attributes. Bettinger and Eerkens (1999) argued that points in that region originally spread as a result of indirect, or model-based, bias, with individuals copying wholesale the design of a single successful model. Hence, differences at the individual level (guided variation vs. indirect bias) can be argued to have generated differences at the population level (uncorrelated attributes vs. correlated attributes).

One inherent limitation in archaeology is that we have access only to population-level historical data. The details of cultural transmission at the level of the individual—who copies what from whom, and how—can only be inferred from these archaeological data, as Bettinger and Eerkens (1999) did, and not directly observed or measured. Mathematical simulations offer one means of addressing this problem, with the results of simple models of cultural transmission matched to archaeological data (e.g., Eerkens et al. 2006). Mathematical models, however, are only as good as their assumptions, in this case assumptions regarding people's propensities to learn socially rather than individually, to conform, to copy the most successful individual, and so on. What are needed are experimental data in order to verify the assumptions and findings of theoretical models.

Some experimental tasks, however, are unrealistically simple. By this we mean that, for example, agents are faced with only two choices, one of which yields a higher payoff. Similarly, it might be assumed that agents exhibit only two traits, one of which has a higher payoff in a particular environment. These scenarios tend to greatly oversimplify real life. For example, even the simplest of human technologies comprise multiple component traits, some of which might be continuous (e.g., projectile-point length) whereas others are discrete but with more than two states (e.g., the shape of a point base). Some traits might be functional (e.g., the thickness or length of points) whereas others are functionless (neutral) (e.g., designs incised on a ceramic vessel). The overall "cultural fitness" of an object is a combination of trait values, each of which interacts with one another as well as with the skill of the manufacturer and user and stochastic factors such as weather conditions (Mesoudi 2014).

Mesoudi and O'Brien (2008a) set out to design a task that was complex enough to yield insights about how people solve real-world technological problems yet simple enough so that the implications of theoretical models of cultural transmission could be tested. Specifically, they tested Bettinger and Eerkens's (1999) hypothesis that the different patterns of projectile-point variation observed in the Great Basin are the result of different cultural transmission processes—guided variation and indirect bias—by experimentally simulating those cultural-transmission processes in the laboratory. They had participants (university students) copy the design of a model after being given information regarding that model's prior success (permitting indirect bias) and then allowed participants to experiment with their point designs in novel selective environments (permitting guided variation). The results matched the patterns of attribute correlation found by Bettinger and Eerkens (1999), with the former points exhibiting highly correlated attributes and the latter points exhibiting less correlated attributes. Hence, more confidence could be placed in Bettinger and Eerkens's (1999) hypothesis that these different archaeological patterns were the result of differences in how projectile-point technology in the two regions was originally transmitted. The finding that cultural transmission was relatively more adaptive when there was a cost to modification suggested a possible explanation for the site differences: Perhaps the prehistoric Nevadan environment was harsher in some respect, imposing a cost on experimentation and necessitating a greater reliance on indirectly biased cultural transmission.

The Mesoudi and O'Brien (2008a) study was designed to also present a more general exploration of cultural transmission in a multimodal adaptive landscape, where point-design attributes are governed by bimodal fitness functions, thus giving multiple locally optimal designs of varying fitness. Mesoudi and O'Brien hypothesized that the divergence in point designs resulting from individual experimentation (the individual-learning component of guided variation) was driven in part by this multimodal adaptive landscape, with different individuals converging by chance on different locally optimal peaks. They argued that indirectly biased horizontal cultural transmission, where individuals search design space and copy the design of the most successful person, allows individuals to escape from local optima and jump to the globally optimal peak, or at least the highest peak found by people in that group (Fig. 9.2) (Lake and Venti 2009; Mesoudi 2008). Mesoudi and O'Brien's experimental results supported this argument, with participants in groups outperforming individual controls when the group participants were permitted to copy each other's point designs. Computer simulations confirmed that this social-learning strategy of "copy-the-successful" was more adaptive than a number of other social-learning strategies, especially in groups of more than 50 people, which have been typical throughout much of human evolution (Dunbar 1992), and showed that the multimodal-adaptive-landscape assumption was key to this advantage.

This latter finding is potentially important, as it demonstrates that the nature of the selective environment will significantly affect aspects of cultural transmission. To reiterate, whereas most previous experiments (e.g., Kameda and Nakanishi 2002, 2003; McElreath et al. 2005) used relatively simple learning tasks that required a participant to

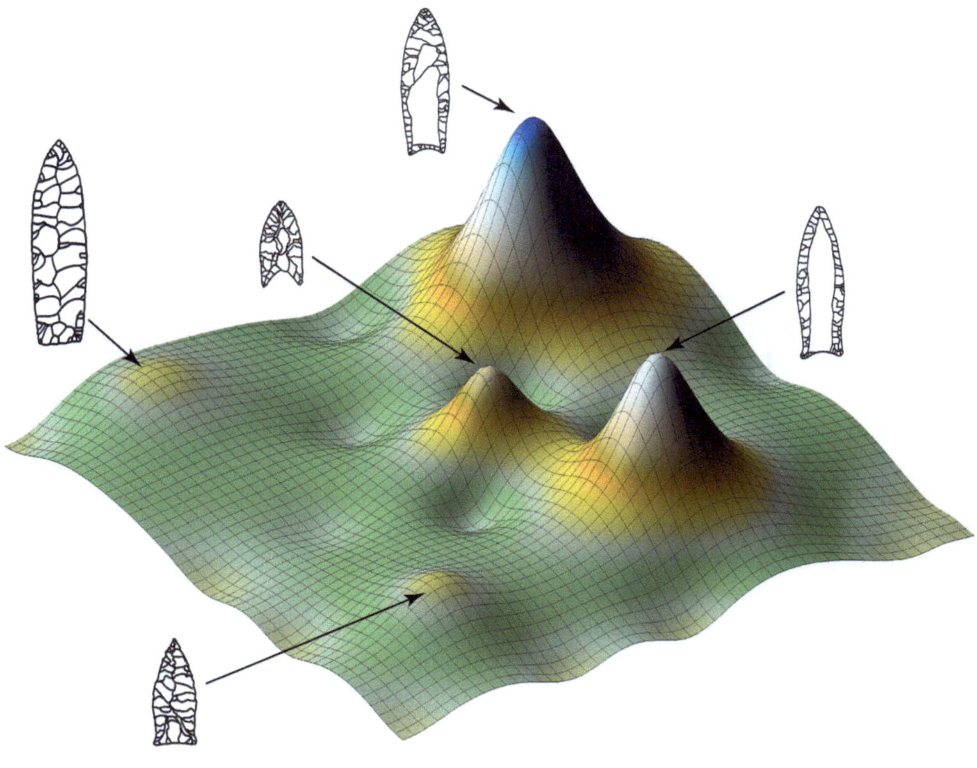

Fig. 9.2 A fictional, and highly simplified, multimodal adaptive landscape of point design. In this design universe, concave-base points have a higher fitness than those with straight bases, and fluted points have a higher fitness than those without flutes. Even among fluted points, however, there are differences. Here, Clovis points are shown as the highest peak—they are globally optimal relative to our design landscape—whereas Cumberland points, which are fluted from the base to the tip, occupy a lower peak—they are suboptimal. Importantly, suboptimality is a relative term. Cumberland points obviously did what they were intended to do in the environment(s) in which they were used

select one of two options (e.g., crops or rabbit locations), Mesoudi and O'Brien used a more complex learning task involving multiple continuous and discrete, functional and adaptively neutral attributes, some of which had bimodal fitness functions. The resulting multimodal adaptive landscape was instrumental in generating and maintaining diversity in the virtual-point designs. They also found in the model that the "copy-the-successful" strategy outperformed the "copy-the-majority" strategy. Indeed, the latter performed no better than individual learning because individuals are just as likely to converge on a local optimum as on a global optimum in the absence of information regarding the success of those individuals, unless individuals at the global optimum outcompete individuals at the local optima and become the majority.

How realistic is this assumption of a multimodal adaptive landscape? Boyd and Richerson (1992) have argued that multimodal adaptive landscapes are likely to be common in cultural evolution and may significantly affect the historical trajectories of artifact lineages, just as population-genetic models suggest that multimodal adaptive landscapes have been important in biological evolution by guiding historical trajectories of biological lineages (Arnold et al. 2001; Lande 1986; Simpson 1944). As we noted earlier, any problems and tasks faced by modern and prehistoric people would have had multiple solutions, some better than others, but all better than nothing. Further, solutions are likely to represent compromises between multiple functions and requirements. With respect to projectile points, for example, Cheshier and Kelly (2006) summarized experimental evidence for tradeoffs in point designs among such factors as accuracy, range, killing power, and durability, noting that "thin, narrow points have greater penetrating power, but wide, thick points create a larger wound that bleeds more easily" (p. 353). Such functional tradeoffs would potentially produce multiple locally optimal point designs, with, for example, one optimal design maximizing penetrating power and another maximizing bleeding.

9.3 The North American Paleolithic and Fluted Points

How might these learning models help us in understanding the cultural landscape of Paleolithic North America after its initial colonization? The exact timing of the colonization is open to question, as is the exact point of entry into the interior of the continent, but what is *not* in question is the point of origin of the early colonists. Despite a few claims to the contrary (e.g., Stanford and Bradley 2012), the overwhelming archaeological and archaeogenetic evidence indicates that humans entered North America by way of Beringia (Goebel et al. 2008; Kemp and Schurr 2010; Morrow 2014; O'Rourke and Raff 2010; Raff and Bolnick 2014; Raff et al. 2010; Rasmussen et al. 2014; Waters and Stafford 2007). Descendants of these migrants moved eastward and then south of the Cordilleran and Laurentide ice sheets, perhaps through an ice-free corridor that ran northwest to southeast through Canada (Catto and Mandryk 1990; Mandryk et al. 2001), and developed a technology known as Clovis (Goebel et al. 2008), which at 13,300–12,800 calBP represents the earliest well-documented archaeological evidence of human occupation of North America.[2]

Clovis is marked by the widespread occurrence of bifacially chipped projectile points that are lanceolate in form, have parallel to slightly convex sides and concave bases, and exhibit a series of flake-removal scars—"flutes"—on one or both faces that extend from the base to about a third of the way to the tip (Bradley 1993; Buchanan and Collard 2010; Buchanan et al. 2012, 2014; Morrow 1995; Sholts et al. 2012; Wormington 1957) (Fig. 9.3). These points were used to tip spears that were thrust and/or thrown. Clovis points were first documented in the American Southwest (Cotter 1937, 1938; Figgins 1927), where they were found alongside the remains of extinct mammals such as mammoth and large bison. They have since been found throughout North America, including Canada and northern Mexico (Anderson and Faught 1998, 2000; Anderson et al. 2010; Buchanan and Collard 2007, 2010; Buchanan et al. 2012; Goebel et al. 2008; Haynes 1964; Holliday 2000; Prasciunas 2011; Sanchez 2001; Sholts et al. 2012; Smallwood 2012; Waters and Stafford 2007).

It has long been suspected that Clovis points originated in the West—the earliest radiocarbon dates (not all of them are universally accepted [e.g., Waters and Stafford 2007]) are from the Aubrey site in northern Texas (ca. 13,450 calBP) and the Sheaman site in Wyoming (13,210 calBP)—but one credible date from the Southeast—Sloth Hole in Florida (Waters and Stafford 2007), at 12,900 calBP—falls inside the 13,300–12,800 calBP date range. With the exception of six radiocarbon dates on hawthorn (*Crataegus sp.*) seeds from Shawnee-Minisink in Pennsylvania (Dent 2007; Gingerich 2007, 2013), at ca. 12,865 calBP, the earliest dates from archaeological sites in the Northeast that have produced large numbers of fluted points—Bull Brook in Massachusetts (Byers 1954; Robinson et al. 2009), Vail in Maine (Gramly 1982), and Debert in Nova Scotia (MacDonald 1968)—consistently fall later than the earliest fluted-point dates in the West (Bradley et al. 2008; Curran 1996; Haynes et al. 1984; Levine 1990; Miller and Gingerich 2013a, b; Robinson et al. 2009).

In the western United States, especially the Plains and Southwest, Clovis points were followed by Folsom points,

[2]Waters and Stafford (2007) use a slightly more conservative span for Clovis, with a maximum span of 13,250–12,800 calBP and a minimum span of 13,125–12,925 calBP.

Fig. 9.3 Clovis points from various North American sites (Photo by Charlotte D. Pevny; courtesy M. R. Waters)

which tend to be smaller in size than Clovis and to have deeper and longer channel flakes (Ahler and Geib 2000; Buchanan and Collard 2010; Collard et al. 2010; Crabtree 1966; Wormington 1957) (Fig. 9.4). Folsom points date to ca. 12,800–11,900 calBP, with the earlier points found in the Northern Plains and the younger ones in the Southern Plains (Collard et al. 2010). In the East, Clovis points were followed by a host of fluted forms such as Gainey/Bull Brook and Crowfield in the Northeast and Great Lakes region; Dalton, Quad, and Cumberland over much of the South and Midsouth; and Simpson and Suwannee in the extreme Southeast (Anderson 1990, 2013; Anderson et al. 1996, 2010; Bradley 1997; Bradley et al. 2008; Brennan, 1982; Bullen 1968; Goodyear 1982; Lewis 1954; MacDonald 1968; Mason 1962; O'Brien et al. 2001; Robinson et al. 2009; Thulman 2007, 2012) (Fig. 9.4).

Not surprisingly, specimens in all these types exhibit variation in size and shape, some more than others. There is, for example, considerable variation among what archaeologists typically would label as Clovis points (Haynes 2013), whereas Folsom points appear to be more standardized in shape, possibly because the Folsom hafting technique had stricter requirements than the Clovis technique (Amick 1995; Buchanan 2006; Judge 1970; Tunnell and Johnson 1991). Focusing on Clovis points, what might account for the variation? Is it the result of drift—that is, is it random—or is there regional patterning that might suggest an adaptive reason? Buchanan et al. (2014) refer to the former as the *continent-wide adaptation hypothesis*. It holds that Clovis groups did not adjust the shape of their points in relation to local environmental conditions (Buchanan and Hamilton 2009; Byers 1954; Haynes 1964; Kelly and Todd 1988; Krieger 1954; Robinson et al. 2009; Sholts et al. 2012; Willey and Phillips 1958) and that variation in shape is the result of drift (Hamilton and Buchanan 2009; Morrow and Morrow 1999). The alternative—the *regional environmental adaptation hypothesis* (Buchanan et al. 2014)—posits that Clovis groups did adapt their hunting equipment to the characteristics of prey and local habitat, which resulted in regional differences in projectile-point shape (Anderson 1990; Meltzer 1988, 1993; Smallwood 2012; Storck and Spiess 1994; Witthoft 1952, 1954).

Fig. 9.4 Tentative chronology of Paleoindian fluted-point types from eastern North America

Buchanan and Hamilton (2009) expanded on the distinction between the competing hypotheses. With respect to projectile points, they defined drift as a measurable change in point form because of neutral stochastic processes caused by sampling effects that occur as the result of cultural transmission in finite, naturally fluctuating populations (Neiman 1995; Shennan and Wilkinson 2001). As a consequence of sampling, drift is amplified in smaller populations, where the number of people from whom to copy, and the number of objects or traits to copy, are limited (Bentley and O'Brien 2011). This process is heightened when populations bud off and become isolated from a parent population (Shennan 2000, 2001). This is known as the "founder effect"—smaller populations retain only a limited set of the cultural variation exhibited among the original population, which is then subject to drift. As Atkinson (2011) points out, the founder effect has been used to explain numerous patterns of variation in cultural replicators, including human material culture (e.g., Diamond 1978; Henrich 2004; Lycett and von Cramon-Taubadel 2008; Rogers et al. 2009).

In contrast to drift, adaptive modifications can be made to improve the functional performance of projectile points in specific environments. Buchanan and Hamilton (2009) proposed that functional innovations made to projectile points are expected to be the result of guided variation, a combination of unbiased cultural transmission and individual learning within specific environmental constraints. For example, in open environments it might be beneficial to have improved aerodynamic capabilities of weapons launched through the air (Lipo et al. 2012), or when hunting prey with comparatively thick hides it might pay to reduce impact-related fractures through the alteration of point shape or hafting arrangements (Cheshier and Kelly 2006; Frison 1989; Hutchings 1997; Musil 1988).

With respect to projectile-point types, how do we explain what appears to be considerably greater regional point diversity in the later portion of the Early Paleoindian period in the East (Fig. 9.4) than what occurred on the Plains (Fig. 9.4), where Folsom was the dominant form for 800–900 years? Was it a continuation of regional adaptation, as Buchanan et al. (2014) proposed for Clovis variation? In terms of learning models, could it reflect continued guided variation? Anderson and Faught (2000; see also Anderson et al. 2011) point out that disruptions in climate and food resources associated with the Younger Dryas (12,890–11,680 calBP), coupled with the disappearance of megafauna (Boulanger and Lyman 2014; Grayson 2007; Meltzer and Mead 1983), could have led to changes in logistic patterns (Boulanger et al. 2015). Large-distance movements may have given way to more-localized movements geared toward a wider range of small animals and plants. Anderson and Faught (2000) propose that the distribution of several projectile-point types—Suwannee and Simpson (Fig. 9.4), for example—within circumscribed ranges in the Southeast might reflect the beginning of that trend.

9.3.1 Studies of Variation in Clovis Points

To place these hypotheses in perspective, we briefly discuss five studies that have attempted to quantify and account for variation in Clovis points. The first four found no direct evidence of regional adaptation whereas the fifth did. The studies are important for what they tell us not only about the Clovis cultural landscape but also about potential limitations of some of the methods used to capture variation in projectile points.

9.3.1.1 Morrow and Morrow (1999)

Using four ratios derived from linear measurements of 449 fluted points from North America, 31 points from Central America, and 61 points from South America, Morrow and Morrow (1999) showed that changes in the form of Early Paleoindian points were positively correlated with latitude, with points becoming more stemmed and shouldered the farther south the sample, culminating in the "Fishtail" points of South America. They considered two possible mechanisms to explain the patterns: (1) point variation was the result of adaptive responses to local environmental conditions across the continents; or (2) variation was a result of drift, which Morrow and Morrow 1999:227) defined as "a process inherent in the ongoing translation of cultural practices from one generation to another under specific geographic and historical circumstances." They discounted adaptation as an explanation for the change in points because they did not detect any correlation between point form and environment. As Buchanan and Hamilton (2009) noted, however, Morrow and Morrow did not specify the environmental parameters that would be necessary to assess the relationship. Although they proposed stylistic drift as the likely mechanism for the change in points, a formal test of their hypothesis was not presented but clearly is warranted before either hypothesis can be rejected.

9.3.1.2 Buchanan and Hamilton (2009)

To test Morrow and Morrow's (1999) hypothesis, Buchanan and Hamilton (2009) generated shape data by measuring 12 interlandmark characters on a sample of 232 points from 26 North American assemblages. They also collected from the literature data on several measures of late Pleistocene regional environmental variation—net primary production, prey availability, prey selection, and prey body size—from eight subregions defined on the basis of physiographic association (Cannon 2004). They then used simple and partial Mantel tests to assess the significance of the correlation between matrices representing point shape and regional measures of environmental variation. They also tested the correlation between point shape and the possible confounding factors of geographic distances among sites, assemblage size, and site type (e.g., kill site versus residential site). Buchanan and Hamilton (2009) found no significant correlations between projectile-point shape and region-specific environmental factors, indicating that variation in shape was not the result of technological adaptive responses to local environmental conditions and therefore was more parsimoniously attributable to drift. They did find evidence of spatial autocorrelation, where regional variation in point shape correlated significantly and positively with geographic distances among sites, as would be expected in situations where populations close in proximity share either cultural phylogenetic histories or extensive horizontal transmission. This is compatible with a scenario of demic splits, which result in regional populations budding off from source populations while maintaining connections through social networks.

9.3.1.3 Hamilton and Buchanan (2009)

Hamilton and Buchanan (2009) used the same 232-point sample as used in the Buchanan and Hamilton (2009) study to examine spatiotemporal gradients in projectile-point size across North America. An earlier study (Hamilton and Buchanan 2007) showed that spatial gradients in Clovis-age radiocarbon dates indicate that the most likely origin of the Clovis colonization of North America was the ice-free corridor. Their analysis demonstrated that the date of the earliest Clovis occupation across the continent decreased linearly with distance from Edmonton, Alberta, traditionally taken to represent the approximate location of the southern exit of the ice-free corridor (Martin 1967; Mosimann and Martin 1975). Thus spatial gradients in Clovis occupations across the continent also reflect temporal gradients.

Hamilton and Buchanan (2009) found that projectile-point size mapped onto the gradient, with size decreasing as sample geographic origin occurred farther away from Edmonton. They also found that the variance in point size was statistically constant over time, which is consistent with biased social-learning practices. They noted that

> It is easily understandable why biased learning strategies would have played an important role in Clovis technologies. Clovis projectile point technology is complex and would have required a significant amount of investment both in terms of time and energy to learn effectively. Under these conditions it is likely that there was a significant amount of variation among the skill-level of flintknappers, such that recognized master flintknappers likely would have held considerable prestige. (Hamilton and Buchanan 2009:67)

Hamilton and Buchanan (2009) further proposed that in a fast-moving and fast-growing population subject to the widespread late Pleistocene environmental changes, conformist bias—copy the majority—would also have been a highly effective strategy for learning, alongside prestige bias—copy the most-skilled point maker. This was their rationale:

> Under circumstances where ecological conditions change on a generational level, the mean trait value is often optimal, leading to frequency-dependent bias, or conformism (Henrich and Boyd 1998). If ecological conditions change much faster than this, social learning will favor trial-and-error learning leading to increased variance. Although the Clovis time period would have seen widespread ecological change over time and space, the rate of this change may not have been experienced within a lifetime (Alroy 2001). As such, Clovis social learning likely involved a combination of both prestige bias and conformism, which had the effect of limiting variance over time.

9.3.1.4 Sholts et al. (2012)

Sholts et al. (2012; see also Gingerich et al. 2014) used laser scanning and Fourier analysis to examine flake-scar patterns on a sample of 34 Clovis points from 7 sites in the Southwest, Southern Plains, and Northern Plains, 5 specimens from the Meekins Neck site in Dorchester County, Maryland (Lowery and Phillips 1994), and 11 modern replicates made by an expert flintknapper. Their analyses suggested that flaking patterns were similar across regions (but not with respect to the replicates), and they concluded that there was a continent-wide standardization of Clovis technology. They tied this to direct transmission from craftsman to craftsman:

> Low flake scar variability among the ancient Clovis points suggests that when the Clovis style swept across the continent, it did not spread via Clovis artisans simply copying finished projectile points or independently developing techniques through trial-and-error. Instead, the similar flake scar patterns suggest that the ancient Clovis points were all created with a very consistent technology. . . . [T]he relative uniformity of flake scar patterns among the geographically diverse Clovis assemblages most likely reflects the Clovis artisans sharing their technical knowledge through direct transmission, i.e. by one knapper showing another the "proper" way to fashion a Clovis-style projectile point.

Sholts et al. (2012) suggested that their scenario was supported by results from modern experimental archaeology. Ferguson (2003), for example, found comparable ranges of variation between points he made and those made by novice knappers whom he had directly assisted as part of an intensive learning process. Conversely, Whittaker (1984) reported that when modern knappers have attempted to copy template points using strategies they acquired on their own or through training, a number of differences between the replicate points and the template points were observed.

9.3.1.5 Buchanan et al. (2014)

Buchanan et al. (2014) re-examined the continent-wide-adaptation versus regional-adaptation hypotheses using the same sample of 241 points employed by both Buchanan and Hamilton (2009) and Hamilton and Buchanan (2009) but expanding it by nine points from four additional assemblages. As opposed to using interlandmark differences to determine shape, as Buchanan and Hamilton (2009) had done, Buchanan et al. (2014) used geometric morphometrics, which creates relative warps, or the principal components of the shape variables. The principal components reflect the major patterns of shape variation within a group of specimens. Figure 9.5 (top) shows the consensus configuration of landmarks, which represents the average shape of all points in the sample. The average point represented in the consensus configuration has a lanceolate-shaped blade and a concave base. The two basal landmarks (landmarks 2 and 3) are the most variable; variation associated with individual landmarks decreases toward the tip (Fig. 9.5 [bottom]).

Figure 9.6 plots the first two relative warps by region. The first relative warp, representing 85 % of the overall variation, is plotted on the X-axis; the second relative warp, representing 4.3 % of the overall variation, is plotted on the Y-axis. Overlap among the regions is evident, but points from the East are more variable than those from the West, particularly along the second relative warp. The wireframes in Fig. 9.6 show deformation from the consensus configuration at the positive and negative ends of each axis to illustrate Clovis shape space. That space is defined along the first relative warp by elliptical blades with deeply concave bases to the left (negative end)—represented by a point from Shoop (Pennsylvania)—and by more linear blades with shallow, rounded concave bases to the right (positive end)—represented by a point from Simon (Idaho). Along the second relative warp, Clovis shape space is defined by lanceolate blades with straight bases at the upper (positive) end—

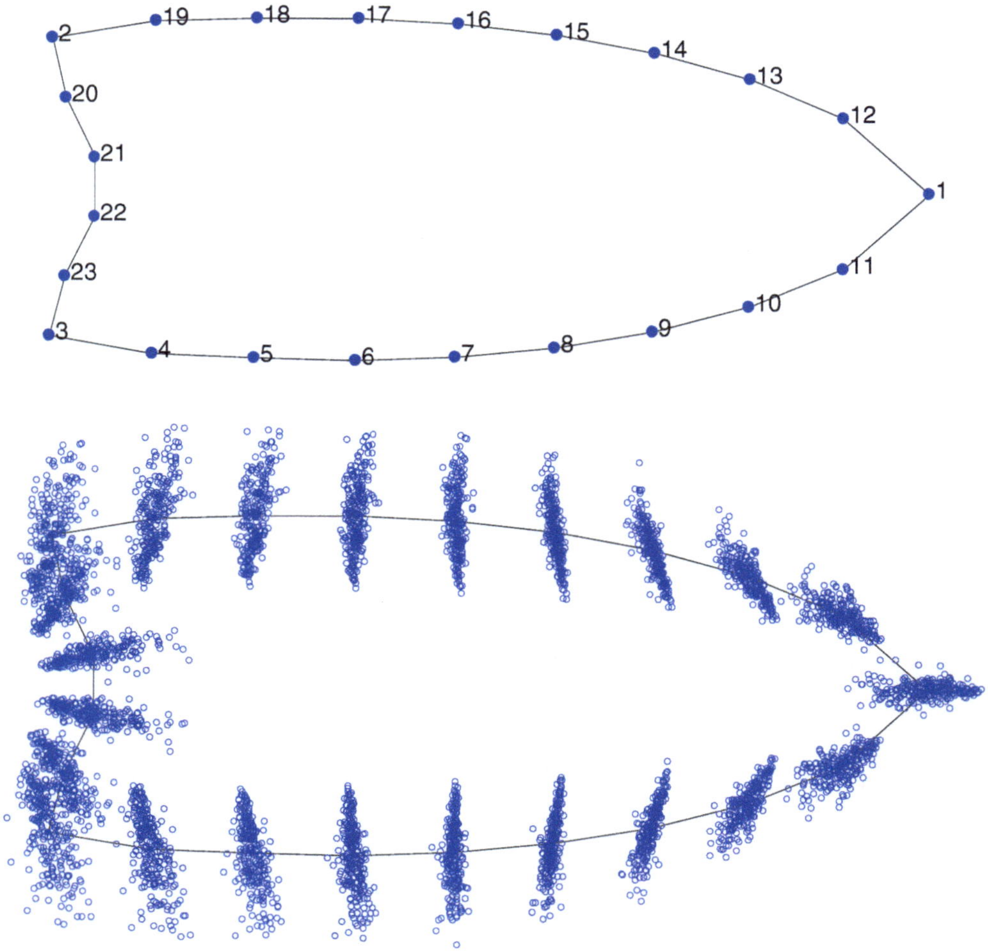

Fig. 9.5 Results of a geometric morphometric shape analysis of 241 Clovis points landmark: *top*, consensus configuration of all landmark configurations; *bottom*, variation in landmark configurations after being translated, scaled, and rotated (From Buchanan et al. 2014)

represented by a point from Murray Springs (Arizona)—and more deltoid blades with deep, concave bases at the lower (negative) end—represented by a point from Vail (Maine). These shape spaces have been casually identified previously (see summary in Gingerich et al. 2014) but not with the precision of the Buchanan et al. (2014) study.

Significance tests showed that among the four subregions in the East, points from the Northeast were significantly different from those from the Midatlantic, Great Lakes, and Midcontinent. In the West, points from the Northwest were significantly different from those from the Southern Plains and Southwest, and Northern Plains points were different from Southern Plains points.

9.3.1.6 Explaining the Interstudy Differences

Why the difference in findings relative to Clovis points? There are at least two reasons, neither of which has to do with the fact that in some studies different samples were used. Buchanan et al. (2014), for example, used virtually the same sample used by Buchanan and Hamilton (2009) and Hamilton and Buchanan (2009), yet came to different conclusions. One reason for the difference probably relates to the different methods used to characterize projectile-point shape. Buchanan and Hamilton (2009) used interlandmark distances to capture point shape, whereas Buchanan et al. (2014) employed geometric morphometrics. The latter approach is known to detect shape similarities and differences better than the former approach (O'Higgins 2000; Slice 2007), and it is likely that the Buchanan et al. (2014) study picked up subtle variation that was undetected by the technique used by Buchanan and Hamilton (2009).

With respect to the Buchanan et al. (2014) study and the Sholts et al. (2012) study, we think there is another reason for the difference: The former examined shape and the latter flake-scar patterning (O'Brien et al. 2014). One clever, and highly significant, aspect of the Sholts et al. (2012; see also Gingerich et al. 2014) study that tells us quite a bit about Clovis-period learning was their inclusion of 11 replicate Clovis points made by Woody Blackwell, who was well known in the knapping world of the 1990s for his ability

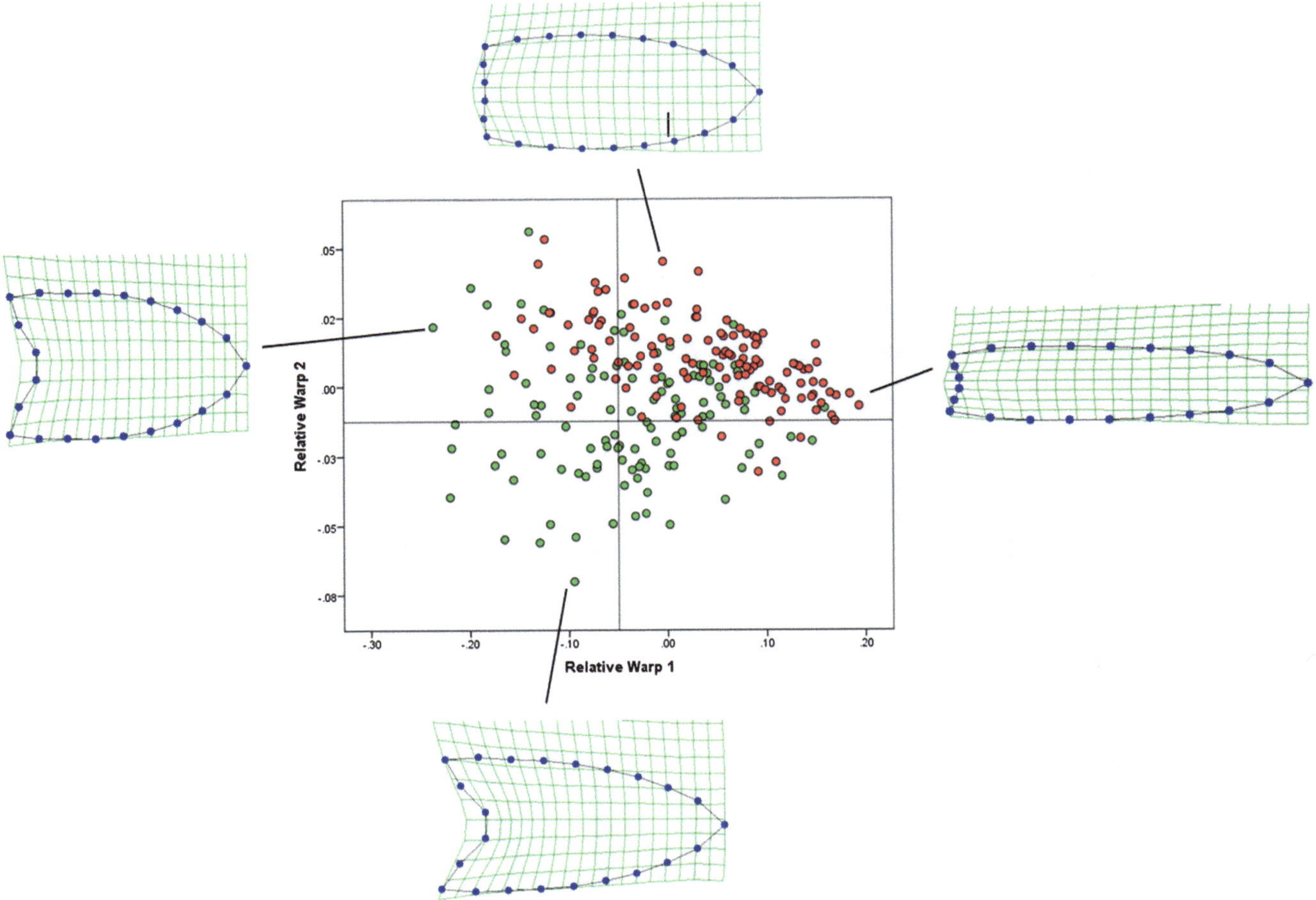

Fig. 9.6 Bivariate plot of relative warp 1 (85 %) against relative warp 2 (4.3 %) for 241 Clovis points (from Buchanan et al. 2014). *Red circles* indicate points from the West and *green circles* indicate points from the East. The four images are deformations from the consensus configurations and display the shape space defined by the first two relative warps. The *upper point* is from Murray Springs (Arizona), the point at the *right* is from Simon (Idaho), the *lower point* is from Vail (Maine), and the point at the *left* is from Shoop (Pennsylvania) (From Buchanan et al. 2014)

to make "superb Clovis points" and "large pieces as thin as anyone could make them" (Whittaker 2004:258). Blackwell copied points from the Drake Clovis cache in Colorado and not only passed them off to a highly knowledgeable collector as authentic but fooled any number of professional archaeologists familiar with Clovis artifacts.

How was Blackwell able to get away with it, at least initially? The answer is, he was a master flintknapper and was able to reverse engineer certain aspects of the Drake points (Preston 1999). Until the study by Sholts et al. (2012), it was widely believed that Blackwell's replicas were all but perfectly executed, and that his mistake, which eventually revealed the points' inauthenticity, was his choice of Brazilian quartz as the raw material for some of the replicas. Sholts et al.'s analysis showed, though, that there was another giveaway: As skilled a knapper as Blackwell was, he could not faithfully copy a Clovis knapper's pattern of flake removal. As Blackwell later said (Preston 1999:85), "I just stopped and looked at this piece and said, 'That really looks like a Drake-style Clovis if I stop right there.' Until then, I had always kept going, cleaning up the edges, making the point smoother, getting the symmetry dead on, and really dressing the thing up. What I'd been losing was its immediacy, its simplicity."

Superimposed front and back flake-scar contours on four points used in Sholts et al.'s (2012) study—one each from the Colby site (Wyoming) and the Drake cache and two of Blackwell's replicates—are shown in Fig. 9.7. Note the difference between the replicates and the authentic points. Figure 9.8 shows the results of a principal components analysis that was carried out to identify shape deviations among the 100 flake-scar contours (front and back) on the sample of 50 points. Most flake-scar contours cluster in the center of the diagram, with the most extreme outlying contours being those of the replicates. In other words, Blackwell could sometimes replicate the flake-removal pattern of a Clovis knapper—note that in terms of the principle components analysis (Fig. 9.8) some of the replicates are indistinguishable from authentic Clovis points—but he was inconsistent in his ability to do so.

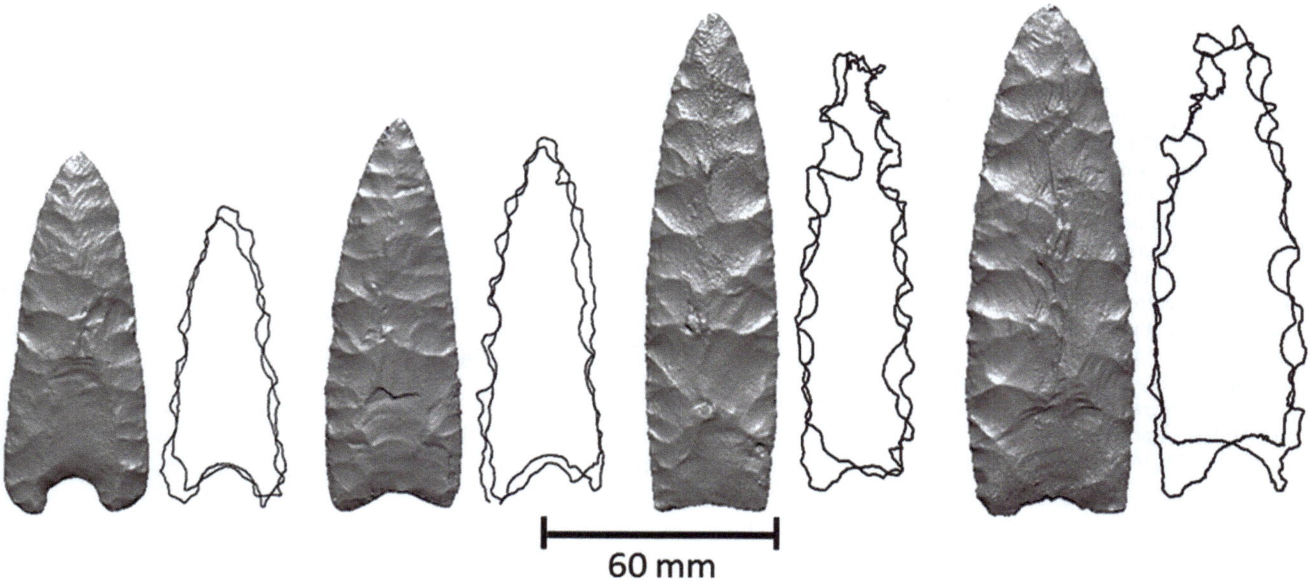

Fig. 9.7 Images of three-dimensional models and overlaid front and back flake-scar contours from prehistoric Clovis points from Colby (*left*) and Drake (*center left*) and two replicate Clovis points (*center right* and *right*) (From Sholts et al. 2012). Despite the markedly different bases on the Colby and Drake points, there is little difference in their flake-scar contours. For the two replicas, their flake-scar contours are more uneven relative to what is seen on prehistoric Clovis points. The replicas also display larger differences between overlaid front and back contours than what is seen on prehistoric specimens

Results of this study support Tankersley's (2004:54) point that biface-manufacture technology is "as much a signature of Clovis as is the morphology of its characteristic projectile point" (Tankersley 2004:54). As Sholts et al. (2012) note, this is especially true for points recovered from the Colby site (Frison and Todd 1986), which have distinctive "C"-shaped bases as opposed to the "typical" Clovis base shape. Despite the odd base shape, they are consistently referred to as Clovis points (Frison 1983). Note that the Colby flake-scar contours are similar to the flake-scar contours of Clovis points from the other assemblages Sholts et al. (2012) examined (Fig. 9.7). In particular, the Drake flake-scar contours show a closer resemblance to those of the Colby points than to the contours of Blackwell's replicates, which he made to mimic the Drake points. Consequently, it appears that the Colby points were manufactured using the same flake-removal process as the other Clovis specimens in the study; they just have a unique base shape. Sholts et al. (2012:3024) believe "this technological uniformity—without evidence for diversification, regional adaptation, or independent innovation—is consistent with Clovis being a short-lived phenomenon."

In summary, taken together the five studies suggest that Clovis learning appears to have been more complicated than any single study demonstrates. The Sholts et al. (2012) study, even with a small sample, indicates there was a standard Clovis lithic-reduction technology that occurred across North America. Whether this standardization was the result of "personal interaction and direct transmission of technological knowledge between Clovis age knappers," as Sholts et al. (2012:3025) propose, is perhaps unknowable, but in terms of learning models, it appears that a good case can be made for some form of biased transmission across the continent (Boulanger et al. 2015; O'Brien et al. 2014). It is understandable why biased learning strategies would have played a key role in fluted-point technologies (Hamilton 2008; Hamilton and Buchanan 2009). The manufacture of a Clovis or Folsom point is a complex procedure that would have required a significant amount of investment both in terms of time and energy to learn effectively (Bradley et al. 2010; Crabtree 1966; Whittaker 2004). Under these conditions, it is likely that there was significant variation among the level of skill exhibited by toolmakers (Bentley and O'Brien 2011; Henrich 2004, 2006)—one does not become a flintknapper, let alone an accomplished one, overnight (Olausson 2008; Pigeot 1990)—such that recognized craftsmen could have held considerable prestige (Hamilton 2008).

Prestige bias—learning from (not simply copying) certain individuals to whom others freely show deference or respect in order to increase the amount and accuracy of information available to the learner (Henrich and Gil-White 2001; Reyes-Garcia et al. 2008)—allows a learner in a novel environment to quickly choose from whom to learn (provided the population is not so large as to "swallow up" highly skilled individuals [Bentley and O'Brien 2011]), thus maximizing his or her chances of acquiring adaptive behavioral solutions to a specific task or enterprise without having to assess directly the adaptiveness of every potential model's behavior

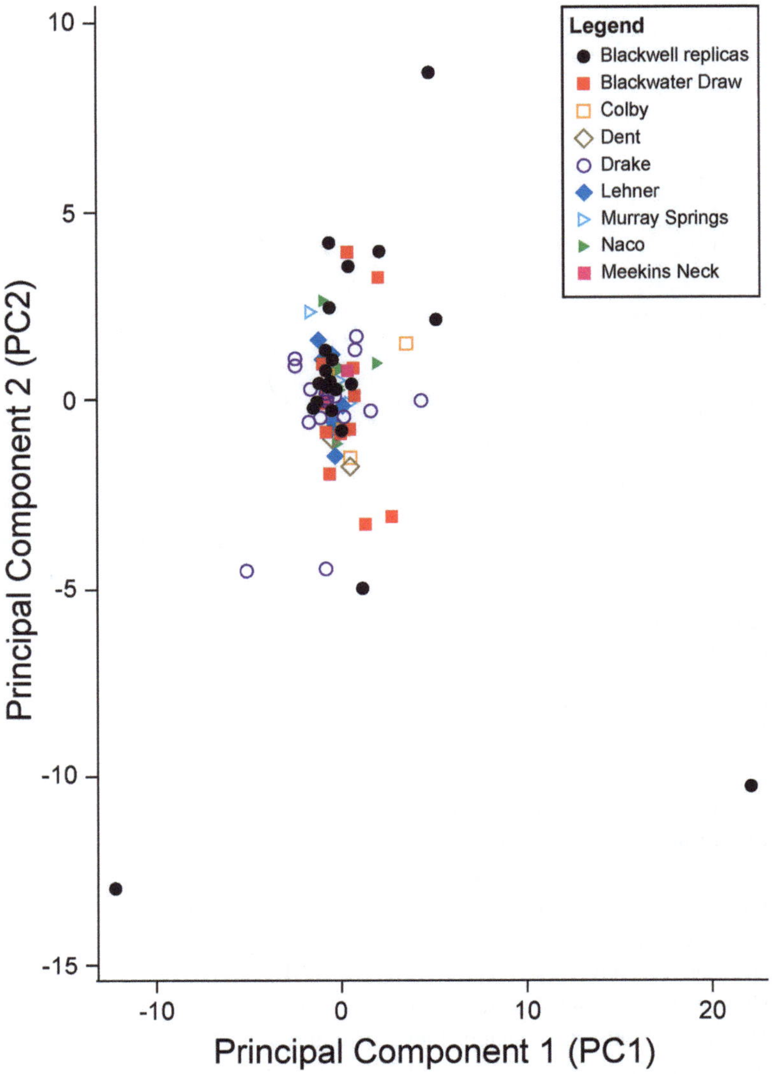

Fig. 9.8 Principal components analysis results for 100 analyzed flake-scar contours on 50 authentic and replicate Clovis points, showing the first principal component (PC 1) versus the second principal component (PC 2) for each contour (From Sholts et al. 2012). Note the outlying *black circles* representing modern replicas for which flake-scar contours deviate from the average shape. Also note that the hollow *orange squares* representing the Colby specimens appear in the center of the diagram, showing that their flake-scar patterns have shapes similar to the other Clovis points

(Atkisson et al. 2012). In a fast-moving and fast-growing population subject to the widespread environmental changes of, say, the North American late Pleistocene landscape, prestige bias would have been a highly effective strategy for social learning (Hamilton 2008) because under circumstances where ecological conditions change, say, on a generational scale, the mean trait value is often optimal, leading to frequency-dependent bias, or conformism (Henrich and Boyd 1998). However, if ecological conditions change faster, social learning may favor individual trial and error or even a combination of the two (Mesoudi 2008; Toelch et al. 2009).

Results of the Buchanan et al. (2014) study—that there is some regional variation in point shape—is in no way at odds with the Sholts et al. (2012) findings of technological uniformity (O'Brien et al. 2014). We propose that patterns of flake removal are less sensitive to adaptive change driven by environmental conditions than is point shape because flaking is less strongly linked to performance than point shape is (Buchanan et al. 2014). In other words, Clovis flintknappers across North America used the same methods to produce points that were similar in flaking pattern yet, where needed, were adapted to different environmental conditions. At the regional level, this takes the appearance of guided variation, with one regional "group" varying its points one way and another regional "group" varying them in an alternative, and oftentimes subtle manner.[3] It is that subtle variation

[3]Recent analysis of Clovis points from one environmentally homogeneous region of the Upper American Midwest demonstrates that although production technique was the same across the sample, differences in shape occur and are highly correlated with the type of chert used to manufacture the points (Eren et al. 2015). These dichotomous results indicate that Clovis foragers engaged in two tiers of social learning. The lower, and more ancestral, tier relates to point flake-scar patterning and can be tied to conformist transmission of ancestral tool-making processes across the Clovis population. The upper, and more-derived, tier relates to point shape. In this case it can be tied to drift that resulted from increased forager interaction at different stone-outcrop hubs. Eren et al. (2015) suspect that we are viewing the very beginnings of a relaxation of social mechanisms that normally

that was just below the visibility threshold in the Buchanan and Hamilton (2009) study but that was picked up in the Buchanan et al. (2014) study. This “structural integrity,” wherein key components are more conservative and therefore less likely to change relative to other components, is also found in other aspects of culture (e.g., Mesoudi and Whiten 2004; Mesoudi et al. 2006; Washburn 2001).

The continent-wide method of point manufacture apparently began to shift immediately following Clovis. In a follow-up study to the one by Sholts et al. (2012), Gingerich et al. (2014) examined flake-removal patterns on specimens of several Early Paleoindian eastern fluted-point types that immediately postdate the height of classic Clovis-point manufacture—for example, Bull Brook (Byers 1954) and Debert/Vail (Gramly 1982; MacDonald 1968) (Fig. 9.4)—and found more variation and bifacial flake-scar asymmetry than what Sholts et al. (2012) found among Clovis points. Gingerich et al. (2014:117) hypothesize that the differences “may represent a time-transgressive shift, where Clovis interaction and the direct transmission of knowledge responsible for consistent reduction techniques is breaking down, causing biface symmetry to become more variable with greater flake scar variation.” They point out that their results may support morphometric studies (e.g., Buchanan and Hamilton 2009) that suggest changes in fluted-point shape resulted from drift and related to a colonization process or a shift in population dynamics. If we had to guess, we would take a shift in population dynamics—that the changes in point form had to do with shifts in the use of space (territories) by Paleoindian groups. Those shifts in turn had implications for how information about point technology and performance was transmitted (O’Brien et al. 2014).

To explore this issue, we undertook a series of phylogenetic analyses aimed at (1) clarifying evolutionary relationships among Paleoindian point forms (Buchanan and Collard 2007, 2010; Collard et al. 2010, 2011; Darwent and O’Brien 2006; O’Brien et al. 2012) and (2) highlighting some of the changes in traits, or characters, of various forms across North America. Several of those studies focused on fluted points from the East and Southeast (O’Brien and Lyman 2000, 2003; O’Brien et al. 2001, 2002, 2013, 2014). Common to those studies was the use of the same eight characters and character states to define projectile-point classes (Fig. 9.9). Because of the nonsystematic manner in which projectile-point types have been created (Anderson 2013; Lyman and O’Brien 2002; Miller and Gingerich 2013b; O’Brien and Lyman 2002; O’Brien et al. 2014), the classes often contain specimens that, in the literature where they were described, were placed in different types.

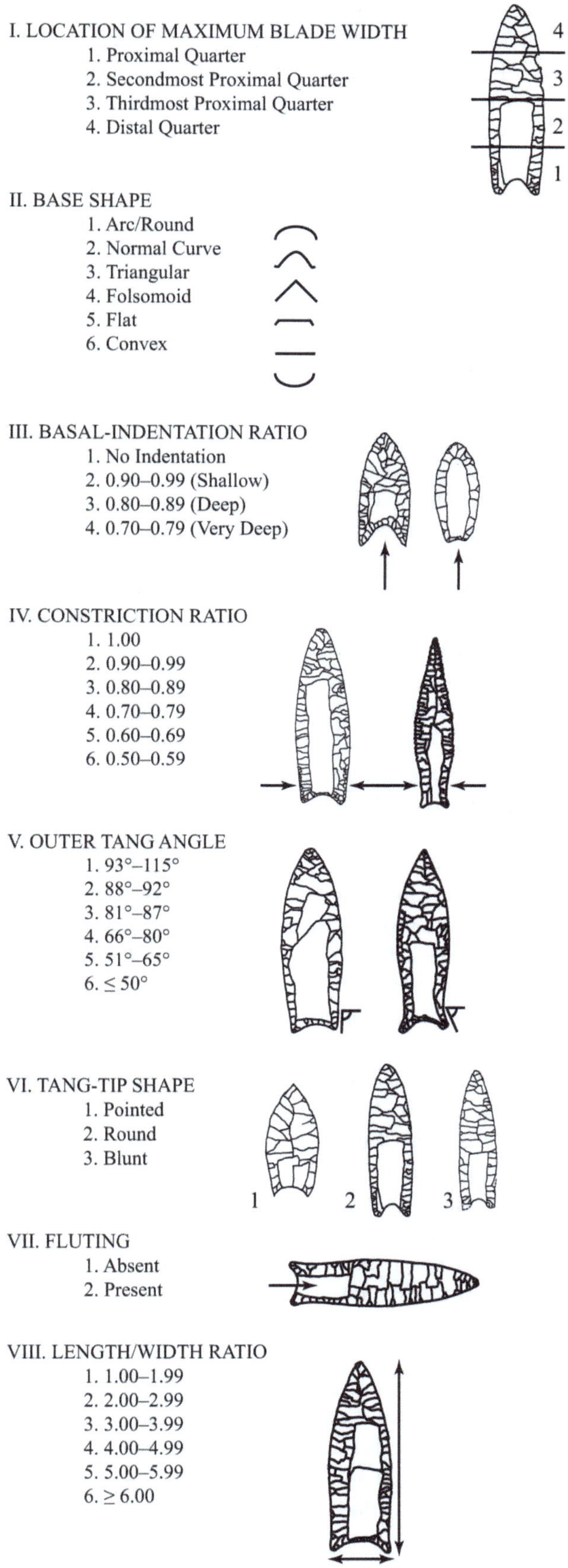

Fig. 9.9 Characters and character states used in the analysis

would act to reinforce ties and a concomitant gradual increase in the diversification of projectile-point shape that will accelerate in the post-Clovis period.

Focusing solely on the latest study (O'Brien et al. 2014), the phylogenetic tree shown in Fig. 9.10, which was built using 218 specimens in 41 classes, contains 48 character-state changes, represented by boxes.[4] Each box is labeled with a Roman numeral indicating the character that has changed; the subscript Arabic numeral indicates the evolved character state (Fig. 9.9). White boxes indicate phylogenetically informative changes—shifts that result from descent with modification as opposed to changes that result from either adaptive convergence (black boxes) or a reversal to ancestral character states (half-shaded boxes). The latter two types of change are not useful in tracing phylogeny, but they do provide information on the kinds of subtle variation present. The tree exhibits numerous clades, defined as two or more related taxa and their common ancestor, some of the larger of which are labeled I–VI.

Projecting the tree into geographic space allows us to observe the significance of the phylogeny in both time and space (Fig. 9.11). Classes in Clade I all contain specimens identified as Clovis points, and all are restricted to the Midwest. Classes in Clade II are skewed toward the Northeast and Middle Atlantic regions. Key constituents of the classes are projectile-point types described as having deep basal indentations—for example, Bull Brook, Debert, and Gainey (Simons et al. 1984) (Fig. 9.4). Several studies have shown that relative depth of the basal indentation varies widely across time and space, with the deepest indentations being in the Northeast and around the Great Lakes (Curran 1996; Ellis 2004; Ellis and Deller 1997; Miller and Gingerich 2013b). Classes in Clade III show a split distribution: One class is restricted to the northern portion of the study area, whereas all other classes in Clade III have distributions in the southern portion. This is not particularly surprising, given that a key constituent of the subclade is Gainey (Fig. 9.4), a point type that occurs primarily along the southern edge of the Great Lakes eastward, although it is found sporadically throughout eastern North America (Morrow and Morrow 2002). Classes in Clade IV occur in a northeast/southwest-trending band from the Tennessee River valley northward, generally following the Ohio River valley. This is also not surprising, given the large number of Cumberland points (Fig. 9.4), a key component of classes in Clade IV, that are found in the Tennessee and central Ohio River valleys (Anderson et al. 2010). Classes in Clade V occur, like those in Clade II, in the Middle Atlantic and Northeast. Classes in Clade VI cluster in the Midwest eastward to the Tennessee River valley. Constituent types include the long, narrow, heavily fluted Cumberland point. Interestingly, Clade VI shows minimal taxonomic diversity and diverges from the superclade comprising the other clades early in the phylogeny.

Of particular interest are the 11 unresolved classes—those that do not fall into one of the six clades—represented in black in Figs. 9.10 and 9.11. In their classic model of Clovis colonization of North America, Kelly and Todd (1988) suggest that the speed of colonization was driven by high rates of residential mobility because of the large foraging areas required of a primarily carnivorous diet. Hamilton and Buchanan (2007) note that Clovis colonists would have moved rapidly through large river systems such as the Missouri and Mississippi drainages, leading to an initially rapid rate of colonization through the midcontinent, which would have then slowed dramatically as diet breadths broadened with the increased biodiversity of the eastern forests (Steele et al. 1998) and as prey size, abundance, and availability changed (Meltzer 1988).

Note the locations of the unresolved classes: They occur in the Upper Midwest near the junction of the Mississippi and Ohio rivers, northeastward along the Ohio River, and southeastward along the Cumberland River. All 11 classes, including the two outgroups, contain specimens identified in the original literature as Clovis points. In some cases, all specimens were identified as Clovis, and in others some were classified as Gainey, Cumberland, Redstone, Debert, and/or Dalton. A working hypothesis based on this distribution would be that the unresolved classes were the products of groups moving rapidly across the landscape—so rapid that there was not enough time for a strong phylogenetic signal to develop. There were technological changes, to be sure—they are what define the classes in the first place—but there were not enough changes to allow much resolution of phylogeny. This conclusion runs parallel to our reasoning for the lack of regional variation in the Buchanan and Hamilton (2009) study of Clovis point shape.

If, as we propose, the unresolved classes are associated primarily with Clovis groups, then a related proposal is that the more-resolved classes, those in clades II–IV, represent later Early Paleoindian points (O'Brien et al. 2014). Figure 9.10 shows the numerous character-state changes that produced those classes. Note that all but one change, the loss of fluting ($VII_2 \rightarrow VII_1$) in Clade III, are either instances of convergence, where knappers or groups of knappers landed on the same adaptive peaks through independent experimentation, or instances of reversal to an ancestral state. This apparent pattern of increased experimentation is what one would expect from the guided-variation model: in the absence of selection, a population will move toward whichever trait is favored by people's individual-learning biases (Mesoudi 2011b; O'Brien et al. 2014). Our proposal of a shift from biased social learning to guided variation accounts for the changes in flake-removal patterns identified by Gingerich et al. (2014) for eastern Paleoindian points compared with the findings of Sholts et al. (2012) for a continent-wide sample of Clovis points. Future work will be directed

[4]See O'Brien et al. (2001, 2013, 2014) for details on how trees were constructed.

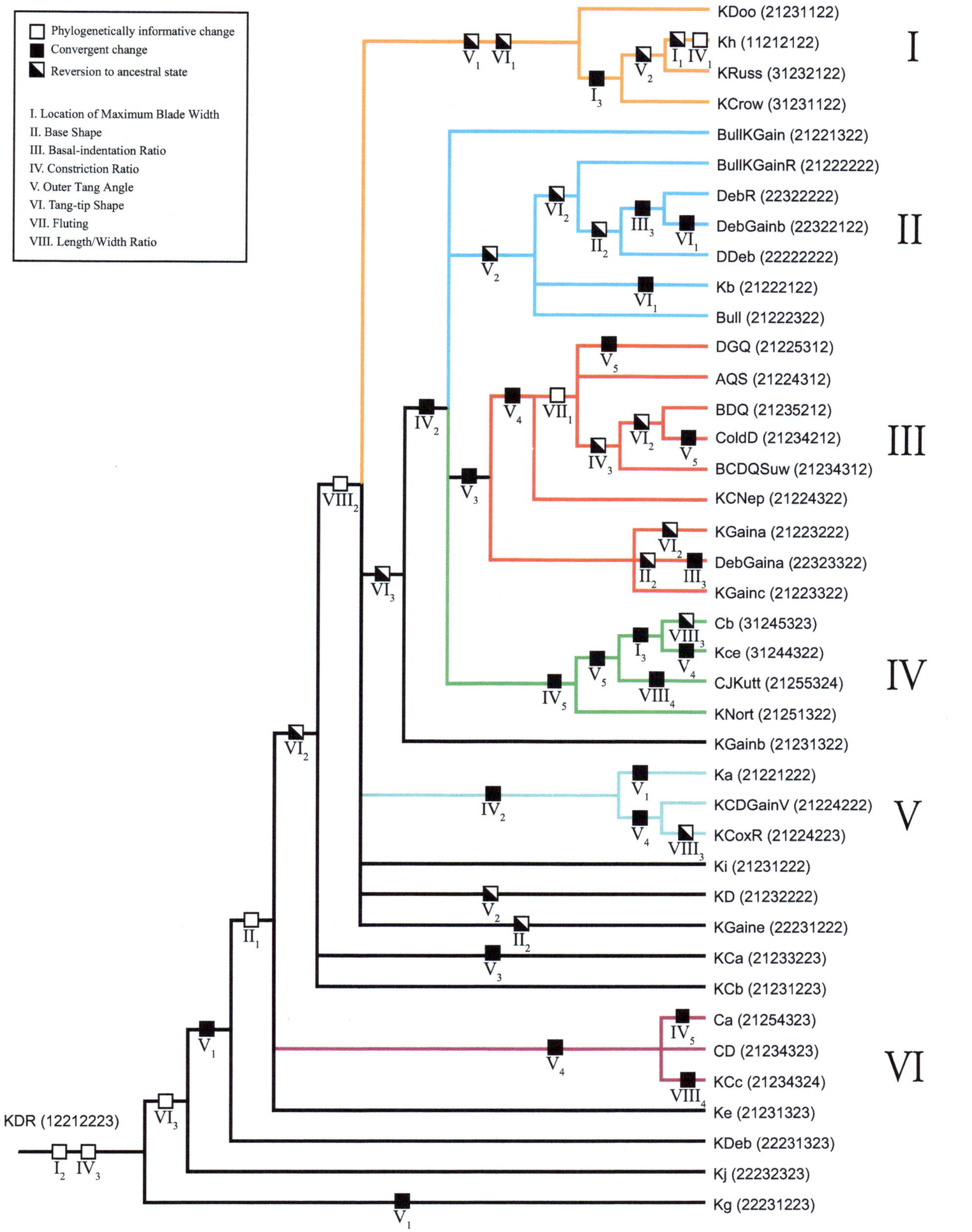

Fig. 9.10 Phylogenetic tree of the 41 classes, with clades shown in different colors (from O'Brien et al. 2014). *Roman numerals* denote characters, and *subscript numbers* denote character states. *Open boxes* indicate phylogenetically informative changes; *shaded boxes* indicate parallel or convergent changes (homoplasy); and *half-shaded boxes* indicate characters that reverted to an ancestral state (The tree is a fifty-percent majority-rule consensus tree based on 100 replicates.)

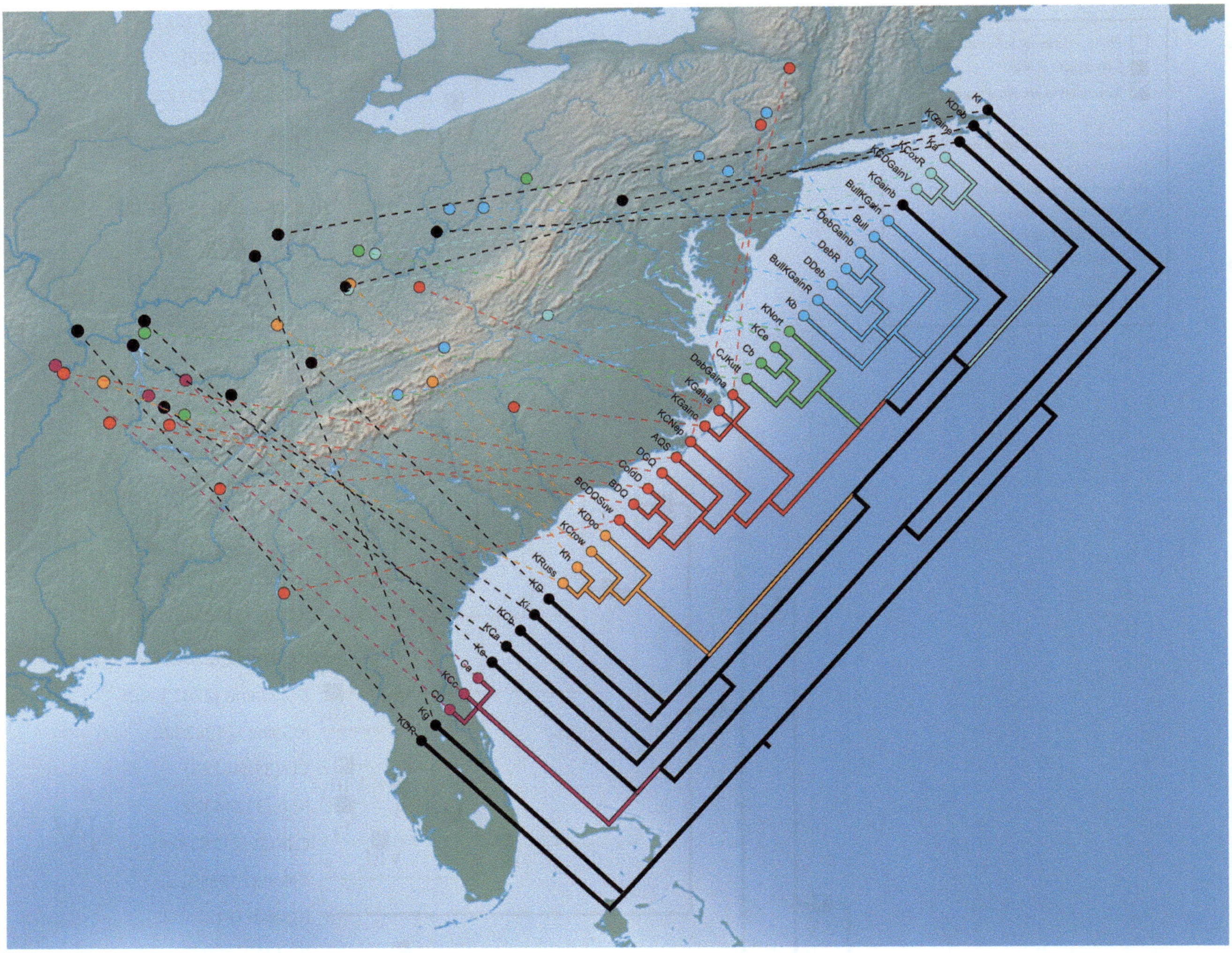

Fig. 9.11 Geographic distribution of clades shown in Fig. 9.10 (from O'Brien et al. 2014)

at developing a better understanding of the "fitness" of character states in terms of performance, which undoubtedly underlies the fact that groups from widely different regions during the late half of the Early Paleoindian period landed on the same fitness peaks in terms of character states—in other words, they found similar solutions to common adaptive problems.

9.4 Conclusion

If nothing else, our discussion should make it clear that in terms of point manufacture, the North American Paleolithic cultural landscape was anything but static. We did not need to conduct the studies reported here in order to reach that conclusion, but taken together, they offer glimpses into just how fluid the landscape was. The studies also underscore the fact that there were myriad subtleties to how information was acquired and transmitted during the Early Paleoindian period—subtleties that in some cases match expectations derived from models of learning. We return to a point made earlier that is difficult to overemphasize: Humans are neither purely social nor purely individual learners. Rather, certain conditions will dictate which kind of learning is used in any particular situation. The analytical tools discussed here, including geometric morphometrics and cladistics, offer a powerful means of moving forward.

Acknowledgments We thank an anonymous reviewer for extremely helpful suggestions for improving the manuscript.

References

Ahler SA, Geib PR (2000) Why flute? Folsom point design and adaptation. J Archaeol Sci 27:799–820

Alroy J (2001) A multispecies overkill simulation of the end-Pleistocene megafaunal mass extinction. Science 292:1893–1896

Amick DS (1995) Patterns of technological variation among Folsom and Midland projectile points in the American Southwest. Plains Anthropol 40:23–38

Anderson DG (1990) The Paleoindian colonization of eastern North America: a view from the southeastern United States. In: Tankersely KB, Isaac BL (eds) Early Paleoindian economies of eastern North America. JAI Press, Greenwich, pp 163–216

Anderson DG (2013) Paleoindian archaeology in eastern North America: current approaches and future directions. In: Gingerich JAM (ed) In the eastern fluted point tradition. University of Utah Press, Salt Lake City, pp 371–403

Anderson DG, Faught MK (1998) The distribution of fluted Paleoindian projectile points: update 1998. Archaeol East N Am 26:163–187

Anderson DG, Faught MK (2000) Palaeoindian artefact distributions: evidence and implications. Antiquity 7:507–513

Anderson DG, O'Steen LD, Sassaman KE (1996) Chronological considerations. In: Anderson DG, Sassaman KE (eds) The Paleoindian and Early Archaic Southeast. University of Alabama Press, Tuscaloosa, pp 3–15

Anderson DG, Miller DS, Yerka SJ, Gillam JC, Johanson DT, Goodyear AC, Smallwood AM (2010) Paleoindian database of the Americas 2010: current status and findings. Archaeol East N Am 38:63–90

Anderson DG, Goodyear AC, Kennett J, West A (2011) Multiple lines of evidence for possible human population decline/settlement reorganization during the early Younger Dryas. Quat Int 242:570–583

Aoki K (2013) Determinants of cultural evolutionary rates. In: Akazawa T, Nishiaki Y, Aoki K (eds) Dynamics of learning in Neanderthals and modern humans. Springer, Tokyo, pp 199–210

Aoki K, Lehmann L, Feldman MW (2011) Rates of cultural change and patterns of cultural accumulation in stochastic models of social transmission. Theor Popul Biol 79:192–202

Aoki K, Wakano JY, Lehmann L (2012) Evolutionarily stable learning schedules and cumulative culture in discrete generation models. Theor Popul Biol 81:300–309

Arnold SJ, Pfrender ME, Jones AG (2001) The adaptive landscape as a conceptual bridge between micro- and macroevolution. Genetica 112–113:9–32

Atkinson QD (2011) Phonemic diversity supports a serial founder effect model of language expansion from Africa. Science 332:346–349

Atkisson C, O'Brien MJ, Mesoudi A (2012) Adult learners in a novel environment use prestige-biased social learning. Evol Psychol 10:519–537

Bar-Yosef O (2013) Neanderthals and modern humans across Eurasia. In: Akazawa T, Nishiaki Y, Aoki K (eds) Dynamics of learning in Neanderthals and modern humans volume 1. Springer, Japan, pp 7–20

Bar-Yosef O, Van Peer P (2009) The chaîne opératoire approach in Middle Paleolithic archaeology. Curr Anthropol 50:103–131

Bentley RA, O'Brien MJ (2011) The selectivity of cultural learning and the tempo of cultural evolution. J Evol Psychol 9:125–141

Bentley RA, Shennan SJ (2003) Cultural transmission and stochastic network growth. Am Antiq 68:459–485

Bentley RA, Hahn MW, Shennan SJ (2004) Random drift and culture change. Proc R Soc Lond Biol 271:1443–1450

Bentley RA, O'Brien MJ, Brock WA (2014) Mapping collective behavior in the big-data era. Behav Brain Sci 37:63–119

Bettinger RL, Eerkens JW (1997) Evolutionary implications of metrical variation in Great Basin projectile points. In: Barton CM, Clark GA (eds) Rediscovering Darwin: evolutionary theory and archeological explanation, Archeological papers of the American Anthropological Association, no. 7. American Anthropological Association, Arlington, pp 177–191

Bettinger RL, Eerkens J (1999) Point typologies, cultural transmission, and the spread of bow-and-arrow technology in the prehistoric Great Basin. Am Antiq 64:231–242

Boas F (1904) The history of anthropology. Science 20:513–524

Boas F (1911) Review of "Methode der ethnologie" by F. Graebner. Science 34:804–810

Boulanger MT, Lyman RL (2014) Northeastern North American Pleistocene megafauna chronologically overlapped minimally with Paleoindians. Quat Sci Rev 85:35–46

Boulanger MT, Buchanan B, O'Brien MJ, Redmond BG, Glascock MD, Eren MI (2015) Neutron activation analysis of 12,900-year-old stone artifacts confirms 450–510+ km Clovis tool-stone acquisition at Paleo Crossing (33ME274), Northeast Ohio, U.S.A. J Archaeol Sci 53:550–558

Boyd R, Richerson PJ (1985) Culture and the evolutionary process. University of Chicago Press, Chicago

Boyd R, Richerson PJ (1992) How microevolutionary processes give rise to history. In: Nitecki MH, Nitecki DV (eds) History and evolution. State University of New York Press, Albany, pp 179–209

Boyd R, Richerson PJ (1996) Why culture is common, but cultural evolution is rare. Proc Br Acad 88:77–94

Boyd R, Richerson PJ (2005) The origin and evolution of culture. Oxford University Press, Oxford

Bradley BA (1993) Paleo-Indian flaked stone technology in the North American High Plains. In: Sofer O, Praslov ND (eds) From Kostenki to Clovis: upper Paleolithic–Paleo-Indian adaptations. Plenum, New York, pp 251–262

Bradley BA (1997) Sloan site biface and projectile point technology. In: Morse DF (ed) Sloan: a Paleoindian Dalton cemetery in Arkansas. Smithsonian Institution Press, Washington, DC, pp 53–57

Bradley JW, Spiess A, Boisvert RA, Boudreau J (2008) What's the point? Modal forms and attributes of Paleoindian bifaces in the New England–Maritimes region. Archaeol East N Am 36:119–172

Bradley BA, Collins MB, Hemmings A (2010) Clovis technology. International Monographs in Prehistory, Ann Arbor, MI

Brennan LA (1982) A compilation of fluted points of eastern North America by count and distribution: an AENA project. Archaeol East N Am 10:27–46

Buchanan B (2006) An analysis of Folsom projectile point resharpening using quantitative comparisons of form and allometry. J Archaeol Sci 33:185–199

Buchanan B, Collard M (2007) Investigating the peopling of North America through cladistic analyses of Early Paleoindian projectile points. J Anthropol Archaeol 26:366–393

Buchanan B, Collard M (2010) A geometric morphometrics-based assessment of blade shape differences among Paleoindian projectile point types from western North America. J Archaeol Sci 37: 350–359

Buchanan B, Hamilton MJ (2009) A formal test of the origin of variation in North American Early Paleoindian projectile points. Am Antiq 74:279–298

Buchanan B, Kilby JD, Huckell BB, O'Brien MJ, Collard M (2012) A morphometric assessment of the intended function of cached Clovis points. PLoS One 7(2):e30530

Buchanan B, O'Brien MJ, Collard M (2014) Continent-wide or region-specific? A geometric-morphometrics-based assessment of variation in Clovis point shape. Archaeol Anthropol Sci 6:145–162

Bullen RP (1968) A guide to the identification of Florida projectile points. Florida State Museum, Gainesville

Byers DS (1954) Bull Brook: a fluted point site in Ipswich, Massachusetts. Am Antiq 19:343–351

Cannon MD (2004) Geographic variability in North American mammal community richness during the terminal Pleistocene. Quat Sci Rev 23:1099–1123

Catto N, Mandryk CAS (1990) Geology of the postulated ice-free corridor. In: Megafauna and man: discovery of America's heartland, Mammoth Site of Hot Springs, South Dakota, Scientific papers, no 1. Mammoth Site of Hot Springs, South Dakota, Hot Springs, pp 80–85

Cavalli-Sforza LL, Feldman MW (1973) Models for cultural inheritance. I. Group mean and within group variation. Theor Popul Biol 4:42–44
Cavalli-Sforza LL, Feldman MW (1981) Cultural transmission and evolution: a quantitative approach. Princeton University Press, Princeton
Cheshier J, Kelly RL (2006) Projectile point shape and durability: the effect of thickness:length. Am Antiq 71:353–363
Collard M, Buchanan B, Hamilton MJ, O'Brien MJ (2010) Spatiotemporal dynamics of the Clovis–Folsom transition. J Archaeol Sci 37:2513–2519
Collard M, Buchanan B, Hamilton MJ, O'Brien MJ (2011) Points and prey: a quantitative test of the hypothesis that prey size influences early Paleoindian projectile point form. J Archaeol Sci 38:852–864
Cotter JL (1937) The occurrence of flints and extinct animals in pluvial deposits near Clovis, New Mexico: Part IV, report on excavation at the gravel pit, 1936. Proc Acad Natl Sci Phila 89:1–16
Cotter JL (1938) The occurrence of flints and extinct animals in pluvial deposits near Clovis, New Mexico: Part VI, report on field season of 1937. Proc Acad Natl Sci Phila 90:113–117
Crabtree DE (1966) A stoneworker's approach to analyzing and replicating the Lindenmeier Folsom. Tebiwa 9:3–39
Curran ML (1996) Palaeoindians in the Northeast: the problem of dating fluted point sites. Rev Archaeol 17:2–11
Darwent J, O'Brien MJ (2006) Using cladistics to construct lineages of projectile points from northeastern Missouri. In: Lipo CP, O'Brien MJ, Collard M, Shennan SJ (eds) Mapping our ancestors: phylogenetic approaches in anthropology and prehistory. Aldine, New York, pp 185–208
Dent RJ (2007) Seed collecting and fishing at the Shawnee Minisink Paleoindian site. In: Walker RB, Driskell BN (eds) Foragers of the terminal Pleistocene in North America. University of Nebraska Press, Lincoln, pp 116–132
Derex M, Beugin MP, Godelle B, Raymond M (2013) Experimental evidence for the influence of group size on cultural complexity. Nature 503:389–391
Diamond J (1978) The Tasmanians: the longest isolation, the simplest technology. Nature 273:185–186
Dunbar RIM (1992) Neocortex size as a constraint on group size in primates. J Hum Evol 22:469–493
Eerkens JW, Lipo CP (2005) Cultural transmission, copying errors, and the generation of variation in material culture and the archaeological record. J Anthropol Archaeol 24:316–334
Eerkens JW, Lipo CP (2007) Cultural transmission theory and the archaeological record: providing context to understanding variation and temporal changes in material culture. J Archaeol Res 15:239–274
Eerkens JW, Bettinger RL, McElreath R (2006) Cultural transmission, phylogenetics, and the archaeological record. In: Lipo CP, O'Brien MJ, Collard M, Shennan SJ (eds) Mapping our ancestors: phylogenetic approaches in anthropology and prehistory. Aldine, New York, pp 169–183
Eerkens JW, Bettinger RL, Richerson PJ (2014) Cultural transmission theory and hunter–gatherer archaeology. In: Cummings V, Jordan P, Zvelebil M (eds) The Oxford handbook of the archaeology and anthropology of hunter–gatherers. Oxford University Press, Oxford
Ehn M, Laland KN (2012) Adaptive strategies for cumulative cultural learning. J Theor Biol 301:103–111
Ellis C (2004) Understanding "Clovis" fluted point variability in the Northeast: a perspective from the Debert site, Nova Scotia. Can J Archaeol 28:205–253
Ellis CJ, Deller DB (1997) Variability in the archaeological record of northeastern early Paleoindians: a view from southern Ontario. Archaeol East N Am 25:1–30
Enquist M, Ghirlanda S, Jarrick A, Wachtmeister CA (2008) Why does human culture increase exponentially? Theor Popul Biol 74:46–55
Enquist M, Ghirlanda S, Eriksson K (2011) Modelling the evolution and diversity of cumulative culture. Philos Trans R Soc B 366:412–423
Eren MI, Lycett SJ (2012) Why Levallois? A morphometric comparison of experimental "preferential" Levallois flakes versus debitage flakes. PLoS One 7(1):e29273
Eren MI, Lycett SJ, Roos C, Sampson CG (2011a) Toolstone constraints and knapping skill: Levallois reduction with two different rock types. J Archaeol Sci 38:2731–2739
Eren MI, Bradley B, Sampson CG (2011b) Middle Paleolithic skill-level and the Individual knapper: an experiment. Am Antiq 76:229–251
Eren MI, Buchanan B, O'Brien MJ (2015) Social learning and technological evolution during the Clovis colonization of the new world. J Hum Evol 80:159–170
Feldman M, Cavalli-Sforza L (1976) Cultural and biological evolutionary processes, selection for a trait under complex transmission. Theor Popul Biol 9:238–259
Ferguson J (2003) An experimental test of the conservation of raw material in flintknapping skill. Lithic Technol 28:113–131
Figgins JD (1927) The antiquity of man in America. Nat Hist 27:229–239
Frison GC (1983) The western plains and mountain region. In: Shutler R Jr (ed) Early man in the New World. Sage, Beverly Hills, pp 109–124
Frison GC (1989) Experimental use of Clovis weaponry and tools on African elephants. Am Antiq 54:766–784
Frison GC, Todd LC (1986) The Colby mammoth site: taphonomy and archaeology of a Clovis kill in northern Wyoming. University of New Mexico Press, Albuquerque
Gingerich JAM (2007) Picking up the pieces: new Paleoindian research in the upper Delaware Valley. Archaeol East N Am 35:117–124
Gingerich JAM (2013) Revisiting Shawnee–Minisink. In: Gingerich JAM (ed) In the eastern fluted point tradition. University of Utah Press, Salt Lake City, pp 218–256
Gingerich AM, Sholts SB, Wärmländer SKTS, Stanford D (2014) Fluted point manufacture in eastern North America: an assessment of form and technology using traditional metrics and 3D morphometrics. World Archaeol 46:101–122
Goebel T, Waters MR, O'Rourke DH (2008) The late Pleistocene dispersal of modern humans in the Americas. Science 319:1497–1502
Goodyear AC (1982) The chronological position of the Dalton horizon in the southeastern United States. Am Antiq 47:382–395
Gramly RM (1982) The Vail site: a Palaeo-Indian encampment in Maine, vol 30, Bulletin of the Buffalo Society of Natural Sciences. Buffalo Society of Natural Sciences, Buffalo, NY
Grayson DK (2007) Deciphering North American Pleistocene extinctions. J Anthropol Res 63:185–213
Hamilton MJ (2008) Quantifying Clovis dynamics: conforming theory with models and data across scales. Ph.D. dissertation, Department of Anthropology, University of New Mexico, Albuquerque
Hamilton MJ, Buchanan B (2007) Spatial gradients in Clovis-age radiocarbon dates across North America suggest rapid colonization from the north. Proc Natl Acad Sci 104:15625–15630
Hamilton MJ, Buchanan B (2009) The accumulation of stochastic copying errors causes drift in culturally transmitted technologies: quantifying Clovis evolutionary dynamics. J Anthropol Archaeol 28: 55–69
Haynes CV (1964) Fluted projectile points: their age and dispersion. Science 145:1408–1413
Haynes G (2013) the weight and meaning of eastern Paleoindian research. In: Gingerich JAM (ed) In the eastern fluted point tradition. University of Utah Press, Salt Lake City, pp 355–370
Haynes CV Jr, Donahue DJ, Jull AJT, Zabel TH (1984) Application of accelerator dating to fluted point Paleoindian sites. Archaeol East N Am 12:184–191
Henrich J (2001) Cultural transmission and the diffusion of innovations: adoption dynamics indicate that biased cultural transmission is the predominant force in behavioral change. Am Anthropol 103: 992–1013

Henrich J (2004) Demography and cultural evolution: why adaptive cultural processes produced maladaptive losses in Tasmania. Am Antiq 69:197–214

Henrich J (2006) Understanding cultural evolutionary models: a reply to Read's critique. Am Antiq 71:771–782

Henrich J (2010) The evolution of innovation-enhancing institutions. In: O'Brien MJ, Shennan SJ (eds) Innovation in cultural systems: contributions from evolutionary anthropology. MIT Press, Cambridge, MA, pp 99–120

Henrich J, Boyd R (1998) The evolution of conformist transmission and the emergence of between-group differences. Evol Hum Behav 19:215–241

Henrich J, Broesch J (2011) On the nature of cultural transmission networks: evidence from Fijian villages for adaptive learning biases. Philos Trans R Soc B 366:1139–1148

Henrich J, Gil-White FJ (2001) The evolution of prestige: freely conferred deference as a mechanism for enhancing the benefits of cultural transmission. Evol Hum Behav 22:165–196

Holliday VT (2000) The evolution of Paleoindian geochronology and typology on the Great Plains. Geoarchaeology 15:227–290

Hoppitt W, Kandler A, Kendal JR, Laland KN (2010) The effect of task structure on diffusion dynamics: implications for diffusion curve and network-based analyses. Learn Behav 38:243–251

Hutchings WK (1997) The Paleoindian fluted point: dart or spear armature? The identification of Paleoindian delivery technology through the analysis of lithic fracture velocity. Ph.D. dissertation, Simon Fraser University, Burnaby, Canada

Judge WJ (1970) Systems analysis and the Folsom–Midland question. Southwest J Anthropol 26:40–51

Kameda T, Nakanishi D (2002) Cost-benefit analysis of social/cultural learning in a nonstationary uncertain environment: an evolutionary simulation and an experiment with human subjects. Evol Hum Behav 23:373–393

Kameda T, Nakanishi D (2003) Does social/cultural learning increase human adaptability? Rogers' question revisited. Evol Hum Behav 24:242–260

Kandler A, Shennan S (2013) A non-equilibrium neutral model for analysing cultural change. J Theor Biol 330:18–25

Kandler A, Steele J (2010) Social learning, economic inequality, and innovation diffusion. In: O'Brien MJ, Shennan SJ (eds) Innovation in cultural systems: contributions from evolutionary anthropology. MIT Press, Cambridge, MA, pp 193–214

Kelly RL, Todd LC (1988) Coming into the country: early Paleoindian hunting and mobility. Am Antiq 53:231–244

Kemp BM, Schurr TG (2010) Ancient and modern genetic variation in the Americas. In: Auerbach BM (ed) Human variation in the Americas. Southern Illinois University Press, Carbondale, pp 12–50

Kempe M, Mesoudi A (2014) An experimental demonstration of the effect of group size on cultural accumulation. Evol Hum Behav 35:285–290

Kempe M, Lycett SJ, Mesoudi A (2012) An experimental test of the accumulated copying error model of cultural mutation for Acheulean handaxe size. PLoS One 7(11):e48333

Kendal J, Giraldeau L-A, Laland KN (2009) The evolution of social learning rules: payoff-based and frequency-dependent biased transmission. J Theor Biol 260:210–219

Kobayashi Y, Aoki K (2012) Innovativeness, population size and cumulative cultural evolution. Theor Popul Biol 82:38–47

Kohler TA, VanBuskirk S, Ruscavage-Barz S (2004) Vessels and villages: evidence for conformist transmission in early village aggregations on the Pajarito Plateau, New Mexico. J Anthropol Archaeol 23:100–118

Krieger AD (1954) A comment on "Fluted point relationships" by John Witthoft. Am Antiq 19:273–275

Kroeber AL (1923) Anthropology. Harcourt, Brace, New York

Kuhn S (2013) Cultural transmission, institutional continuity and the persistence of the Mousterian. In: Akazawa T, Nishiaki Y, Aoki K (eds) Dynamics of learning in Neanderthals and modern humans. Springer, Tokyo, pp 105–113

Lake MW, Venti J (2009) Quantitative analysis of macroevolutionary patterning in technological evolution: bicycle design from 1800 to 2000. In: Shennan SJ (ed) Pattern and process in cultural evolution. University of California Press, Berkeley, pp 147–161

Laland KN (2004) Social learning strategies. Learn Behav 32:4–14

Laland K, Brown G (2011) Sense and nonsense: evolutionary perspectives on human behavior, 2nd edn. Oxford University Press, Oxford

Lande R (1986) The dynamics of peak shifts and the pattern of morphological evolution. Paleobiology 12:343–354

Levine MA (1990) Accommodating age: radiocarbon results and fluted point sites in northeastern North America. Archaeol East N Am 18:33–63

Lewis TMN (1954) The Cumberland point. Okla Anthropol Soc Bull 2:7–8

Lipo CP, Madsen ME, Dunnell RC, Hunt T (1997) Population structure, cultural transmission, and frequency seriation. J Anthropol Archaeol 16:301–333

Lipo CP, Dunnell RC, O'Brien MJ, Harper V, Dudgeon J (2012) Beveled projectile points and ballistics technology. Am Antiq 77:774–788

Lowery DL, Phillips T (1994) The Meekins Neck Paleoindian site complex, Dorchester County, Maryland: a development of a Paleoindian settlement model for the Delmarva Peninsula. Md Archaeol 30(2):29–36

Lycett SJ, von Cramon-Taubadel N (2008) Acheulean variability and hominin dispersals: a model-bound approach. J Archaeol Sci 35:553–562

Lyman RL (2008) Cultural transmission in North American anthropology and archaeology, ca. 1895–1965. In: O'Brien MJ (ed) Cultural transmission and archaeology: issues and case studies. Society for American Archaeology Press, Washington, DC, pp 10–20

Lyman RL, O'Brien MJ (1997) The concept of evolution in early twentieth-century Americanist archaeology. In: Barton CM, Clark GA (eds) Rediscovering Darwin: evolutionary theory and archeological explanation, Archeological papers of the American Anthropological Association, no. 7. American Anthropological Association, Arlington, VA, pp 21–48

Lyman RL, O'Brien MJ (2002) Classification. In: Hart JP, Terrell JE (eds) Darwin and Archaeology: a handbook of key concepts. Bergin & Garvey, Westport, CT, pp 69–88

Lyman RL, O'Brien MJ (2003) Cultural traits: units of analysis in early twentieth-century anthropology. J Anthropol Res 59:225–250

MacDonald GF (1968) Debert: a Palaeo-Indian site in central Nova Scotia. Nat Mus Man, Anthropol Paper, no. 16, Ottawa

Mandryk CAS, Josenhans H, Fedje DW, Mathewes RW (2001) Late quaternary paleoenvironments of northwestern North America: implications for inland versus coastal migration routes. Quat Sci Rev 20:301–314

Martin PS (1967) Prehistoric overkill. In: Martin PS, Wright HE (eds) Pleistocene extinctions: the search for a cause. Yale University Press, New Haven, CT, pp 75–120

Marwick B (2005) What can archaeology do with Boyd and Richerson's cultural evolutionary program? Rev Archaeol 26:30–40

Mason OT (1895) Similarities in culture. Am Anthropol 8:101–117

Mason RJ (1962) The Paleo-Indian tradition in eastern North America. Curr Anthropol 3:227–278

McElreath R, Lubell M, Richerson PJ, Waring TM, Baum W, Edsten E, Efferson C, Paciotti B (2005) Applying evolutionary models to the laboratory study of social learning. Evol Hum Behav 26:483–508

Meltzer DJ (1988) Late Pleistocene human adaptations in eastern North America. J World Prehist 2:1–52

Meltzer DJ (1993) Is there a Clovis adaptation? In: Soffer O, Praslov ND (eds) From Kostenki to Clovis: upper Paleolithic–Paleo-Indian adaptations. Plenum, New York, pp 293–310

Meltzer DJ, Mead JI (1983) The timing of Late Pleistocene mammalian extinctions in North America. Quat Res 19:130–135

Mesoudi A (2008) An experimental simulation of the "copy-successful-individuals" cultural learning strategy: adaptive landscapes, producer–scrounger dynamics, and informational access costs. Evol Hum Behav 29:350–363

Mesoudi A (2010) The experimental study of cultural innovation. In: O'Brien MJ, Shennan SJ (eds) Innovation in cultural systems: contributions from evolutionary anthropology. MIT Press, Cambridge, MA, pp 175–191

Mesoudi A (2011a) An experimental comparison of human social learning strategies: payoff-biased social learning is adaptive but underused. Evol Hum Behav 32:334–342

Mesoudi A (2011b) Cultural evolution: how Darwinian theory can explain human culture and synthesize the social sciences. University of Chicago Press, Chicago

Mesoudi A (2011c) Variable acquisition costs constrain cumulative cultural evolution. PLoS One 6(3):e18239

Mesoudi A, Lycett SJ (2009) Random copying, frequency-dependent copying and culture change. Evol Hum Behav 30:41–48

Mesoudi A (2013) Studying cultural transmission within an interdisciplinary cultural evolutionary framework. In: Ellen R, Lycett SJ, Johns SJ (eds) Cultural transmission: a critical anthropological synthesis. Oxford University Press, Oxford, pp 131–147

Mesoudi A, O'Brien MJ (2008a) The cultural transmission of Great Basin projectile-point technology I: an experimental simulation. Am Antiq 73:3–28

Mesoudi A, O'Brien MJ (2008b) The cultural transmission of Great Basin projectile-point technology II: an agent-based computer simulation. Am Antiq 73:627–644

Mesoudi A, O'Brien MJ (2008c) The learning and transmission of hierarchical cultural recipes. Biol Theory 3:63–72

Mesoudi A, Whiten A (2004) The hierarchical transformation of event knowledge in human cultural transmission. J Cogn Cult 4:1–24

Mesoudi A, Whiten A, Dunbar R (2006) A bias for social information in human cultural transmission. Br J Psychol 97:405–423

Miller DS, Gingerich JAM (2013a) Regional variation in the terminal Pleistocene and early Holocene radiocarbon record of eastern North America. Quat Res 79:175–188

Miller DS, Gingerich JAM (2013b) Paleoindian chronology and the eastern fluted point tradition. In: Gingerich JAM (ed) In the eastern fluted point tradition. University of Utah Press, Salt Lake City, pp 9–37

Morrow JE (1995) Fluted point manufacture: a perspective from the Ready Lincoln Hills site, 11JY46, Jersey County, Illinois. Midcont J Archaeol 20:167–191

Morrow JE (2014) Review of "Across Atlantic ice: the origin of America's Clovis culture", by Stanford DJ, Bradley, BA. Lithic Technol 39:76–78

Morrow JE, Morrow TA (1999) Geographic variation in fluted projectile points: a hemispheric perspective. Am Antiq 64:215–231

Morrow JE, Morrow TA (2002) Exploring the Clovis–Gainey–Folsom continuum: technological and morphological variation in midwestern fluted points. In: Clark JE, Collins MB (eds) Folsom technology and lifeways, vol 4, Lithic technology, special publication. Dept. of Anthropology, University of Tulsa, Tulsa, OK, pp 141–157

Mosimann JE, Martin PS (1975) Simulating overkill by Paleoindians. Am Sci 63:304–313

Musil RR (1988) Functional efficiency and technological change: a hafting tradition model for prehistoric North America. In: Willig JA, Aikens CM, Fagan JL (eds) Early human occupation in far western North America: the Clovis–Archaic interface, Nevada State Museum, Anthropological papers, no. 21. Nevada State Museum, Carson City, pp 373–387

Nakahashi W (2013) Cultural evolution and learning strategies in hominids. In: Akazawa T, Nishiaki Y, Aoki K (eds) Dynamics of learning in Neanderthals and modern humans. Springer, Tokyo, pp 245–254

Neiman FD (1995) Stylistic variation in evolutionary perspective: inferences from decorative diversity and interassemblage distance in Illinois Woodland ceramic assemblages. Am Antiq 60:7–36

O'Brien MJ, Bentley RA (2011) Stimulated variation and cascades: two processes in the evolution of complex technological systems. J Archaeol Method Theory 18:309–335

O'Brien MJ, Lyman RL (2000) Applying evolutionary archaeology: a systematic approach. Kluwer Academic/Plenum, New York

O'Brien MJ, Lyman RL (2002) The epistemological nature of archaeological units. Anthropol Theory 2:37–57

O'Brien MJ, Lyman RL (2003) Cladistics and archaeology. University of Utah Press, Salt Lake City

O'Brien MJ, Darwent J, Lyman RL (2001) Cladistics is useful for reconstructing archaeological phylogenies: Palaeoindian points from the southeastern United States. J Archaeol Sci 28:1115–1136

O'Brien MJ, Lyman RL, Saab Y, Saab E, Darwent J, Glover DS (2002) Two issues in archaeological phylogenetics: taxon construction and outgroup selection. J Theor Biol 215:133–150

O'Brien MJ, Buchanan B, Collard M, Boulanger MT (2012) Cultural cladistics and the early history of North America. In: Pontarotti P (ed) Evolutionary biology: mechanisms and trends. Springer, New York, pp 23–42

O'Brien MJ, Collard M, Buchanan B, Boulanger MT (2013) Trees, thickets, or something in between? Recent theoretical and empirical work in cultural phylogeny. Isr J Ecol Evol 59(2):49–61

O'Brien MJ, Boulanger MT, Buchanan B, Collard M, Lyman RL, Darwent J (2014) Innovation and cultural transmission in the American Paleolithic: phylogenetic analysis of eastern Paleoindian projectile-point classes. J Anthropol Archaeol 34:100–119

O'Higgins P (2000) Quantitative approaches to the study of craniofacial growth and evolution: advances in morphometric techniques. In: O'Higgins P, Cohn M (eds) Vertebrate ontogeny and phylogeny: implications for the study of hominid skeletal evolution. Academic Press, London, pp 163–185

O'Rourke DH, Raff JA (2010) The human genetic history of the Americas: the final frontier. Curr Biol 20:R202–R207

Olausson D (2008) Does practice make perfect? Craft expertise as a factor in aggrandizer strategies. J Archaeol Method Theory 15: 28–50

Pigeot N (1990) Technical and social actors: flint knapping specialists and apprentices at Magdalenian Etiolles. Archaeol Rev Camb 9:126–141

Powell A, Shennan S, Thomas MG (2009) Late Pleistocene demography and the appearance of modern human behavior. Science 324:1298–1301

Prasciunas MM (2011) Mapping Clovis: projectile points, behavior, and bias. Am Antiq 76:107–126

Premo L (2012) Local extinctions, connectedness, and cultural evolution in structured populations. Adv Complex Syst 15(1150002):1–18

Premo LS (2014) Cultural transmission and diversity in time-averaged assemblages. Curr Anthropol 55:105–114

Premo LS, Scholnick JB (2011) The spatial scale of social learning affects cultural diversity. Am Antiq 76:163–176

Preston DD (1999) Woody's dream. New Yorker 75(34):80–87

Putt SS, Woods AD, Franciscus RG (2014) The role of verbal interaction during experimental bifacial stone tool manufacture. Lithic Technol 39:96–112

Raff JA, Bolnick DA (2014) Palaeogenomics: genetic roots of the first Americans. Nature 506:162–163

Raff JA, Bolnick DA, Tackney J, O'Rourke DH (2010) Ancient DNA perspectives on American colonization and population history. Am J Phys Anthropol 146:503–514

Rasmussen M, Anzick SL, Waters MR, Skoglund P, DeGiorgio M, Stafford TW Jr, Rasmussen S et al (2014) The genome of a late Pleistocene human from a Clovis burial site in western Montana. Nature 506:225–229

Reader SM, Laland KN (2002) Social intelligence, innovation and enhanced brain size in primates. Proc Natl Acad Sci 99: 4436–4441

Rendell L, Boyd R, Cownden D, Enquist M, Eriksson K, Feldman MW, Fogarty L, Ghirlanda S, Lillicrap T, Laland KN (2010) Why copy others? Insights from the social learning strategies tournament. Science 328:208–213

Rendell L, Boyd R, Enquist M, Feldman MW, Fogarty L, Laland KN (2011a) How copying affects the amount, evenness and persistence of cultural knowledge: insights from the social learning strategies tournament. Philos Trans R Soc B 366:1118–1128

Rendell L, Fogarty L, Hoppitt WJE, Morgan TJH, Webster MM, Laland KN (2011b) Cognitive culture: theoretical and empirical insights into social learning strategies. Trends Cogn Sci 15:68–76

Reyes-Garcia V, Molina JL, Broesch J, Calvet L, Huanca T, Saus J, Tanner S, Leonard WR, McDade TW (2008) Do the aged and knowledgeable men enjoy more prestige? A test of predictions from the prestige-bias model of cultural transmission. Evol Hum Behav 29:275–281

Richerson PJ, Boyd R (2005) Not by genes alone: how culture transformed human evolution. University of Chicago Press, Chicago

Robinson BS, Ort JC, Eldridge WA, Burke AL, Pelletier BG (2009) Paleoindian aggregation and social context at Bull Brook. Am Antiq 74:423–447

Rogers DS, Feldman MW, Ehrlich PR (2009) Inferring population histories using cultural data. Proc Biol Sci 276:3835–3843

Sanchez MG (2001) A synopsis of Paleo-Indian archaeology in Mexico. Kiva 67:119–136

Sapir E (1916) Time perspective in aboriginal American culture: a study in method, vol 90, Memoir (Geological Survey of Canada). Government Printing Bureau, Ottawa

Schillinger K, Mesoudi A, Lycett SJ (2014) Copying error and the cultural evolution of "additive" vs. "reductive" material traditions: an experimental assessment. Am Antiq 79:128–143

Sharon G (2009) Acheulian giant-core technology. Curr Anthropol 50:335–367

Shennan SJ (2000) Population, culture history and the dynamics of culture change. Curr Anthropol 41:811–835

Shennan SJ (2001) Demography and cultural innovation: a model and some implications for the emergence of modern human culture. Camb Archaeol J 11:5–16

Shennan SJ, Wilkinson JR (2001) Ceramic style change and neutral evolution: a case study from Neolithic Europe. Am Antiq 66:577–593

Sholts SB, Stanford DJ, Flores LM, Wärmländer SKTS (2012) Flake scar patterns of Clovis points analyzed with a new digital morphometrics approach: evidence for direct transmission of technological knowledge across early North America. J Archaeol Sci 39:3018–3026

Simons DB, Shott MJ, Wright HT (1984) The Gainey site: variability in a Great Lakes Paleo-Indian assemblage. Archaeol East N Am 12:266–279

Simpson GG (1944) Tempo and mode in evolution. Columbia University Press, New York

Slice DE (2007) Geometric morphometrics. Annu Rev Anthropol 36:261–281

Smallwood AM (2012) Clovis technology and settlement in the American Southeast: using biface analysis to evaluate dispersal models. Am Antiq 77:689–713

Stanford DJ, Bradley BA (2012) Across Atlantic ice: the origin of America's Clovis culture. University of California Press, Berkeley

Steele J, Adams J, Sluckin T (1998) Modelling Paleoindian dispersals. World Archaeol 30:286–305

Steele J, Glatz C, Kandler A (2010) Ceramic diversity, random copying, and tests for selectivity in ceramic production. J Archaeol Sci 37:1348–1358

Storck PL, Spiess AE (1994) The significance of new faunal identifications attributed to an early Paleoindian (Gainey complex) occupation at the Udora site, Ontario. Am Antiq 59: 121–142

Tankersley KB (2004) The concept of Clovis and the peopling of North America. In: Barton CM, Clark GA, Yesner DR, Pearson GA (eds) The American continents: a multidisciplinary approach in human biogeography. University of Arizona Press, Tucson, pp 49–63

Thulman DK (2007) A typology of fluted points from Florida. Fla Anthropol 60(4):63–75

Thulman DK (2012) Discriminating Paleoindian point types from Florida using landmark geometric morphometrics. J Archaeol Sci 39:1599–1607

Toelch U, van Delft MJ, Bruce MJ, Donders R, Meeus MTH, Reader SM (2009) Decreased environmental variability induces a bias for social information use in humans. Evol Hum Behav 30: 32–40

Tomasello M, Kruger AC, Ratner HH (1993) Cultural learning. Behav Brain Sci 16:495–511

Tunnell C, Johnson L (1991) Comparing dimensions for Folsom points and their byproducts from the Adair–Steadman and Lindenmeier sites and other localities. Texas Hist Comm Archeol Rep, no. 1. Austin

Tylor EB (1871) Primitive culture. Murray, London

Washburn DK (2001) Remembering things seen: experimental approaches to the process of information transmittal. J Archaeol Method Theory 8:67–99

Waters MR, Stafford TW Jr (2007) Redefining the age of Clovis: implications for the peopling of the Americas. Science 315:1122–1126

Whiten A, Hinde RA, Laland KN, Stringer CB (2011) Culture evolves. Philos Trans R Soc B 366:938–948

Whittaker JC (1984) Arrowheads and artisans: stone tool manufacture and individual variation at Grasshopper Pueblo. PhD dissertation, Department of Anthropology, University of Arizona, Tucson

Whittaker JC (2004) American flintknappers: stone age art in the age of computers. University of Texas Press, Austin

Willey GR, Phillips P (1958) Method and theory in American archaeology. University of Chicago Press, Chicago

Witthoft J (1952) A Paleo-Indian site in eastern Pennsylvania: an early hunting culture. Proc Am Phil Soc 96:464–495

Witthoft J (1954) A note on fluted point relationships. Am Antiq 19:271–273

Wormington HM (1957) Ancient man in North America, Popular series (Denver Museum of Natural History), no. 4. Denver Museum of Natural History, Denver

Wynn T, Coolidge FL (2010) How Levallois reduction is similar to, and not similar to, playing chess. In: Nowell A, Davidson I (eds) Stone tools and the evolution of human cognition. University of Colorado Press, Boulder, pp 83–103

Experimental Studies of Cumulative Culture in Modern Humans: What Are the Requirements of the Ratchet? 10

Christine A. Caldwell

Abstract

The success of *Homo sapiens* as a species may be explained, at least in part, by their learning abilities. The archaeological record suggests that the material culture of humans during the Palaeolithic was fluid and diverse. Social learning abilities may therefore have allowed *Homo sapiens* to adapt rapidly to novel or changeable environmental conditions. A capacity for cumulative cultural evolution is certainly apparent in all contemporary human societies, whereas it appears either absent or extremely rare in other extant species. Here I review laboratory studies of cumulative culture in modern adult humans, designed to shed light on the social information required for this type of learning to occur. Although it has been suggested that cumulative culture may depend on a capacity for imitation, we found that imitation (at least in the narrow sense of action copying) was not necessary for human participants to exhibit ratchet-like effects of improvement over learner generations. We discuss the need for high fidelity reproduction in cumulative culture (independent of action copying).

Keywords

Cultural evolution • Cumulative culture • Emulation • Imitation • Ratchet effect

10.1 Introduction

Homo sapiens dominate the planet in the present day, but the success of the species can be traced back much further. Hill et al. (2009) have suggested that even if the hunter-gatherer lifestyle of humans had never been superseded by agriculture and industry, the worldwide population would still have reached over 70 million. The early success of *H. sapiens* contrasts strikingly with the fate of *Homo neanderthalensis* however, as *H. neanderthalensis* probably went extinct just a few thousand years after *H. sapiens* dispersed from Africa and colonised Europe.

This extreme discrepancy in the relative success of the two species appears not to be attributable to physical characteristics, because in this respect the Neanderthals should have had the advantage, having evolved in Europe with physiological adaptations to the northern environment (Mithen 2014). However, what we know about the behaviour of the two species suggests that cognitive differences are a possibility. While Neanderthal artefacts appear to have remained relatively unchanged for over 300,000 years (e.g. Mithen 2005; Mellars 1999), human artefacts from around the time of the Neanderthal extinction show huge diversity, and were clearly used for particular specialised functions (Diamond 1989).

Therefore there seems to be fairly clear evidence that, at least from the Upper Palaeolithic onwards, humans exhibited cumulative cultural evolution. Cumulative cultural evolution refers to cultural change that accumulates over many generations leading to the evolution of behaviours that no individual could invent (Boyd and Richerson 1996).

C.A. Caldwell (✉)
Psychology, School of Natural Sciences, University of Stirling, Stirling FK9 4LA, UK
e-mail: c.a.caldwell@stir.ac.uk

A. Mesoudi and K. Aoki (eds.) *Learning Strategies and Cultural Evolution during the Palaeolithic*, Replacement of Neanderthals by Modern Humans Series, DOI 10.1007/978-4-431-55363-2_10,

In human culture this phenomenon has been described as producing a "ratchet"-like effect (Tomasello 1999), in that beneficial modifications are preserved without backwards slippage until further improvements are discovered.

Some recent proposals have suggested that cumulative culture became apparent at this point in human history as a consequence of the size and structure of human populations (e.g. Henrich 2004; Kline and Boyd 2010; Powell et al. 2009; Shennan 2001). Using simulation models, Powell et al. (2009) found that high population densities and high migration rates between subpopulations resulted in accumulation of modifications and increased complexity in technologies. On this basis they concluded that human population dynamics were a key driver in the acceleration of rates of cultural change. However, these models rest on the assumption that any cognitive requirements for cumulative culture were already in place (making these changes more fundamental). Therefore, although it remains possible that cognitive capacities for cumulative culture may have been masked by demographic factors in Neanderthals, it is nonetheless of major interest to know whether there are particular aspects of the cognition and/or learning of modern humans which are implicated in cumulative culture, which may not have been present in other hominin species, including ancestors of *H. sapiens* as well as Neanderthals. Any such capacities might offer a plausible explanation for the replacement of *H. neanderthalensis* by *H. sapiens*, since the technological advances made possible by cumulative culture could have represented an important competitive advantage.

10.2 Laboratory Studies of Cumulative Culture in Humans

In recent years, members of my research group have been pursuing a related question, concerning the uniqueness of human cumulative culture amongst currently extant species. Although there is evidence of cultural variation in chimpanzees (Whiten et al. 1999), as well as some other nonhuman primate species (e.g. Panger et al. 2002; Perry et al. 2003; van Schaik et al. 2003), few if any of the behaviours described are suggestive of cumulative culture. In fact, there is currently little evidence of anything resembling cumulative cultural evolution in any nonhuman species (Dean et al. 2014). The few compelling examples of complex, multistage tool use that exist in chimpanzees (e.g. Sanz et al. 2009), are accepted to be very rare, as well as falling far short of the elaborateness of human culture (Whiten 2011).

This evolutionary anomaly therefore presents something of a puzzle. The capacity for cumulative culture has allowed modern humans to dominate the planet (Hill et al. 2009), so it would appear to represent a uniquely powerful mechanism for adapting to novel and changing environments. Nonetheless, humans remain the only species to have developed this particular facility.

Various hypotheses have been put forward to explain this evolutionary discontinuity. Until the mid-1990s, the apparent absence of evidence of visuo-motor imitation in nonhuman species (e.g. monkeys: Visalberghi and Fragaszy 1990; apes: Tomasello et al. 1987, 1994) offered a conveniently neat explanation, and prominent theorists in the field cited such evidence in their accounts of the uniqueness of human cumulative culture. For example, Boyd and Richerson (1996) noted that simpler social learning processes were unlikely to result in cumulative cultural evolution. The simple social learning processes they referred to included local enhancement (where the attention of an observer is drawn to a particular location) and stimulus enhancement (attention drawn to a class of objects). In such cases, the behaviour is learned by trial and error, with the presence of another individual simply making that learning more likely. Boyd and Richerson (1996) argued that this sort of learning cannot generate cumulative cultural evolution, which by definition must allow learners to proceed from a more advanced starting point than was possible for previous generations. In contrast, learning by local or stimulus enhancement effectively necessitates that each new learner has to start from scratch, wiping out any useful innovations which may have been chanced upon by others.

However, there is now a wealth of evidence showing copying of specific techniques (using "two-action" experimental designs) in chimpanzees (e.g. Whiten et al. 1996, 2005; Whiten 1998) and other nonhuman primates (e.g. Voelkl and Huber 2000; Dindo et al. 2008). Such evidence calls into question the idea that other species do not exhibit cumulative culture because of a lack of imitative ability.

More recently, Tomasello and colleagues (e.g. Tomasello 1999; Tennie et al. 2009) have been even more specific about the types of social learning expected to generate cumulative culture. Tomasello (1999) has stated that, "cumulative cultural evolution depends on imitative learning, and perhaps active instruction on the part of adults, and cannot be brought about by means of "weaker" forms of social learning such as local enhancement, emulation learning, ontogenetic ritualization, or any form of individual learning." (p. 39). The specific reference to emulation learning as a mechanism not capable of supporting cumulative culture is noteworthy here. Emulation learning has been defined in a number of different ways in the literature on social learning (e.g. see Whiten et al. 2009 for a review). However, its usage derives originally from a chapter by Wood (1989), referring to "instances where children achieve common goals to those modelled, but do so by using idiosyncratic means that were never observed" (p. 72). Tomasello (1990) then used the term to describe chimpanzees' responses following a demonstration of tool use (in Tomasello et al.'s 1987 study), as the chimpanzees had shown increased manipulation of the tool, without copying the exact method of the demonstrator.

Since then, however, the term "emulation" has become used in rather different contexts, typically referring to situations in which the learner could have (or indeed has) achieved the exact same outcome as the demonstrator, but without necessarily copying bodily movements (e.g. through attending to how the tool or object itself moves). On this basis, Tennie et al. (2009) have criticised Whiten and colleagues' claims of imitation from two-action studies, asserting that the chimpanzees, "were not learning about actions but rather about how or where the box works—without attending much or at all to the behavioural techniques used (emulation learning)." (Tennie et al. 2009, p. 2407).

Tennie et al. (2009) therefore characterised chimpanzees as "product-oriented" (focussing on effects and end products), in contrast with "process-oriented" human learners, who focus on bodily actions. There does seem to be some evidence to suggest that, in comparative studies, human children may be more focussed on copying actions, with chimpanzees being more goal-oriented (e.g. Horner and Whiten 2005; Tennie et al. 2010). However, in relation to the question of whether these types of learning can support cumulative culture, this debate raises the intriguing question of whether cumulative culture is still possible for humans, even when restricted to learning from "product", rather than "process" information. It was this question that led my research group into our initial attempts to study human cumulative culture under laboratory conditions.

Time (minutes: seconds)	Participants present in test group (black indicates building, grey indicates observing)
0:00	**1** 2 3
2:30	**1** **2** 3 4
5:00	**2** **3** 4 5
7:30	**3** **4** 5 6
10:00	**4** **5** 6 7
12:30	**5** **6** 7 8
15:00	**6** **7** 8 9
17:30	**7** **8** 9 10
20:00	**8** **9** 10
22:30	**9** **10**
25:00	**10**
27:30	

Fig. 10.1 Group composition over time in the microsociety design of Caldwell and Millen (2008a, b). Generational succession was simulated through the repeated removal of experienced participants and introduction of naïve participants. Each row of the table shows the group composition at any given time, made up of observing participants and participants actually engaged in the task. Participants were randomly assigned the positions 1–10

10.2.1 Study 1: Establishing the Methods for Studying Cumulative Culture in the Laboratory

Our initial goal was simply to establish that it was possible to study cumulative cultural evolution under laboratory conditions. We therefore wanted to identify tasks that participants could be asked to complete, which could in principle be used to show a learning-like effect of improved performance over successive attempts, but crucially, involving different individuals making each of these attempts. The tasks we chose also needed to be ones which participants could complete in a relatively short time period, in order to make the studies feasible. We also required tasks that would allow a clear objective measure of performance, so that we could track improvements (in terms of the prescribed goal measure) over learner generations.

We ran our initial experiments (reported in full in Caldwell and Millen 2008b) using two different tasks. In the first task, participants were asked to build a paper aeroplane from a single sheet of paper. The goal measure was the flight distance of their plane (which we recorded by giving them the best distance of three throws, to allow for mis-throws). In the second task, participants were asked to build a tower from raw spaghetti and a small amount of modelling clay. The goal measure for this task was the height of the tower, so participants were to aim to build their tower to be as tall as possible.

For both tasks, individual participants were given a total of 5 min to build their own artefact. They were also given 5 min of observation time prior to this, during which they could observe other participants engaged in the task. Our design therefore assumed that participants would most benefit from social information early on in their experimental "lifespan", and that their own opportunities for innovation should follow this, in order to allow cumulative culture to develop. Using formal mathematical models, Aoki et al. (2012) have shown that such an ordering (social learning followed by individual learning) is evolutionarily adaptive under certain conditions. It is therefore plausible that this represents an evolved learning strategy in humans.

For each task, we ran ten chains each composed of ten participants, who took part in the task one after the other, each beginning their own attempt at 2½ min intervals. Figure 10.1 displays a schematic illustrating the role of each of the ten participants in any given chain at any point during testing. Running an entire chain of ten participants therefore took a total of 27½ min, as can be seen from the schematic. Participants' completed artefacts were retained in the test area for inspection by their successors in the chain.

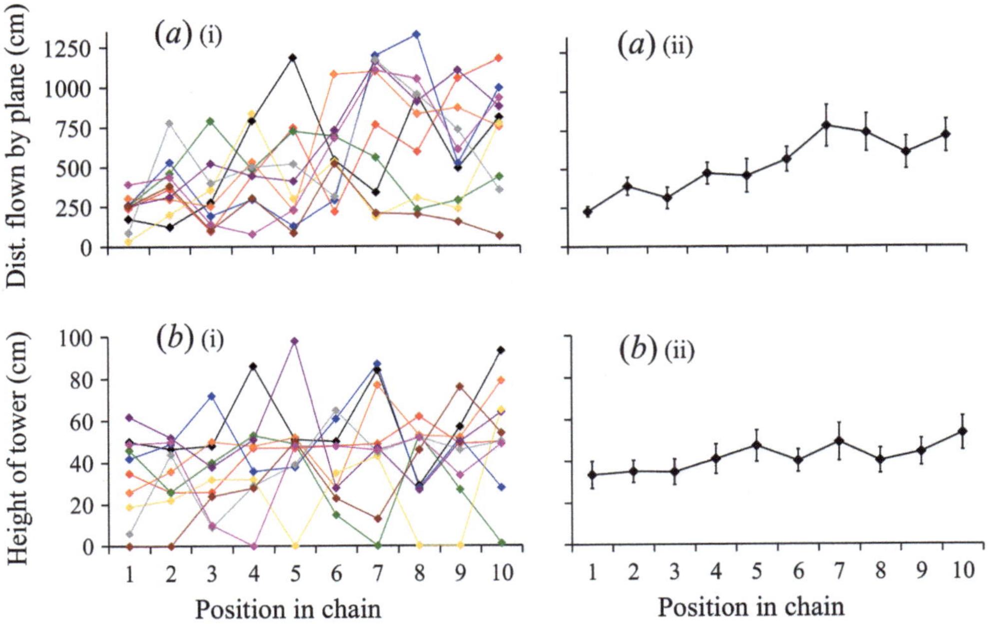

Fig. 10.2 Tower height and flight distance data from Caldwell and Millen (2008a, b). Panel (**a**)(**i**) displays the data for the ten chains of paper aeroplanes, with panel (**a**)(**ii**) showing the mean distance flown for each position in the chains (error bars indicate +/− SEM). Panel (**b**)(**i**) displays the data for the ten chains of spaghetti towers, with panel (**b**)(**ii**) showing the mean height for each position in the chains (error bars indicate +/− SEM)

Therefore, although participants could only observe their two immediate predecessors actually building their artefacts, they could in theory attempt to reproduce the design of any of the earlier attempts from the finished product.

Our predictions for these initial experiments were that participants' scores on the goal measures (flight distance of planes and height of towers) would increase over learner generations, such that later participants were performing better. We expected to find this because we believed that participants would be able to benefit from the earlier learning efforts of other participants, through observation of both successful and unsuccessful approaches of others they were able to observe, and through copying of apparently effective completed designs. We also expected to find some evidence for cultural traditions in the artefacts produced by the participants in our chains, as a result of participants copying other designs. We therefore expected that pairs of designs taken from the same chain would be more similar to each other than pairs of designs taken from different chains.

With regard to our prediction that scores would be higher for later generation participants, we found this to be the case for both tasks. Figure 10.2 displays the data for both the paper aeroplane task and the spaghetti tower task, with data from individual chains displayed in the panels on the left and averaged data in the panels on the right. Despite a high level of variability in success on both tasks (apparent from the individual chain data) there were significant trends in the direction of higher scores in later generations. This aspect of our findings supported our basic expectation of a "ratchet-like" effect across multiple successive learners.

In terms of our prediction that designs would be more similar within chains, compared with across them, this was also upheld. Naïve coders rated the similarity of pairs of designs, using photographs taken during testing, and as expected we found that pairs from the same chain were typically rated as being more similar to each other, compared with pairs from different chains. This offered additional insights into the effects that we were seeing, as it was clear from this finding that participants were indeed copying elements of others' designs. However, since designs also tended to improve with transmission, this must have involved more than just random copying, probably entailing selective copying of effective designs and/or innovative modification of previous examples.

Figure 10.3 shows examples of towers built by participants in two chains of one of our experiments. It should be noted, however, that the towers depicted in Fig. 10.3 were taken from Caldwell and Millen (2010), from an experimental condition found to show particularly strong evidence of copying. Differences between chains (and similarities within them) are rarely as striking as those seen here.

10.2.2 Study 2: Comparing Learning Mechanisms

Having established that these methods could be used to tap into ratchet effects in social learning, we then wanted to manipulate the learning opportunities available to our participants in order to determine whether any particular type of social information was necessary for this to occur (Caldwell and Millen 2009). We separated the different sources of social information into information about: (a) actions (corresponding to imitation, according to most definitions in the literature, e.g. Whiten and Ham 1992; Heyes 1993), (b) results, in the form of both finished products and feedback

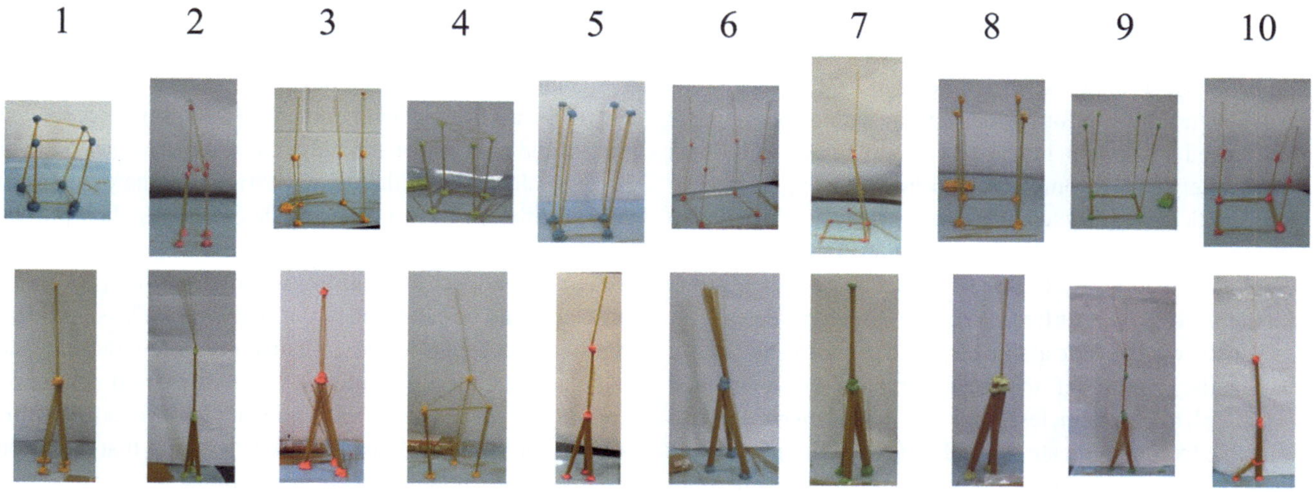

Fig. 10.3 Examples of spaghetti towers created by participants in Caldwell and Millen (2010). Each row displays the complete set of towers that were produced by one chain of participants. The towers are ordered from *left* to *right*, in the order in which they were produced (i.e. the first participant's tower is on the far *left*, and the tenth on the far *right*)

about performance (corresponding to emulation according to most definitions, e.g. Whiten et al. 2004; Call et al. 2005), and (c) teaching, in the form of verbal communication between participants. By combining these different sources of information into their various possible permutations, we would be able to establish which of these were necessary for obtaining a cumulative learning effect over multiple individuals.

We used the paper aeroplane task because with this task it was possible to separate out information about actions from information about results, allowing us to run a "pure" actions-only condition involving no feedback on performance. In our spaghetti tower task, observation of another's actions during construction inevitably also provided information about results in that it was possible to see the height of the tower as it was built. However, with the paper aeroplane task, performance (flight) was separated from construction, and we could therefore exploit this to separate out these sources of information within our experimental design. It should however be noted that, even for the paper aeroplane task, it was not possible to completely separate out results in the form of folds made in the paper, since observing another's actions meant that participants were also able to see the effects of those actions on the material. Close inspection of completed planes could however be separated from observation of others' actions. Further details are provided below, regarding how information from actions, results and teaching were manipulated.

We manipulated the availability of information by making alterations to the experimental set-up and the membership of the test group. So, for example, information in the form of actions was made available by having participants in the test group in full view of one another so that they could observe others building, as well as having participants enter the test group 5 min before starting their own build (in order to benefit from dedicated observation time). Conditions with actions information available were therefore, in this respect, similar to the design of our previous study (described in the previous section). In conditions with no information about actions, participants who were engaged in construction were screened off from one another in the test group so that they could not see each other building planes. The 5 min observation period was also removed, so participants simply entered the test group and began building their own plane straight away.

In the conditions in which information about results was available, participants were provided with the two most recently completed planes so that they could inspect these finished products. They were also given explicit information about the flight distances of these planes, and could observe other participants throwing their planes at the end of their build. Conditions with results information available were therefore largely similar to our previous research using this task, although it should be noted that the availability of finished products was more carefully controlled in this study (with only the two most recently completed planes available), compared with the previous one discussed (in which all completed artefacts were retained for inspection). In this particular study it was considered important to maximise consistency in the amount of information available to participants, for reasons of comparability with the other learning conditions. For conditions in which results information was unavailable, participants in the test group were screened off from those flying their planes, completed planes were not given to the test group members to inspect, and flight distances were not fed back by the experimenter.

In order to vary the presence or otherwise of teaching, efforts had to be made to ensure the availability, or unavailability as appropriate, of such information. For conditions in which teaching information was available, participants were asked to remain in the test group for an additional 5 min, after having completed and flown their plane. They were explicitly instructed that their role was to help the other participants with the task, and that they should pass on any information that they believed might help their group members. This was different to our previous design, as previously we had not assigned participants to the role of teacher; they simply left the test group once their own build was completed. Nonetheless, we knew that when participants were able to communicate about the task, useful information could be (and often was) exchanged between members of the test group. Hence for the conditions without information from teaching, we instructed participants not to engage in any form of verbal communication during their participation in the experiment. Although complete adherence to the no talking rule could not be guaranteed, any participants who began to discuss the task were reminded of the rule by the experimenter (who was always present). All sessions were recorded and a subset from each condition were transcribed to confirm that verbal instruction relating to the task was, or was not, being exchanged, as appropriate to the condition.

The various different combinations of these sources of information gave rise to a total of seven different experimental conditions: actions, results & teaching; actions & results; actions & teaching; actions only; results & teaching; results only; and teaching only. For each condition, within-chain analyses provided the measure of the value of the social information. The first participants in any chain had no access any form of social information, whereas participants in the last few generations not only had access to social information, but had access to social information from others who had themselves received opportunities for social learning from experienced models (see Figs. 10.1 and 10.2).

Further details of these conditions, including diagrams of the experimental set-up, can be found in Caldwell and Millen (2009), with full procedural information for each condition provided as supplementary information. In each of these conditions, ten chains each consisting of ten participants were run, following the design of our previous study.

All of these conditions were analysed individually for evidence of cumulative culture (as evidenced by a trend for improved performance over learner generations). Interestingly, we found that all seven of the conditions showed this effect. This suggested that any one of the sources of information was sufficient, independently, for our participants to benefit from the earlier efforts of the members of their chain. This was noteworthy, particularly with regard to the results-only condition.

Essentially, our results-only condition represented our test for the possibility of cumulative culture occurring on the basis of emulation learning alone. In this condition participants had information about end products in the form of completed planes, and also information about outcomes in the shape of the flight distances of these planes. However, participants in this condition could not observe others building their planes, and furthermore they were not permitted to talk to their test group partner, so no information about the process of building could be conveyed in this way either. The significant improvement found in these chains therefore indicated that cumulative cultural evolution was possible from emulation learning only, at least for this particular task. We return to this issue in the Sect. 10.3, as there is little doubt that imitation and teaching are very likely to facilitate transmission of many other skills. However it is clear that it is not imitative ability, in and of itself, which gives humans the capacity for cumulative cultural evolution.

10.2.3 Study 3: High Fidelity Copying and Cumulative Cultural Evolution

One of the reasons that imitation has been proposed to be crucial for cumulative cultural evolution is that it is believed to result in high fidelity transmission (e.g. Tennie et al. 2009). However, Caldwell and Millen's (2009) finding, detailed above, suggests that cumulative cultural evolution can occur in the absence of action copying, purely from information about end products and results. In relation to Tennie et al.'s (2009) argument, this means that either high fidelity transmission is possible without imitation, or perhaps that high fidelity transmission is not necessary for cumulative cultural evolution.

The former possibility, that high fidelity transmission can occur without imitation, seems very plausible based on our experiments using the paper aeroplane and spaghetti tower tasks. These tasks produce fairly simple end product designs, which can be reproduced with reasonable accuracy even if one has not observed the process of building.

Nonetheless, this raised the additional question of how closely participants copied end products in our tasks. Caldwell et al. (2012) set out to quantify the degree of matching exhibited by participants who were presented only with finished products. On this occasion we used the spaghetti tower task, for which we had generally found two main design types arising in our participants' solutions (Caldwell and Millen 2008b, 2010). Figure 10.3 displays the towers built by two complete chains of participants in Caldwell and Millen's (2010) study, and these illustrate these two main design types. One design type, which we have labelled "cubic", typically begins with a square base of spaghetti, with four parallel upright spaghetti columns, often finished

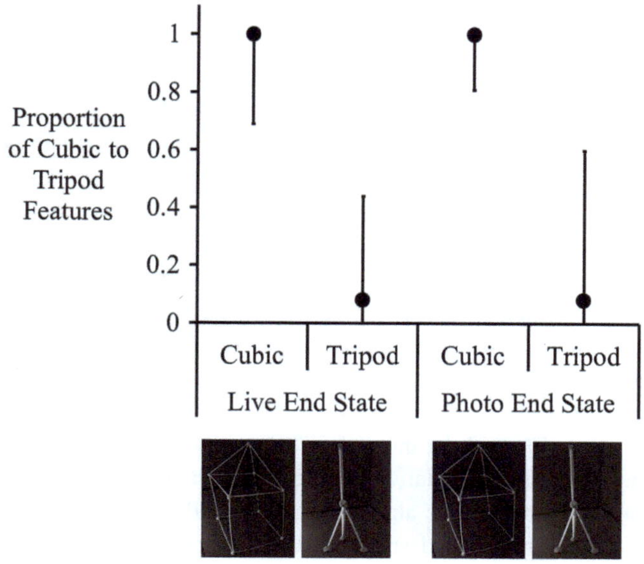

Fig. 10.4 Median proportion of cubic to tripod tower type features in participants' towers in live and photo end-state conditions of Caldwell et al. (2012) (error bars display quartiles)

with a square pyramid upper section. The other design type, which we have labelled "tripod", typically begins with modelling clay feet supporting three (occasionally more, or fewer) spaghetti columns which converge to a single point, with upper levels as single columns of spaghetti.

These two design types offered the possibility of exposing participants to experimentally manipulated contrasting stimuli, which allowed us to objectively code, and therefore quantify, the similarity of participants' solutions in relation the two possible alternatives. As in our previous experiments however, the goal was the same, i.e. the participants were instructed simply to build their tower to be as tall as possible. However, in this version, Caldwell et al. (2012) presented finished towers as examples of previous participants' efforts, in place of the face-to-face interaction involved in the previous studies detailed above. We therefore exposed participants to example towers of either the cubic or tripod type, and then coded the towers they produced in terms of the features in common with each of the two designs.

We found that participants showed a strong tendency to match the design type that they had been exposed to, and this was the case regardless of whether the completed towers were physically present, or shown as photographs (see Fig. 10.4). The finding led us to conclude that although relatively high fidelity copying does appear to be associated with the sort of transmission that can generate cumulative cultural evolution, such high fidelity copying is nonetheless possible on the basis of information about end products, without necessarily requiring observation of another individual's actions. Once again, an important caveat should be mentioned, as the task in question makes it easy to reproduce the solutions from inspecting the end products alone. For many tasks, especially more complex ones, this will not necessarily be the case, and we will return to this issue in the following section.

10.3 Cognitive Requirements of Cumulative Culture?

10.3.1 Action Copying

As noted above, for these two tasks it appears that observation of other's actions is not necessary either for accurate reproduction, or for cumulative cultural evolution. However, this is not to say that action copying is never required, since it inevitably will be for other behaviours. As we have discussed elsewhere (e.g. Caldwell and Millen 2009), imitation and/or teaching will always be needed in order to learn behaviours that leave no physical trace (e.g. for communicative gestures, where the bodily movement itself is the goal).

Furthermore, any sequences of behaviour that are "cognitively opaque" to the potential learner (e.g. Gergely and Csibra 2006) will be transmitted much more readily through either faithful imitation or instruction from a skilled partner. Cognitive opacity, as Gergely and Csibra have defined it, refers to behaviours for which it is unclear *how* the required actions bring about the desired goal, and therefore this will inevitably pose problems for a learner restricted to emulation and other non-imitative forms of social learning.

Boyd et al. (2011) provide multiple examples of skills necessary for survival in Arctic conditions, used by the local Inuit populations, which provide excellent examples of cognitive opacity. The most effective clothing for the winter climate, for example, is created from Caribou skin, harvested in autumn when the fur is at just the right thickness. The hide must then undergo a lengthy process of stretching, scraping and moistening, followed by further stretching, in order to produce material that is sufficiently soft and workable. The garment must then be constructed in a very specific way in order to create a shape which retains heat but releases moisture (Otak 2005, cited in Boyd et al. 2011). Details such as the importance of harvesting at a particular time of year, the steps involved in processing, and the relevance of the garment's shape, would all qualify as cognitively opaque, since their functional significance would not be obvious to a naïve observer. Gergely and Csibra (2006) propose that imitative learning is an important means by which such behaviours can be transmitted without loss of important functional details such as these.

In contrast, the paper aeroplane and spaghetti tower tasks which we used were relatively cognitively transparent from this point of view. Furthermore, even if participants were unaware of the functional significance of particular paper folds, or precise placement of modelling clay, they would nonetheless generally be able to infer the actions required to reproduce these fairly accurately, based on the finished product alone.

Interestingly however, despite emphasising the significance of cognitive opacity in relation to imitative abilities, Gergely and Csibra (2006) do not seem to share Tomasello's (1999) view on the relationship between cumulative culture and imitation. They state: "We argue that the cognitive opacity of cultural products in early hominid cultural environments represented evolutionary pressure for the selection of a new type of social-cognitive learning mechanism to solve this learnability problem and to ensure fast and efficient transmission of culturally relevant knowledge.". The implication of this proposal is that any human-unique social learning mechanisms are perhaps better viewed as a consequence, rather than cause, of complex human culture. If this is indeed the case, then researchers may need to look elsewhere for their explanations of how cumulative culture arose in the first place.

Certainly, it appears that action-copying abilities alone cannot explain the species distribution of cumulative culture. Our own findings have illustrated what is possible for adult humans using emulation learning, and there is increasing evidence of action copying abilities in nonhuman species (e.g. Hopper et al. 2007).

10.3.2 High Fidelity Copying

The interest in imitation specifically, in relation to cumulative culture, may derive in part from a general confusion within the literature over definitions of imitation (see Caldwell and Whiten 2002, for a review). Within the literature on comparative psychology, a particular focus on action copying dates back to Thorndike's (1898) interest in imitation, as defined as the ability to, "from an act witnessed learn to do the act" (p. 50). Action copying has typically been assumed to be particularly cognitively challenging due to the cross-modal mapping required between observation of another's actions and one's own performance of the same action. On this basis, vocal imitation (such as that documented in birds) is typically dismissed as cognitively undemanding (e.g. Whiten and Ham 1992; Heyes 1993; Byrne 2002). Emulation learning is viewed in the same way, since reproducing an effect on the environment requires no understanding of the correspondence between one's own body and another's.

However, it seems clear that this cognitively complex cross-modal mapping is not what Boyd and Richerson (1996) had in mind when proposing "true imitation" as necessary for cumulative culture, since they specifically mention bird song traditions as an exceptional case. Boyd and Richerson (1996) were probably more interested in faithful reproduction in a more general sense, without being concerned with the precise cognitive mechanisms involved.

More recently, Galef (2013) has drawn specific attention to the relative unimportance of Thorndike-inspired definitions (and tests) of imitation in relation to our understanding of cumulative culture, noting that, "the kind of imitation that might lead to faithful copying and cumulative culture is not Thorndike's imitation and is not illustrated by the two-action method." (p. 126). Galef (2013) proposes instead that cumulative culture is much more likely to be dependent on imitation in its common-usage sense, involving using another's performance as a template for refinement of one's own efforts.

Therefore, whilst the ability to copy actions may not be necessary for cumulative culture, a motivation to reproduce others' performance almost certainly is. Participants in our paper aeroplane and spaghetti tower experiments were no doubt motivated to reproduce what they believed to be effective designs and, for these tasks, this was perfectly possible even in the absence of information about actions.

Such an interpretation would also be consistent with Wasielewski's (2014) recent argument regarding the importance of imitation for cumulative culture involving cognitively opaque behaviour. Wasielewski (2014) highlighted Acerbi et al.'s (2011) insight that behaviours can vary in the extent to which solutions close to the optimal design result in payoffs that are also close to optimal. For such behaviours, trial-and-error feedback can allow the learner to find the optimal solution even when transmission is relatively low fidelity. In contrast, for behaviours which may be superficially similar to others with very different payoffs, high fidelity transmission would be required. Wasielewski (2014) points out that low resolution learning is sufficient to explain most of the evidence for action copying and cultural variation in nonhumans, as a result of the possibility of behavioural refinement guided by feedback from non-optimal solutions.

However, although high fidelity reproduction may be necessary for cumulative culture to occur, this does not allow us to conclude that this is the key difference between humans and other species which allowed humans to develop cumulative culture. Returning to comparisons with Neanderthals and other extinct hominins is enlightening in this regard, since it would appear that a proclivity for high fidelity reproduction may not have been unique to humans during the Palaeolithic. It has already been mentioned that the culture of Neanderthals remained relatively static for a matter of hundreds of thousands of years. However, the strong technological traditions observed are suggestive of very high fidelity transmission (Mithen 1994). This suggests that mechanisms facilitating faithful social transmission were present in hominin species which exhibited little evidence of cumulative culture. Some other factor may therefore have accounted for the unusual complexity of human culture.

It is possible that the critical development was one which, operating alongside high fidelity reproduction, tended to push change in the direction of increased complexity and improvements in utility. Any one of a number of factors might thus have been influential. Advances in individual learning capacities (e.g. Aoki 2013), for example, might have allowed individuals to modify their culturally inherited repertoire of skills in predominantly beneficial ways. Alternatively, particular social learning strategies (Laland 2004) could have been responsible for generating shifts towards increasing utility through cultural selection (e.g. see Mesoudi's 2011 study of payoff bias). Finally, as noted in the very first section of this chapter, it remains possible that the cognitive capacities required for cumulative culture were already in place well in advance of the Upper Palaeolithic, but that the rapid accumulation and diversification of technology was only manifest in populations of certain size and structure. The remaining puzzle associated with these interpretations is that, in contrast to high fidelity social transmission, none of them appear to be unique to humans. Animals have been shown to have capacities for capacities for innovation and to exhibit success-biased and payoff-biased social learning (e.g. see Dean et al. 2014 for a review of explanations of human cumulative culture). Likewise, humans are not unique in the size of their communities, or their rates of migration. Perhaps there are quantitative or qualitative differences between humans and other species in terms of how innovation and/or social learning biases operate, although if so, these have yet to be elucidated. A further possibility is that high fidelity copying, innovation, and biased transmission, in combination with certain demographic catalysts, are all critical components for the expression of cumulative culture, and that modern humans are the only species to date for which all of these ingredients have combined in quite this way.

10.3.3 Concluding Remarks

The potential for rapid adaptation permitted by cumulative culture may well provide an explanation for the replacement of the Neanderthals by modern humans. The cognitive capacities underlying this process therefore hold significant scientific importance. Although it remains unclear exactly what caused the explosion of cumulative culture in modern humans during the Palaeolithic, our studies of cumulative culture in contemporary humans, reviewed in the current chapter, have provided some insights into the related debate over the uniqueness of human culture compared with other extant species.

Our findings have primarily helped to shed some light on the issue of the relevance (or otherwise) of action copying. As noted above, there is also good reason to remain cautious over similar explanations of cumulative culture in terms of high fidelity copying, even though there is better reason to believe that this may be a prerequisite. Investigations into the specific social learning strategies employed by humans, and the nature of human innovation, may have the potential to offer further insights. Such investigations might, in due course, find evidence with converges with other proposals contending that there were qualitative differences in cognition between humans and Neanderthals (e.g. Mithen 2014; Wynn and Coolidge 2004). However, irrespective of the eventual explanation, it is likely that experimental studies of social learning in contemporary humans (particularly when combined with comparative data from nonhuman species) will continue to offer one of the most critical sources of evidence in settling this vigorous and stimulating debate.

Acknowledgements CC was supported by ESRC grants RES-061-23-0072 and RES-062-23-1634.

References

Acerbi A, Tennie C, Nunn CL (2011) Modeling imitation and emulation in constrained search spaces. Learn Behav 39:104–114

Aoki K (2013) Determinants of cultural evolutionary rates. In: Akazawa T, Nishiaki Y, Aoki K (eds) Dynamics of learning in Neanderthals and modern humans volume 1: cultural perspectives. Springer, Dordrecht, pp 199–210

Aoki K, Wakano JY, Lehmann L (2012) Evolutionarily stable learning schedules and cumulative culture in discrete generation models. Theor Popul Biol 81:300–309

Boyd R, Richerson PJ (1996) Why culture is common, but cultural evolution is rare. Proc Br Acad 88:77–93

Boyd R, Richerson PJ, Henrich J (2011) The cultural niche: why social learning is essential for human adaptation. Proc Natl Acad Sci U S A 108:10918–10925

Byrne RW (2002) Imitation of novel complex actions: what does the evidence from animals mean? Adv Stud Behav 31:77–105

Caldwell CA, Millen AE (2008a) Studying cumulative cultural evolution in the laboratory. Philos T Roy Soc B 363:3529–3539

Caldwell CA, Millen AE (2008b) Experimental models for testing hypotheses about cumulative cultural evolution. Evol Hum Behav 29:165–171

Caldwell CA, Millen AE (2009) Social learning mechanisms and cumulative cultural evolution: is imitation necessary? Psychol Sci 20:1478–1483

Caldwell CA, Millen AE (2010) Conservatism in laboratory microsocieties: unpredictable payoffs accentuate group-specific traditions. Evol Hum Behav 31:123–130

Caldwell CA, Whiten A (2002) Evolutionary perspectives on imitation: is a comparative psychology of social learning possible? Anim Cogn 5:193–208

Caldwell CA, Schillinger K, Evans CL, Hopper LM (2012) End state copying by humans (*Homo sapiens*): implications for a comparative perspective on cumulative culture. J Comp Psychol 126:161–169

Call J, Carpenter M, Tomasello M (2005) Copying results and copying actions in the process of social learning: chimpanzees (*Pan troglodytes*) and human children (*Homo sapiens*). Anim Cogn 8:151–163

Dean LG, Vale GL, Laland KN, Flynn EG, Kendal RL (2014) Human cumulative culture: a comparative perspective. Biol Rev 89: 284–301

Diamond J (1989) The great leap forward. Discover 10:50–60
Dindo M, Thierry B, Whiten A (2008) Social diffusion of novel foraging methods in brown capuchin monkeys (*Cebus apella*). Proc R Soc B 275:187–193
Galef BG (2013) Imitation and local enhancement: detrimental effects of consensus definitions on analyses of social learning in animals. Behav Process 100:123–130
Gergely G, Csibra G (2006) Sylvia's recipe: the role of imitation and pedagogy in the transmission of human culture. In: Enfield NJ, Levinson SC (eds) Roots of human sociality: culture, cognition, and human interaction. Berg Publisher, Oxford, pp 229–255
Henrich J (2004) Demography and cultural evolution: how adaptive cultural processes can produce maladaptive losses: The Tasmanian case. Am Antiq 69:197–214
Heyes CM (1993) Imitation, culture and cognition. Anim Behav 46:999–1010
Hill K, Barton M, Hurtado AM (2009) The emergence of human uniqueness: characters underlying behavioral modernity. Evol Anthropol 18:187–200
Hopper LM, Spiteri A, Lambeth SP, Schapiro SJ, Horner V, Whiten A (2007) Experimental studies of traditions and underlying transmission processes in chimpanzees. Anim Behav 73:1021–1032
Horner V, Whiten A (2005) Causal knowledge and imitation/emulation switching in chimpanzees (*Pan troglodytes*) and children (*Homo sapiens*). Anim Cogn 8:164–181
Kline MA, Boyd R (2010) Population size predicts technological complexity in Oceania. Proc Roy Soc B 277:2559–2564
Laland KN (2004) Social learning strategies. Learn Behav 32:4–14
Mellars P (1999) The Neanderthal problem continued. Curr Anthropol 40:341–350
Mesoudi A (2011) An experimental comparison of human social learning strategies: payoff-biased social learning is adaptive but underused. Evol Hum Behav 32:334–342
Mithen SJ (1994) Technology and society during the middle Pleistocene: hominid group size, social learning and industrial variability. Cam Archaeol J 3:3–36
Mithen SJ (2005) The singing Neanderthals: the origins of music, language, mind and body. Weidenfeld & Nicolson, London
Mithen SJ (2014) The cognition of Homo neanderthanlensis and H. sapiens: does the use of pigment necessarily imply symbolic thought? In: Akazawa T, Ogihara N, Hiroki C. Tanabe HC, Terashima H (eds) Dynamics of learning in Neanderthals and modern humans volume 2: cognitive and physical perspectives. Springer, Tokyo, Japan, pp 7–15
Otak LA (2005) Iniqsimajug: Caribou-skin preparation in Igloolik, Nunavut. In: King JC, Pauksztat B, Storrie R (eds) Arctic clothing. McGill-Queens University Press, Montreal
Panger MA, Perry S, Rose L, Gros-Louis J, Vogel E, MacKinnon KC, Baker M (2002) Cross-site differences in foraging behaviour of white-faced capuchins (*Cebus capucinus*). Am J Phys Anthropol 119:52–66
Perry S, Baker M, Fedigan L, Gros-Louis J, Jack K, MacKinnon KC, Manson JH, Panger MA, Pyle K, Rose L (2003) Social conventions in wild white-faced capuchin monkeys: evidence for traditions in a Neotropical primate. Curr Anthropol 44:241–268
Powell A, Shennan S, Thomas MG (2009) Late Pleistocene demography and the appearance of modern human behavior. Science 324:1298–1301
Sanz C, Call J, Morgan D (2009) Design complexity in termite-fishing tools of chimpanzees (*Pan troglodytes*). Biol Lett 5:293–296
Shennan S (2001) Demography and cultural innovation: a model and its implications for the emergence of modern human culture. Camb Archaeol J 11:5–16
Tennie C, Call J, Tomasello M (2009) Ratcheting up on the ratchet: on the evolution of cumulative culture. Philos Trans Royal Soc B 364:2405–2415
Tennie C, Greve K, Gretscher H, Call J (2010) Two-year-old children copy more reliably and more often than nonhuman great apes in multiple observational learning tasks. Primates 51:337–351
Thorndike EL (1898) Animal intelligence: an experimental study of the associative process in animals. Psychol Rev Monogr 8:1–109
Tomasello M (1990) Cultural transmission in the tool use and communicatory signaling of chimpanzees? In: Parker ST, Gibson KR (eds) "Language" and intelligence in monkeys and apes. Cambridge University Press, Cambridge, UK, pp 274–311
Tomasello M (1999) The cultural origins of human cognition. Harvard University Press, Cambridge, MA
Tomasello M, Davis-Dasilva M, Camak L, Bard K (1987) Observational learning of tool use by young chimpanzees. Hum Evol 2:175–183
Tomasello M, Call J, Nagell K, Olguin R, Carpenter M (1994) The learning and use of gestural signals by young chimpanzees: a trans-generational study. Primates 35:137–154
van Schaik CP, Ancrenaz M, Borgen G, Galdikas B, Knott CD, Singleton I, Suzuki A, Utami SS, Merrill M (2003) Orangutan cultures and the evolution of material culture. Science 299:102–105
Visalberghi E, Fragaszy DM (1990) Do monkeys ape? In: Parker ST, Gibson KR (eds) "Language" and intelligence in monkeys and apes. Cambridge University Press, Cambridge, UK, pp 247–273
Voelkl B, Huber L (2000) True imitation in marmosets. Anim Behav 60:195–202
Wasielewski H (2014) Imitation is necessary for cumulative cultural evolution in an unfamiliar, opaque task. Hum Nat 25: 161–179
Whiten A (1998) Imitation of the sequential structure of actions by chimpanzees (*Pan troglodytes*). J Comp Psychol 112:270–281
Whiten A (2011) The scope of culture in chimpanzees, humans and ancestral apes. Philos T Roy Soc B 366:935–1187
Whiten A, Ham R (1992) On the nature and evolution of imitation in the animal kingdom: reappraisal of a century of research. Adv Stud Behav 21:239–283
Whiten A, Custance DM, Gomez JC, Texidor P, Bard KA (1996) Imitative learning of artificial fruit processing in children (*Homo sapiens*) and chimpanzees (*Pan troglodytes*). J Comp Psychol 110:3–14
Whiten A, Goodall J, Mcgrew WC, Nishida T, Reynolds V, Sugiyama Y, Tutin C, Wrangham R, Boesch C (1999) Cultures in chimpanzees. Nature 399:682–685
Whiten A, Horner V, Litchfield CA, Marshall-Pescini S (2004) How do apes ape? Learn Behav 32:36–52
Whiten A, Horner V, de Waal FBM (2005) Conformity to cultural norms of tool use in chimpanzees. Nature 437:737–740
Whiten A, McGuigan N, Marshall-Pescini S, Hopper LM (2009) Emulation, imitation, over-imitation and the scope of culture for child and chimpanzee. Philos Trans Roy Soc B 364:2417–2428
Wood D (1989) Social interaction as tutoring. In: Bornstein MH, Bruner JS (eds) Interaction in human development. Lawrence Erlbaum, Hillsdale, pp 59–80
Wynn T, Coolidge FL (2004) The expert Neandertal mind. J Hum Evol 46:467–487

11 Learning in the Acheulean: Experimental Insights Using Handaxe Form as a 'Model Organism'

Stephen J. Lycett, Kerstin Schillinger, Marius Kempe, and Alex Mesoudi

Abstract

'Learning' is a process by which an individual gains new information. In the case of 'social learning', this process occurs because at least one individual has undertaken an activity that results in another individual learning something new. For an individual, therefore, 'learning' is an active process that takes place *in vivo*. For those faced with the challenge of studying learning in the Palaeolithic, however, all we are left with are inert objects (e.g., stone artefacts). Yet, understanding social learning during the Palaeolithic is a fundamental step toward understanding matters of our cultural evolution. Evolutionary biology, faced with similar problems, has made substantial progress in understanding matters of transmission, the effects of transmission on phenotypic variation, rates of mutation, etc. via the use of laboratory experiments, especially through the use of so-called 'model organisms'. Here, we describe two experiments that use handaxe form in the manner of a 'model organism' in order to understand the effects of copying error. We go on to discuss why understanding these micro-evolutionary effects can ultimately lead to a greater understanding of learning dynamics in handaxe-making hominin populations. These experiments illustrate that the characteristic size and shape parameters of handaxe traditions will have been inherently unstable. In the case of shape, in particular, this suggests that a learning mechanism other than pure observation of others' artefacts was used. Individual (trial-and-error) learning could conceivably constrain variation somewhat, but costs associated with knapping would encourage the adoption of social learning mechanisms that would countermand the inevitable effects of copying error with reduced risk to tool manufacturers.

Keywords

Acheulean handaxes • Copying error • Cultural evolution • Cultural mutation • Social learning

11.1 Introduction

One of the enduring challenges facing Palaeolithic archaeology is to understand its basic database—i.e., knapped artefacts—in terms of behavioural, social, and cognitive implications, especially as these relate to wider questions of human evolution over the longer term (e.g., Isaac 1972; Gowlett 2010). It is, therefore, the hope of everyone working with the Palaeolithic record that key insights into these factors can be provided by examining, at various scales

S.J. Lycett (✉)
Department of Anthropology, University at Buffalo SUNY, 380 MFAC-Ellicott Complex, Buffalo, NY 14261, USA
e-mail: sjlycett@buffalo.edu

K. Schillinger
Department of Anthropology, University of Kent, Canterbury CT2 7NR, UK

M. Kempe • A. Mesoudi
Department of Anthropology and Centre for the Coevolution of Biology and Culture, Durham University, Durham DH1 3LE, UK

A. Mesoudi and K. Aoki (eds.) *Learning Strategies and Cultural Evolution during the Palaeolithic*, Replacement of Neanderthals by Modern Humans Series, DOI 10.1007/978-4-431-55363-2_11,

of analysis, patterns observable in knapped stone artefacts, especially as recorded in terms of their variable attributes, form, and/or character (Clarke 1968; O'Brien and Lyman 2000; Lycett and Chauhan 2010). Major changes or trends in the Palaeolithic record inevitably draw particular attention, perhaps none more so than the appearance and temporal durability of so-called "handaxes" (Gowlett 2011).

The appearance of bifacial handaxes is often taken as evidence of a major behavioural shift in the production of stone tools by hominins, particularly in comparison with the preceding Oldowan (e.g., Roche 2005; Gowlett 2006; Lepre et al. 2011; Beyene et al. 2013). This is fuelled by the fact that their manufacture, either through the production and shaping of large flake blanks or via the reduction of stone nodules, involved not merely the production of discrete flake tools as was the case with the Oldowan, but serial knapping events strung together to produce the resultant form (Edwards 2001; Roche 2005). Formally, 'handaxes' are defined by the imposition of a long axis by means of invasive bifacial knapping to produce a relatively large (generally >10 cm in length) elongated, bilaterally-organized form, with a sharp edge that extends around a major portion of their extremities (Roe 1976; Isaac 1977; Schick and Toth 1993; Gowlett 2006). Evidence from experiments, residue analysis, usewear, design theory, cut-marks, and archaeological context has led many to contend that the form of such artefacts was driven, at least in part, by functional requirements relating to demands imposed by their use as cutting and/or chopping tools (e.g., Jones 1980; Keeley 1980; Roberts and Parfitt 1999; Domínguez-Rodrigo et al. 2001; Simão 2002; Gowlett 2006; Bello et al. 2009; Yravedra et al. 2010). Such artefacts first appear in the archaeological record of Africa ~1.75–1.5 MYA (Lepre et al. 2011; Beyene et al. 2013), but they subsequently appear in Western Europe and large parts of Asia, and remain a persistent feature of the archaeological record for over one million years (Clark 1994; Lycett and Gowlett 2008; Gowlett 2011). In specific terms, production of these artefacts represents a shift from the manufacture of relatively simple cutting tools (flakes) produced by bouts of knapping not necessarily directed toward the production of deliberate core forms (Toth 1985), to a situation where knapping events were strategically oriented toward *shaping* the residual block of stone (Roche 2005; Gowlett 2006).

Models of cultural evolution highlight the importance of understanding the mechanisms that underlie particular historical trends seen in the archaeological record (e.g., Cavalli-Sforza and Feldman 1981; Boyd and Richerson 1985; O'Brien and Lyman 2000; Henrich and McElreath 2003; Mesoudi and O'Brien 2008; Mesoudi 2011; Shennan 2011; Lycett and von Cramon-Taubadel 2015). Following Darwin (1859), these models emphasize the co-existence of three essential factors in bringing about historical change via a process of 'descent with modification': (1) a mechanism of inheritance, or 'learning' (2) the existence of variation in inherited properties, and (3) the differential representation of inherited variants through time (Mesoudi et al. 2004). What do we know about learning in prehistoric hominins from the production of artefacts such as handaxes? 'Social learning', in any form, is effectively a means by which certain information about a particular behavioural 'pattern' (however defined) is transmitted from one individual to another such that the behavioural pattern is, in turn, repeated (Heyes 1994; Byrne and Russon 1998; Whiten et al. 2004). It is sometimes (e.g., Mithen 1999) assumed that the repeated production of handaxe artefacts over time and space alone implies not only social learning, but specific forms of social learning such as imitation (i.e. copying of behavioural factors involved in their production, in addition to information gleaned from seeing the artefact alone). Such assumptions are certainly consistent with the macro-scale phenomenon of the 'Acheulean' technocomplex, whereby handaxes were repeatedly produced over large swathes of time and space (see above). Inevitably, however, any statement about learning from temporal and spatial patterns alone is limited in potency and would benefit from strengthening by other lines of enquiry. It may be worth considering in greater detail *why* the repeated production of broadly similar artefacts might be telling us something about social learning, especially in terms of the specific parameters that characterize the artefacts referred to as 'handaxes', using an evolutionary framework that moves from the micro to macro perspective.

Binford (e.g., 1983) noted some time ago that archaeology faces a fundamental dilemma in terms of trying to understand dynamic processes (human activities) from 'static' data (i.e. artefacts). Anthropologists, of course, can say things about people by studying *people*; archaeologists on the other hand, are forced to say things about people by studying *objects* people in the past left behind. This is the both the wonder and the curse of archaeology (Clarke 1968). In terms of understanding cultural evolutionary factors in the Palaeolithic, the basic problem is that we want to know something about learning and transmission from artefacts such as handaxes. In terms of their dislocation from the population dynamics and social processes that produced them, however, handaxes dug out of the ground are as dead as the fossils that palaeontologists use to reconstruct dynamic biological processes of transmission, the generation of variation, and the processes responsible for the sorting of that variation over time (i.e. evolution).

The analogy between 'dead' fossils and 'dead' artefacts in terms of understanding dynamic evolutionary processes is potentially an interesting one in terms of looking for prospective solutions. Over several decades, evolutionary biology has made substantial progress in understanding matters of transmission, the effects of transmission on phenotypic variation, rates of mutation, etc. via the use of laboratory experiments (Bataillon et al. 2013). What is needed are experiments that attempt to understand how the dynamics of micro-evolutionary processes affect artefactual variation, in order to better understand what the macro patterns seen in the archaeological record imply on a more secure basis. Experiments have a long history in the study of archaeological artefacts, including stone tools (Johnson 1978). Surprisingly, however, there are only a small number of experimental studies that have expressly studied microevolutionary effects in artefacts in order to derive their macroevolutionary implications for the study of culture as seen in the archaeological record (Eerkens 2000; Mesoudi and O'Brien 2008; Kempe et al. 2012; Schillinger et al. 2014).

Interestingly, biologists frequently use so-called 'model organisms' in experimental studies of evolutionary processes. 'Model' organisms are used because they enable a more secure understanding of phenomena of wide interest, from seemingly discrete, even trivial, laboratory experiments. Commonly used model organisms, such as fruit flies (*Drosophila spp.*), tend to have a variety of characteristics that make them particularly suitable for such experiments, including economy, speed of replication, and controllability (e.g., Ashburner and Novitski 1976; Greenspan 2004; Ashburner et al. 2005). The most suitable model organisms thus display some of the complexities of the phenomenon of interest, yet are generally not so complex that they are unwieldy in experimental settings. Elsewhere (Schillinger et al. 2014), we have argued that in regard to the study of cultural evolutionary phenomena, simple experiments that replicate certain aspects of handaxe form (e.g., their size and/or shape) make a particularly useful subject of study.

Here, we describe two experiments that we have undertaken using handaxe form in the manner of a 'model organism' in order to understand the effects of copying errors, or 'mutation' in cultural evolutionary models. We go on to discuss these experiments in terms of why understanding micro-evolutionary effects, such as copy error, can ultimately lead to a greater understanding regarding issues of 'learning' in handaxe-making hominin populations of the Palaeolithic.

11.2 Experiment 1: Considering the Effects of Size Mutation in the Acheulean

In the context of artefactual variation, the study of variation generation at a microevolutionary level is the equivalent of studying genetic mutation in biology (Cavalli-Sforza and Feldman 1981; Eerkens and Lipo 2005). In principle, a number of potential mechanisms (e.g., deliberate embellishment) might lead to the generation of new cultural variants, and it is important to note that the deliberate, intentional introduction of variation does not invalidate an evolutionary theory of cultural change (Mesoudi 2008). However, it is also recognized that unintentional copying errors (i.e. imperfect replication) during the manufacture of artefacts can lead to the introduction of novel variation in material traditions (Clarke 1968: 161; Eerkens and Lipo 2005; Hamilton and Buchanan 2009).

In our first experiment (Kempe et al. 2012) we were interested in testing the 'accumulated copying error' (or 'ACE') model proposed by Eerkens and Lipo (2005), in which random error in a quantitative artefact dimension (e.g., handaxe 'length') is generated by the physiological limitations of the hominin perceptual system. Eerkens and Lipo drew on experimental findings suggesting that the accuracy of human perception has physiological limits, which influences our ability to perceive differences between objects (e.g., Coren et al. 1994). If the difference in size between two objects is below some threshold, then this size difference will tend to be imperceptible. Such error thresholds are always relative to the size of the object, rather than absolute. The perceptual threshold below which humans fail to discriminate variation in the size of different objects is termed the 'Weber fraction' and is now established at a level of ~3 % difference for a dimensional variable such as 'length' (Eerkens 2000; Kempe et al. 2012). In other words, if a person is presented with two objects and the difference in their length is less than 3 %, they will generally fail to perceive this difference. Such insights provide a basis for comparing patterns of size variation in artefactual assemblages (Eerkens 2000; Eerkens and Bettinger 2001; Kempe et al. 2012).

Eerkens and Lipo (2005) applied this basic principle to the repeated cultural transmission of artefacts. They assumed that when attempting to copy the morphology of an artefact as faithfully as possible, and in the absence of formal measurement aids (e.g., scaled rules), the manufacturer will inevitably make copying errors that are imperceptible due to the aforementioned perception thresholds. If that person's copied artefact is in turn copied by another person, and so on along a transmission chain, then copying errors will compound over time, possibly creating significant morphological change compared to the original artefact. Moreover, if multiple such transmission chains evolve independently, then the variation between these diverging chains is likely to become substantial and to increase over time. Note that this process will take place regardless of whether any other cultural evolutionary forces are at work. Eerkens and Lipo presented a simple simulation model of this process in which a continuous trait value is transmitted over successive generations of individuals with a 3 % random normal error rate, and with 10 independently evolving chains. Their simulation showed that, as expected, the independent chains diverged over time as some became larger and others became smaller. Due to the randomness of the error, the overall mean value did not change over time, while the between-chain variation did increase over time.

The aim of our experiment was to provide an explicit empirical test of Eerkens and Lipo's (2005) ACE model of artefact transmission. Although their assumption that 3 % is the perceptual threshold below which humans fail to discriminate variation in the size of different objects is based on previous experimental findings from psychophysics, it is unclear (1) whether this 3 % error threshold is uniform across a large population of individuals, given that psychophysicists typically obtain their estimates from just one or a handful of experimental participants; (2) whether this 3 % threshold, originally obtained for simple lines or abstract geometric shapes, also applies to more realistic artefact shapes, such as handaxes; and (3), whether it is valid to simply extrapolate a single individual's perceptual error along successive transmission episodes or whether there are, for example, unexpected dynamics introduced by the compounding of individual errors.

In order to address these three points, we (Kempe et al. 2012) first derived the ACE model within an explicit mathematical framework and simulated its effects. The model operated on the basis that an individual copies a continuously-valued artefactual trait (e.g., handaxe 'length') with a set amount of copying error, drawn randomly from a normal distribution. Given that copy-error is scaled proportionally to that of the attribute being copied, the resultant value is obtained by multiplying the original (starting) value by the randomly sampled copying error. Full details are provided in Kempe et al. (2012). Simulating this model in 10 independently evolving chains over 400 generations of copying demonstrated that mean artefact size (across all 10 chains) will remain stable (Fig. 11.1). This is because, although most chains become smaller across time, the few transmission chains that get larger become increasingly large because error is scaled relatively to the attribute being copied. Equally, chains that tend toward the production of smaller artefacts deviate less overall from the original value in absolute terms. Hence, perhaps counter intuitively, the average artefact value across independently evolving chains will remain relatively stable across time, even though *variance* across the different transmission chains increases. These results were, therefore, broadly in agreement with Eerkens and Lipo's (2005) original formulation of the effects of the ACE model.

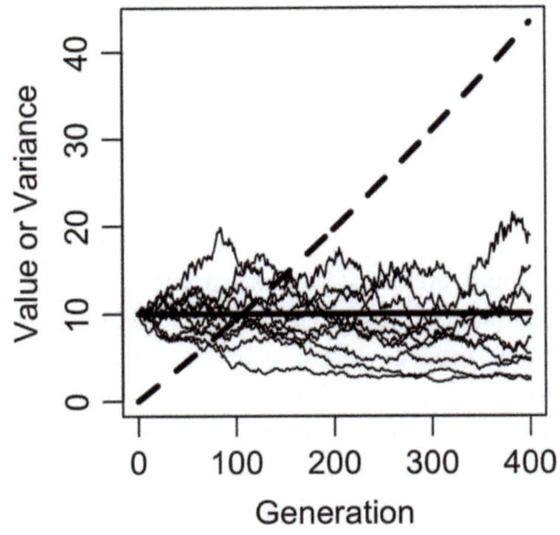

Fig. 11.1 Results of simulation based on copy error model. Output shows 10 chains evolving over 400 generations (*black lines*) and theoretically predicted mean (*thick black line*) and variance (*thick dashed line*)

Fig. 11.2 Example of the resizing experiment using electronic touchscreen. Each participant was asked to resize the handaxe image on the *right* so as to match the size of the previous participant's as closely as possible, which is shown on the *left*. Participants pressed the tick mark to complete the copying task

The experimental portion of the study aimed to determine more accurately the variance of the distribution seen in such copying errors and also to test whether mean and sample variance values obtained from experimental data matched their model-predicted values. The experiment consisted of individual volunteers, randomly allocated to independent transmission chains, resizing an image of a handaxe using an iPad (Fig. 11.2). Each participant was shown the handaxe of the previous person in their chain and then asked to resize a second handaxe to match the size of the previous person's handaxe as closely as possible. This experimental set-up, as with the use of 'model organisms' referred to earlier, has the advantage of procedural simplicity, in that it required only application of a pinching gesture on the electronic screen and no time limit was imposed. Hence, perceptual error was emphasized over manufacturing error, as is necessary given the aims of our study. A video demonstration of the experiment is provided in the online supplementary information (open access) of Kempe et al. (2012).

A total of 200 participants were asked to faithfully copy the size of the previous participant's handaxe image using this experimental set up, divided into 20 independent transmission chains of 10 participants each. Two alternative conditions were implemented. In the first condition, the size of the image to be rescaled began at the maximum possible, which in this instance was 14.4 cm, given the height of the electronic screen. The second condition involved the starting size of the image to be rescaled being set at 1/3 of maximum (i.e. 4.8 cm). The transmission chains divided equally between these 'start-larger' and 'start-smaller' conditions. Copying errors were found to be normally distributed and on the order of 2.69 % for the larger condition and 3.99 % for the smaller condition, with an overall mean of 3.43 %, which corresponds well to the 3 % value reported in the psychophysiological literature. The experimental findings also supported the model's prediction that between-chain variance should increase over time and did so in a manner quantitatively in line with the model. However, when the initial size of the image that the participants resized was larger than the size of the image they were copying, subjects tended to increase the size of the image, resulting in the mean size increasing rather than staying constant, as the raw model would predict. This latter observation represents a novel and unanticipated empirical finding that, to our knowledge, has no precedent in psychology, illustrating the value of experimental simulations of cultural transmission over multiple generations rather than inferring long-term dynamics from the characteristics of single individuals.

The final part of the study involved comparing our results to a dataset of genuine Acheulean handaxes. This dataset provided quantitative data for 2061 complete handaxes from 21 different sites located in a total of five countries (Israel, Morocco, South Africa, Tanzania, United Kingdom) covering an age range of some 1.2 million years (Marshall et al. 2003). The coefficient of variation (CV) for length in this sample was 0.30 and for breadth was 0.23. We then simulated the output of the ACE model, setting copy-error rates to the 3.43 % determined by our experiments. These simulations determined that CV values of greater than 0.30 will be produced by copying error in less than 200 copying generations. Conservatively setting each generation at 20 years would imply that in $\leq$4,000 years of copying, variation would exceed that seen in the genuine Acheulean sample. Indeed, copying error would need to be only on the order of 0.17 % (i.e. 20 times smaller than the 3.43 % seen in our experiments) in order to produce the level of total variation seen in the 1.2 million years covered by the handaxes concerned. Clearly, what these results imply is that certain factors must be countermanding the inevitable effects of size copying error and lineage mutation implied by the model.

Further experiments and models might examine such factors, as well as test the effect of relaxing our simplifying assumptions. For example, we assume in the model and experiment that transmission is one-to-one, with individuals acquiring the trait from a single individual. Ethnographic evidence (e.g. Stout 2005) suggests that the transmission of artefacts may often be many-to-one, with individuals acquiring the trait from multiple cultural parents, which our design could easily be modified to explore (see Kempe and Mesoudi 2014 for an experimental methodology that incorporates multiple cultural parents).

11.3 Experiment 2: Considering the Effects of Shape Mutation in the Acheulean

It bears emphasizing that our first experiment, as described above, considers only mutation in the *size* (i.e. scale) of handaxe attributes, and the patterns of variation that will be produced through inevitable copying errors induced by physiological limits to perceive such differences. Although the analyses we describe suggest that Acheulean handaxes exhibited a definite range of size variability (see also Crompton and Gowlett 1993; Gowlett 2009), a fundamental distinguishing feature of handaxes are their characteristic *shape* properties (Roe 1976; Wynn 1995; Roche 2005). Shape properties of artefacts, independently of their size, may have specific functional or aesthetic significance (Lycett 2008; Winter-Livneh et al. 2013). Indeed, although size and shape are often conflated ('form' = size + shape), both conceptually and empirically, size and shape are fundamentally distinct (Bookstein 1989; Jungers et al. 1995). While the size of an object is a univariate property and can therefore be described quantitatively by a single measure of scale such as volume, shape is inherently a multivariate property. A quantitative concept of shape, therefore, relies not on the appreciation of a single variable such as 'length', but on the relative relationships between multiple aspects of morphometric variation in a given object. As Gowlett (2006) has noted, the deliberate manufacture of handaxe shape requires—minimally—the interrelated manipulation of the *relative* length variable (s), width (s) and aspects of thickness variability on the part of their manufacturer, and control of those interrelated properties during the knapping process.

The potential importance of shape copying errors in the case of handaxe production is particularly emphasized given that it has been proposed that the production of artefacts via knapping may be a particularly 'error prone' process. For instance, some time ago, Deetz (1967) noted that in the case of a 'reductive' process of manufacture, such as the knapping of stone artefacts, errors are not easily reversed. As Baumler (1995: 11) put it more recently, in the case of stone tool manufacture "each [flake] removal is irrevocable and its consequences are permanent". Conversely, Deetz (1967) contended that in the case of more readily reversible (or 'additive') processes of manufacture, such as basketry or pottery, errors are readily corrected due to the ease by

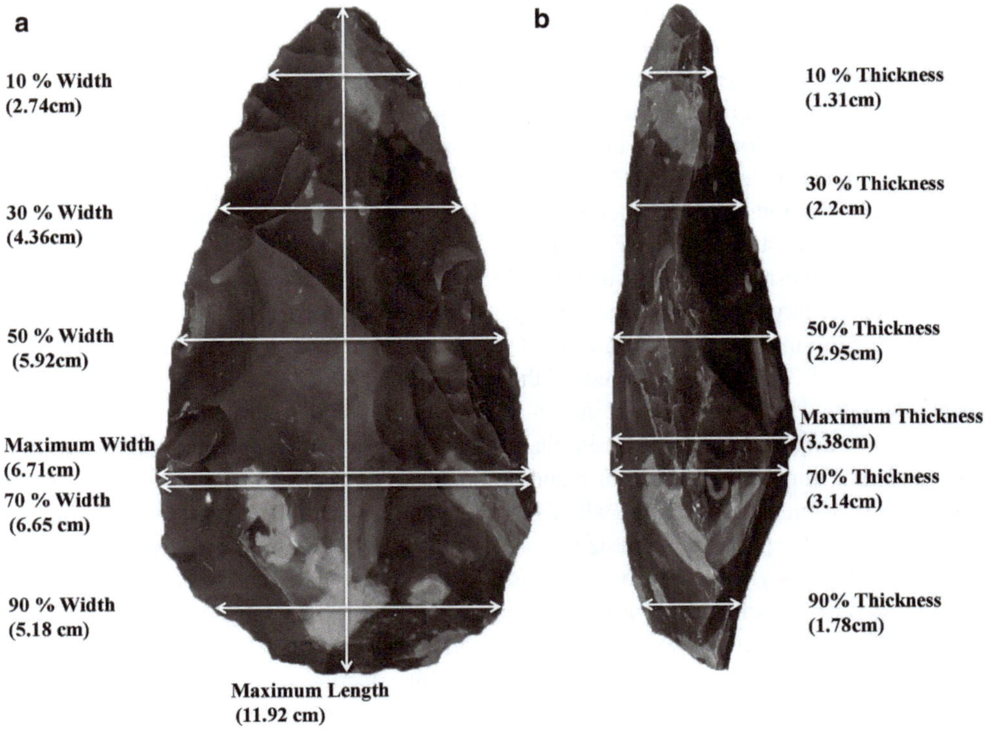

Fig. 11.3 Flint replica handaxe provided to participants as the 'target' model during the copying task. Participants in each condition were asked to copy the shape of this handaxe from standardized plasticine blocks using a steel table knife (Replica knapped by SJL.)

which material can be either added or removed. According to Deetz, differences between such alternative manufacturing processes would lead inevitably to greater levels of variation in non-reversible manufacturing traditions, such as those used in handaxe production.

Testing predictions of this form through studies of the archaeological record would be fraught with difficulty given the differing situational conditions under which alternative sets of artefacts (e.g., pots versus baskets) might be made, even within the same community. Moreover, comparing variation in artefacts across differing raw materials may be problematic given that the medium of manufacture itself (e.g., stone versus clay) might influence variation patterns in particular ways. A further specific problem in the case of artefacts such as handaxes is that their production requires skilled behaviour built over months, if not years, of practice (Edwards 2001) and is an activity that can even prove dangerous (Whittaker 1994). Such issues provide serious challenges to the implementation of an experimental approach that requires ready-recruitment of participants in numbers amenable to the implementation of statistical analysis. What is needed is a 'model organism' approach that enables implementation of fundamental controls such that the key contrasts between the two alternative manufacturing conditions are emphasized, while replicating the essential (i.e. 'additive' and 'reductive') features of the processes of interest under safe conditions.

Given these considerations, we (Schillinger et al. 2014) implemented an experimental procedure that consisted of a simple copying task. Participants were asked to copy the shape of a replica Acheulean handaxe (Fig. 11.3) as accurately as possible using a standardized block of plasticine and a stainless steel table knife. Following Deetz (1967), the central prediction that we tested is that the implementation of reductive manufacturing processes, where material can be removed but not added, automatically leads to an overall higher rate in copying error for shape than under reversible manufacturing conditions. In the context of this hypothesis, we specifically targeted the statistical effects of copying error on shape attributes, using a dataset of size-adjusted morphometric variables. It should be noted, therefore, that in contrast to the type of experiment described earlier, this experiment is not so much aimed at the issue of perceptual bias (in terms of establishing a baseline error rate), as procedural bias (i.e. additive versus reductive manufacturing processes) and establishing whether one procedure has intrinsically greater error rates than the other.

A total of 60 participants were recruited to take part in this experiment. Of these, 30 were female (mean age = 26, s.d. = 5.4, age range = 18–44 years) and 30 were male (mean age = 28, s.d. = 9.8, age range = 18–64 years). Equal numbers of males and females were employed deliberately in order to control for any potential confound in terms of sex differences in visuo-spatial abilities (see e.g., Wynn et al.

1996). The participants were divided equally between two experimental conditions: an 'additive-reductive' condition, whereupon participants were instructed that they were free to both remove and add plasticine during the manufacture of their replica, and a 'reductive-only' condition in which participants were strictly required only to remove material in producing their copy. In order to control for memory effects (see e.g., Eerkens 2000), participants were permitted to examine the target handaxe for 1 min prior to beginning the copying task, and were free to compare their model to the target at any point during the 30 min allotted for completion of the task. All participants were able to complete the task within this timeframe.

Upon completion of the task, each participant's model was photographed in plan- and profile-views according to a standardized orientation protocol (see Schillinger et al. 2014 for full details). Thereafter, measurements were obtained digitally for 42 standardized variables (28 plan-view and 14 profile-view) from each plasticine handaxe using the freely available morphometrics software tpsDig v2.16 (Rohlf 2010). Given that our analysis specifically focused on monitoring shape-related changes, the data were size-adjusted via use of the geometric mean method (Jungers et al. 1995; Lycett et al. 2006). This method of size-adjustment effectively removes size (scaling) variation between specimens by equalizing their volumes, yet retains their relevant shape data (Jungers et al. 1995). The geometric mean of a series of *n* variables ($a_1, a_2, a_3 \ldots a_n$) is equivalent to $\sqrt[n]{a_1 \times a_2 \times a_3 \times \cdots \times a_n}$. Simply, the geometric mean is the *n*th root of the product of all *n* variables (Sokal and Rohlf 1995: 43). The method proceeds on a specimen-by-specimen basis, dividing each variable in turn by the geometric mean of the variables to be size-adjusted. Hence, to implement the method, the geometric mean of each handaxe replica was calculated separately and thereafter each of the 42 morphometric variables for each specimen were divided by the geometric mean for that particular specimen. The size-adjusted values of the 42 morphometric variables for each of the 60 replicas were subtracted from the equivalent 42 variables of the target flint replica. Thereafter, mean shape error was computed for each of the 42 morphometric variables across the 30 replicas obtained in each experimental condition.

In terms of results, the additive-reductive condition had an overall mean copying error rate of 0.115 (s.d. = 0.040) across all variables. The reductive-only condition had a mean of 0.134 (s.d. = 0.053) across all variables. The results of a conservative Mann-Whitney *U* test demonstrated that copying error rates for the 42 variables in the reductive-only condition were statistically greater than in the additive-reductive condition ($U = 621.5$; asymptotic $p = 0.0191$; Monte Carlo $p = 0.0199$). Figure 11.4 shows the overall distribution of the copy errors from the two conditions in the form of 25–75 percentile box plots.

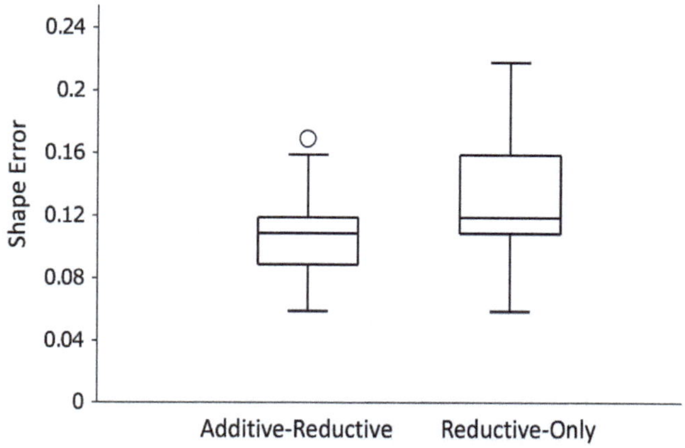

Fig. 11.4 Box plots showing shape error distribution in each of the two experimental conditions. *Horizontal lines* indicate the median error in each case, boxes depict the 25–75 percentile, while whiskers mark the largest datum point ≤1.5× box range

Overall, the results of these analyses were, therefore, entirely consistent with the proposition of Deetz (1967) that copying errors, at least in terms of shape, will be higher in artefacts produced via processes of irreversible reduction, than in artefacts produced via reversible processes of manufacture. In other words, 'mutation' rates in the shape attributes of artefacts produced under irreversible, or 'reductive', conditions—such as handaxes produced via stone knapping—are intrinsically greater than those produced via alternative means.

11.4 Discussion

Here, we have reviewed the results of two experiments designed to examine mutation rates produced by copying error in cultural evolutionary models. At this juncture, however, the reader may well be asking what any of this implies for questions concerning the learning of handaxe production in the hominin societies that made them.

The value of these experiments in approaching questions of this type, we would contend, lies in their microevolutionary perspective. Evolution is a process in which inherited variants are replicated differentially, and imperfectly, through time. Variants are replicated differentially because certain variants fail to be reproduced either due to 'selective' factors (i.e. differences between variants lead to a differential likelihood of successful replication in given circumstances) or due to some random (chance) factor, which results in 'drift' of characteristics within populations. The replication process in any genuine evolutionary system is, however, always imperfect (even if only in very subtle ways), thus ensuring that new variants may appear across time. These new variants, of course, become the engine for yet further evolutionary change. In the case of artefacts, this was recognised some time ago by Harrison (1930: 111) who noted:

> it is clear that the size and form of any one-piece artifact, or of any such component of a compound artifact, may be altered very considerably by the cumulative effect of a number of changes each small in itself ... In some instances, variational modifications arise through the copying from other artifacts of features of form, with resultant changes in shape and proportions; ... It also plays a part in mutational progress.

The results of the first experiment we describe (Kempe et al. 2012), considered the role of size mutation in these terms. We simulated the assumptions of the accumulated copying error model under realistic size copying-error rates established via empirical experiment. These analyses were able to show that drift alone would lead to size variation exceeding that seen in archaeological samples of handaxes in, conservatively, as little as 4,000 years. In other words, it seems that in the case of a learning model in which individuals simply copied the size of handaxes made by those nearby, mutation effects over time would inevitably lead relatively quickly to unrealistic levels of size variation.

Simple learning (size copying) plus a level of random variation does not, therefore, explain the relatively constrained levels of size variation seen across the samples of handaxes considered in our study (Kempe et al. 2012). What this plausibly suggests is that functionally-related cultural selection, such as the need to fit into the hand comfortably (Gowlett 2006), was constraining variation in ways that do not fit the simple drift-copying model. Such constraints may have been inducted by individual ('trial-and-error') learning, whereby an individual hominin 'scaled' their handaxes to what worked well in a functional capacity given their own physical size, strength, etc. Indeed, the importance of 'feedback' mechanisms between considerations of this nature and the attributes of artefacts such as handaxes was mentioned some time ago by Clarke (1968: 181–182, 649). However, within any socially-mediated context of observation of, and learning about, handaxe production and usage, some notion of suitable size parameters is also likely to have been inducted in novice handaxe producers through direct observation of others' tools (Fragaszy et al. 2013), essentially 'resetting' the drift clock with each generation, especially in terms of the population mean.

Our second experimental analysis (Schillinger et al. 2014), considered copying error in terms of *shape* mutation, independently of size factors. The participants engaged in one of two alternative conditions; one representing an irreversible ('reductive-only') manufacturing process, and the other representing a reversible ('additive-reductive') manufacturing process. Participants in each condition were asked to copy the shape of a target form (a flint replica handaxe) as accurately as possible utilizing a standardized block of plasticine and a stainless steel table knife. Our analysis found that replicas produced in the reductive-only condition displayed statistically greater levels of shape-copying error than those produced in the additive-reductive condition. In other words, mutation rates in the shape properties of material traditions produced under reductive conditions (such as stone knapping) are intrinsically greater than those produced via alternative means. Hence, shape mutation rates are process dependent, and in the case of traditions produced through knapping will be inherently prone to copying error, especially via an imitative (i.e. goal copying) form of learning. This point is especially notable given that the characteristic shape of handaxes was imposed by hominins through the knapping sequence itself, and is not easily determined merely by selection of starting form as might more easily be the case with 'size' parameters. Indeed, handaxes were produced on tabular material, rounded nodules and cobbles, and flake blanks, the latter of which sometimes involved the instigation of regionally distinctive knapping routines that produced such flake 'blanks' (Sharon 2009). Hence, the characteristic form of handaxes seen in the archaeological record was imposed on materials comprised of a variety of differing 'start' points. Indeed, at least one study suggests that handaxes made of stone and bone at the same locality show no statistical differences in outline shape, despite the differing nature of these materials (Costa 2010).

Elsewhere, we have noted that the appearance rate of new cultural variants may conceptually be linked to potential for evolutionary change (Schillinger et al. 2014: 137) akin to the concept of 'evolvability' in biology (Ridley 2004: 587). It must be stressed, of course, that while 'evolvability' in these terms might be used to describe the potential for change brought about by selective factors (either natural or cultural), it can also be used to describe potential for the degradation of culturally transmitted traits, leading eventually to their extinction, or cultural 'collapse' of a particular tradition. Indeed, although variation is required for selection to operate, and is therefore a prerequisite of cumulative cultural evolution, equally it has been known for some time in biology that 'mutation load' is a factor that may ultimately prove fatal to population viability (Simpson 1953). Hence, in the case of items of material culture, such as handaxes produced via reductive processes, their potential for evolutionary 'corruption' in terms of shape would be higher than artefactual traditions produced via processes in which errors are more easily corrected. A visual demonstration of this effect is shown in Fig. 11.5, which illustrates merely 15 'generations' of different experimental participants copying the previous participant's copy in the form of a transmission chain, when the initial shape starts out in a form similar to that of a handaxe. In this (albeit anecdotal) example, it is readily visible that erosion due to effects of copying error can

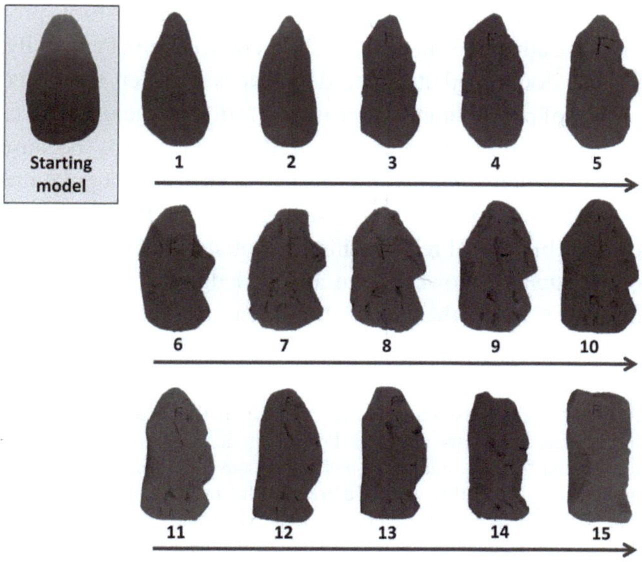

Fig. 11.5 Transmission chain produced by 15 participants copying the 3D shape of the 'artefact' of the previous participant (starting initial 'target' shown *top left*). Each of these 3D models was carved from standardized foam blocks (22.3 × 11 × 7.8 cm) using a plastic table knife. It should be noted, therefore, that the production of these shapes requires no specialized skills or knowledge, and they are produced on an easily malleable material

potentially have potent effects on the integrity of 'handaxe' shape within a limited number of generations.

Given these findings, what does this imply for the learning of 'shape' in handaxe-producing communities of the Palaeolithic? One possibility is that individual (trial-and-error) learning helped to mitigate the relatively strong effects of shape mutation in handaxe traditions, at least once the 'concept' of handaxe tools had been instigated via other means (e.g., stimulus enhancement learning and emulation). There are reasons, however, to be cautious that individual learning alone explains phenomena of the scale attested in the archaeological record. One pertinent factor to consider here is the inherently dangerous nature of flintknapping involving the percussive removal of razor sharp stone flakes, which inevitably also leads to small flakes and sharp chips of stone being thrown into the air. Ethnographically and historically recorded comments on such injury risks are attested in the literature (e.g., Pope 1918: 117; Kroeber 1961: 184; Hampton 1999: 267). Today, many flintknappers who produce stone artefacts for academic purposes or recreation deliberately wear protective gear in the form of safety glasses, gloves and thick padding. One contemporary academic flintknapper is known to have severed a tendon with a small flake (~5 × 20 mm) requiring corrective surgery and causing permanent debilitation of movement in his hand (Whittaker 1994: 3–4). Painful open wounds, blood loss, risk of infection to injuries, eye damage/loss are noted risks, in addition to damaged ligaments that might be caused by incorrect form. These factors are important when we consider that there is widespread agreement that social learning strategies will be favourable to asocial learning strategies (i.e. trial-and-error learning) whenever the activity to be learned is costly or hazardous (Boyd and Richerson 1985; Feldman et al. 1996). This is strongly supported by empirical evidence from studies of learning in non-human animals, which indicate that social learning will be favoured wherever asocial learning is more costly (e.g., Mineka and Cook 1988; Chivers and Smith 1995; Kelley et al. 2003), as well as in contemporary humans using similar computer-based learning tasks as those described above (Mesoudi and O'Brien 2008).

One social learning mechanism potentially available to handaxe-producing hominins, beyond pure copying of artefact shape (i.e. emulation), is imitation. This would imply some copying of not only the form of the object (the goal, or 'end state') but also some fidelity in terms of copying the actual behaviours used by others in the manufacture of their handaxes. Subtle differences in the details of manufacture, or 'process controls' (sensu Patten 2005) that aim to increase the likelihood of intended outcomes during the knapping process would make obvious targets for such bouts of imitation. Such possibilities would also explain why metric studies of handaxe form have persistently indicated *statistical* differences between assemblages of handaxes from different regions or sites (e.g., Wynn and Tierson 1990; Lycett and Gowlett 2008; Lycett and von Cramon-Taubadel 2015), which remain difficult to explain solely on the basis of reduction and/or raw material factors (Sharon 2008; Eren et al. 2014; Lycett and von Cramon-Taubadel 2015). However, further support for the instigation of imitative learning must be further substantiated via independent means.

11.5 Conclusions

Assertions of social learning are often invoked for handaxe production, sometimes even invoking specific mechanisms of social learning such as imitation (e.g., Mithen 1999). These assertions are made largely on the strength that these mechanisms most plausibly explain the repeated pattern of handaxe production over considerable swathes of time and space. Given the scale of the phenomenon under consideration, these assertions are not necessarily unreasonable. This is especially the case given what we have learned in recent decades about the role of social learning in the acquisition of tool use behaviours of our closest living primate relatives (e.g., Lonsdorf et al. 2004; Whiten and Mesoudi 2008; Horner and de Waal 2009; Humle et al. 2009; Biro et al. 2010), and what we know from the ethnographic record in terms of the learning of stone artefacts that are similar to handaxes (e.g., Stout 2005). However, such assertions are problematic in the case of prehistoric handaxes without considering the role, scale, incidence and potential magnitude of cumulative copying errors in precise terms. In essence, what our experiments show is that inevitable copying error in size *and* shape factors would relatively quickly lead to the disintegration of handaxe traditions attested empirically in the archaeological record. Handaxe traditions are not inherently stable; in fact, a combination of factors relating to copying error in both size and shape factors would make them inherently *unstable*. In the case of shape, in particular, this suggests that a learning mechanism other than *pure* observation of others' artefacts (i.e. goal emulation, or 'end-state' copying), was used in the learning of handaxe manufacture. Individual (trial-and-error) learning could conceivably constrain variation somewhat, but the costs associated with knapping would encourage the adoption of social learning mechanisms that would countermand the inevitable effects of copying error that we describe with reduced risk to tool manufacturers. Determining what such mechanisms are—in precise terms—will, however, require further work. Hence, do we still have a long way to go in order to fully understand the dynamics of learning in the Acheulean? Absolutely. What the preceding discussion should demonstrate, however, is the importance of an evolutionary perspective that can work from a micro- to macro-scale level, and importantly, is informed by data derived from experimental work of the type we describe.

11.6 Final Remarks in the Context of the RNMH Project Objectives

As Nishiaki (2013: 173) noted recently, a primary objective of the RNMH research project is to investigate potential factors in the replacement of hominin populations referred to as 'Neanderthals' by those commonly referred to as 'Modern Humans'. If the points we have made here regarding imitation capacities in handaxe producing hominin populations are correct (and we re-emphasize the need for further corroboration), then imitative capacities were a plesiomorphic feature, present in the last common ancestor of both groups of populations. Hence, if we are correct, differences in imitative capacities are unlikely to have played a major role in the replacement process. This, it should be noted, is despite the fact that imitation may be a key variable in the fidelity of social learning systems, which a priori, may make it a potential explanative candidate. If modal differences in the innate (i.e. genetically determined) social learning capacities of these two hominin populations played any role in the replacement of one by another, this leaves only two potential candidates: (1) that capacities for more sophisticated social learning mechanisms, such as teaching (Tehrani and Riede 2008), were different or (2) that one population possessed greater capacity for innovation than the other, which in turn, provided an advantage in a competitive environment. Personally, we are doubtful that innate differences between these two groups of populations will provide a *fully* satisfactory answer to the question of population replacement, partly because of the demonstrable resilience of Neanderthal populations over millennia prior to their disappearance, and the proximity of their biological relationship to contemporary populations of *H. sapiens*. However, we acknowledge the considerable amount of work that needs to be done to fully support our position.

Acknowledgements The research discussed in this chapter was funded by a Leverhulme Trust Research Project Grant (F/07 476/AR) awarded to AM and SJL. For numerous invaluable conversations and comments while undertaking this work we are grateful to Noreen von Cramon-Taubadel. We are also grateful to Ken Aoki and Jamie Tehrani for helpful comments on a previous version of this paper.

References

Ashburner M, Novitski E (1976) The genetics and biology of *drosophila* 1A. Academic, London
Ashburner M, Golic K, Hawley RS (2005) *Drosophila*: a laboratory handbook, 2nd edn. Cold Spring Harbor Laboratory Press, New York
Bataillon T, Joyce P, Sniegowski P (2013) As it happens: current directions in experimental evolution. Biol Lett 9:20120945
Baumler MF (1995) Principles and properties of lithic core reduction: implications for levallois technology. In: Dibble HL, Bar-Yosef O (eds) The definition and interpretation of levallois technology. Prehistory Press, Madison, pp 11–23
Bello SM, Parfitt SA, Stringer C (2009) Quantitative micromorphological analyses of cut marks produced by ancient and modern handaxes. J Archaeol Sci 36:1869–1880
Beyene Y, Katoh S, WoldeGabriel G, Hart WK, Uto K, Sudo M et al (2013) The characteristics and chronology of the earliest Acheulean at konso, Ethiopia. Proc Natl Acad Sci U S A 110:1584–1591
Binford LR (1983) In pursuit of the past. Thames & Hudson, London
Biro B, Carvalho S, Matsuzawa T (2010) Tools, traditions, and technologies: interdisciplinary approaches to chimpanzee nutcracking. In: Lonsdorf EV, Ross SR, Matsuzawa T (eds) The mind of the chimpanzee: ecological and experimental perspectives. Chicago University Press, Chicago, pp 141–155
Bookstein FL (1989) 'Size and shape': a comment on semantics. Syst Zool 38:173–180
Boyd R, Richerson P (1985) Culture and the evolutionary process. University of Chicago Press, Chicago
Byrne RW, Russon AE (1998) Learning by imitation: a hierarchical approach. Behav Brain Sci 21:667–721
Cavalli-Sforza LL, Feldman MW (1981) Cultural transmission and evolution: a quantitative approach. Princeton University Press, Princeton
Chivers DP, Smith RJF (1995) Chemical recognition of risky habitats is culturally transmitted among fathead minnows, *Pimephales promelas* (osteichthyes, cyprinidae). Ethol 99:286–296
Clark JD (1994) The Acheulian industrial complex in Africa and elsewhere. In: Corruccini RS, Ciochon RL (eds) Integrative paths to the past. Prentice Hall, Englewood Cliffs, pp 451–469
Clarke DL (1968) Analytical archaeology. Methuen, London
Coren S, Ward LM, Enns JT (1994) Sensation and perception. Harcourt Brace, San Diego

Costa AG (2010) A geometric morphometric assessment of plan shape in bone and stone Acheulean bifaces from the middle Pleistocene site of Castel di Guido, Latium, Italy. In: Lycett SJ, Chauhan PR (eds) New perspectives on old stones: analytical approaches to paleolithic technologies. Springer, New York, pp 23–41

Crompton RH, Gowlett JAJ (1993) Allometry and multidimensional form in Acheulean bifaces from kilombe, Kenya. J Hum Evol 25:175–199

Darwin C (1859) On the origin of species. John Murray, London

Deetz J (1967) Invitation to archaeology. The Natural History Press, Garden City

Domínguez-Rodrigo M, Serralllonga J, Juan-Tresserras J, Alcala L, Luque L (2001) Woodworking activities by early humans: a plant residue analysis on Acheulian stone tools from Peninj (Tanzania). J Hum Evol 40:289–299

Edwards SW (2001) A modern knapper's assessment of the technical skills of the late Acheulean biface workers at Kalambo falls. In: Clark JD (ed) Kalambo falls prehistoric site: volume III. Cambridge University Press, Cambridge, pp 605–611

Eerkens JW (2000) Practice makes within 5 % of perfect: visual perception, motor skills, and memory in artifact variation. Curr Anthropol 41:663–668

Eerkens JW, Bettinger RL (2001) Techniques for assessing standardization in artifact assemblages: can we scale variability? Am Antiq 66:493–504

Eerkens JW, Lipo CP (2005) Cultural transmission, copying errors, and the generation of variation in material culture and the archaeological record. J Anthropol Archaeol 24:316–334

Eren MI, Roos CI, Story B, von Cramon-Taubadel N, Lycett SJ (2014) The role of raw material differences in stone tool shape variation: an experimental assessment. J Archaeol Sci 49:472–487

Feldman MW, Aoki K, Kumm J (1996) Individual versus social learning: evolutionary analysis in a fluctuating environment. Anthropol Sci 104:209–232

Fragaszy DM, Biro D, Eshchar Y, Humle T, Izar P, Resende B, Visalberghi E (2013) The fourth dimension of tool use: temporally enduring artefacts aid primates learning to use tools. Philos Trans R Soc B 368:20120410

Gowlett JAJ (2006) The elements of design form in Acheulian bifaces: modes, modalities, rules and language. In: Goren-Inbar N, Sharon G (eds) Axe age: Acheulian tool-making from quarry to discard. Equinox, London, pp 203–221

Gowlett JAJ (2009) Artefacts of apes, humans, and others: towards comparative assessment and analysis. J Hum Evol 57:401–410

Gowlett JAJ (2010) The future of lithic analysis in palaeolithic archaeology: a view from the Old World. In: Lycett SJ, Chauhan PR (eds) New perspectives on old stones: analytical approaches to paleolithic technologies. Springer, New York, pp 295–309

Gowlett JAJ (2011) The empire of the Acheulean strikes back. In: Sept J, Pilbeam D (eds) Casting the net wide: papers in honor of Glynn Isaac and his approach to human origins research. American School of Prehistoric Research, Cambridge, MA, pp 93–114

Greenspan RJ (2004) Fly pushing: the theory and practice of drosophila genetics, 2nd edn. Cold Springer Harbor Laboratory Press, New York

Hamilton MJ, Buchanan B (2009) The accumulation of stochastic copying errors causes drift in culturally transmitted technologies: quantifying Clovis evolutionary dynamics. J Anthropol Archaeol 2:55–69

Hampton OW (1999) Culture of stone: sacred and profane uses of stone among the Dani. Texas A&M, College Station

Harrison HS (1930) Opportunism and the factors of invention. Am Anthropol 32:106–125

Henrich J, McElreath R (2003) The evolution of cultural evolution. Evol Anthropol 12:123–135

Heyes CM (1994) Social learning in animals: categories and mechanisms. Biol Rev 69:207–231

Horner V, de Waal FBM (2009) Controlled studies of chimpanzee cultural transmission. Prog Brain Res 178:3–15

Humle T, Snowdon CT, Matsuzawa T (2009) Social influences on ant-dipping acquisition in the wild chimpanzees (*Pan troglodytes verus*) of bossou, guinea, West Africa. Anim Cogn 12(S1):37–48

Isaac GL (1972) Early phases of human behaviour: models in lower palaeolithic archaeology. In: Clarke DL (ed) Models in archaeology. Methuen, London, pp 167–199

Isaac GL (1977) Olorgesailie: archaeological studies of a middle Pleistocene lake basin in Kenya. University of Chicago Press, Chicago

Johnson LL (1978) A history of flint-knapping experimentation, 1838–1976. Curr Anthropol 19:337–372

Jones PR (1980) Experimental butchery with modern stone tools and its relevance for palaeolithic archaeology. World Archaeol 12:153–165

Jungers WL, Falsetti AB, Wall CE (1995) Shape, relative size, and size adjustments in morphometrics. Yearb Phys Anthropol 38: 137–161

Keeley LH (1980) Experimental determination of stone tool uses: a microwear analysis. University of Chicago Press, Chicago

Kelley JL, Evans JP, Ramnarine IW, Magurran AE (2003) Back to school: can antipredator behaviour in guppies be enhanced through social learning? Anim Behav 65:655–662

Kempe M, Mesoudi A (2014) An experimental demonstration of the effect of group size on cultural accumulation. Evol Hum Behav 35:285–290

Kempe M, Lycett SJ, Mesoudi A (2012) An experimental test of the accumulated copying error model of cultural mutation for Acheulean handaxe size. PLoS One 7(11):e48333

Kroeber T (1961) Ishi in two worlds. University of California Press, Berkeley

Lepre CJ, Roche H, Kent DV, Harmand S, Quinn RL, Brugal J-P et al (2011) An earlier origin for the Acheulian. Nature 477:82–85

Lonsdorf EV, Eberly LE, Pusey AE (2004) Sex differences in learning in chimpanzees. Nature 428:715–716

Lycett SJ (2008) Acheulean variation and selection: does handaxe symmetry fit neutral expectations? J Archaeol Sci 35:2640–2648

Lycett SJ, Chauhan PR (2010) Analytical approaches to palaeolithic technologies: an introduction. In: Lycett SJ, Chauhan PR (eds) New perspectives on old stones: analytical approaches to palaeolithic technologies. Springer, New York, pp 1–22

Lycett SJ, Gowlett JAJ (2008) On questions surrounding the Acheulean 'tradition'. World Archaeol 40:295–315

Lycett SJ, von Cramon-Taubadel N (2015) Toward a "quantitative genetic" approach to lithic variation. J Archaeol Method Theory. doi: 10.1007/s10816-013-9200-9

Lycett SJ, von Cramon-Taubadel N, Foley RA (2006) A crossbeam co-ordinate caliper for the morphometric analysis of lithic nuclei: a description, test and empirical examples of application. J Archaeol Sci 33:847–861

Marshall G, Gamble C, Roe D (2003) The Acheulian biface project: a digital archive for teaching and research. Antiquity 77 (online)

Mesoudi A (2008) Foresight in cultural evolution. Biol Philos 23: 243–255

Mesoudi A (2011) Cultural evolution: how Darwinian theory can explain culture and synthesize the social sciences. Chicago University Press, Chicago

Mesoudi A, O'Brien MJ (2008) The cultural transmission of great basin projectile point technology I: an experimental simulation. Am Antiq 73:3–28

Mesoudi A, Whiten A, Laland KN (2004) Is human cultural evolution Darwinian? evidence reviewed from the perspective of *The origin of species*. Evolution 58:1–11

Mineka S, Cook M (1988) Social learning and the acquisition of snake fear in monkeys. In: Galef BG, Zentall TR (eds) Social learning: psychological and biological perspectives. Erlbaum, Hillsdale, pp 51–73

Mithen S (1999) Imitation and cultural change: a view from the stone age, with specific reference to the manufacture of handaxes. In: Box HO, Gibson KR (eds) Mammalian social learning: comparative and ecological perspectives. Cambridge University Press, Cambridge, pp 389–399

Nishiaki Y (2013) 'Gifting' as a means of cultural transmission. In: Akazawa T, Nishiaki Y, Aoki K (eds) Dynamics of learning in Neanderthals and Modern Humans, vol 1. Springer, Tokyo, pp 173–185

O'Brien MJ, Lyman RL (2000) Applying evolutionary archaeology: a systematic approach. Kluwer Academic/Plenum, New York

Patten B (2005) Peoples of the flute: a study in anthropolithic forensics. Stone Dagger Publications, Denver

Pope ST (1918) Yahi archery. University of California Press, Berkeley

Ridley M (2004) Evolution, 3rd edn. Blackwell, Oxford

Roberts MB, Parfitt S (eds) (1999) Boxgrove: a middle Pleistocene hominid site at eartham quarry, boxgrove, West Sussex. English Heritage, London

Roche H (2005) From simple flaking to shaping: stone-knapping evolution among early hominins. In: Roux V, Bril B (eds) Stone knapping: the necessary conditions for a uniquely hominin behaviour. McDonald Institute, Cambridge, pp 35–48

Roe D (1976) Typology and the trouble with handaxes. In: Sieveking G de G, Longworth IH, Wilson KE (eds) Problems in economic and social archaeology. Duckworth, London, pp 61–70

Rohlf FR (2010) TpsDig2 v2.16. Department of Ecology and Evolution, State University of New York, Stony Brook. http://life.bio.sunysb.edu/morph/

Schick KD, Toth N (1993) Making silent stones speak: human evolution and the dawn of human technology. Weidenfeld and Nicolson, London

Schillinger K, Mesoudi A, Lycett SJ (2014) Copying error and the cultural evolution of "additive" versus "reductive" material traditions: an experimental assessment. Am Antiq 79:128–143

Sharon G (2008) The impact of raw material on Acheulian large flake production. J Archaeol Sci 35:1329–1344

Sharon G (2009) Acheulian giant-core technology. Curr Anthropol 50:335–367

Shennan S (2011) Descent with modification and the archaeological record. Philos Trans R Soc B 366:1070–1079

Simão J (2002) Tools evolve: the artificial selection and evolution of paleolithic stone tools. Behav Brain Sci 25:419

Simpson GG (1953) The major features of evolution. Columbia University Press, New York

Sokal RR, Rohlf FJ (1995) Biometry, 3rd edn. W.H. Freeman, New York

Stout D (2005) The social and cultural context of stone knapping skill acquisition. In: Roux V, Bril B (eds) Stone knapping: the necessary conditions for a uniquely hominin behaviour. McDonald Institute, Cambridge, pp 331–340

Tehrani JJ, Riede F (2008) Towards an archaeology of pedagogy: learning, teaching and the generation of material culture traditions. World Archaeol 40(3):316–331

Toth N (1985) The Oldowan reassessed: a close look at early stone artefacts. J Archaeol Sci 12:101–120

Whiten A, Mesoudi A (2008) Establishing an experimental science of culture: animal social diffusion experiments. Philos Trans R Soc B 363:3477–3488

Whiten A, Horner V, Litchfield CA, Marshall-Pescini S (2004) How do apes ape? Learn Behav 32:36–52

Whittaker JC (1994) Flintknapping: making and understanding stone tools. University of Texas Press, Austin

Winter-Livneh R, Svoray T, Gilead I (2013) Shape reproducibility and architectural symmetry during the chalcolithic period in the southern Levant. J Archaeol Sci 40:1340–1353

Wynn T (1995) Handaxe enigmas. World Archaeol 27:10–24

Wynn T, Tierson F (1990) Regional comparison of the shapes of later Acheulean handaxes. Am Anthropol 92:73–84

Wynn TG, Tierson FD, Palmer CT (1996) Evolution of sex differences in spatial cognition. Yearb Phys Anthropol 39:11–42

Yravedra J, Domínguez-Rodrigo M, Santonja M, Pérez-González A, Panera J, Rubio-Jara S et al (2010) Cut marks on the middle pleistocene elephant carcass of Áridos 2 (Madrid, Spain). J Archaeol Sci 37:2469–2476

Index

A. Mesoudi and K. Aoki (eds.) *Learning Strategies and Cultural Evolution during the Palaeolithic*, Replacement of Neanderthals by Modern Humans Series, DOI 10.1007/978-4-431-55363-2,

MIX
Papier aus verantwortungsvollen Quellen
Paper from responsible sources
FSC® C105338

If you have any concerns about our products,
you can contact us on
ProductSafety@springernature.com

In case Publisher is established outside the EU,
the EU authorized representative is:
Springer Nature Customer Service Center GmbH
Europaplatz 3, 69115 Heidelberg, Germany

Printed by Libri Plureos GmbH
in Hamburg, Germany